Applied GIS and Spatial Analysis

Applied GIS and Spatial Analysis

Editors

JOHN STILLWELL
GRAHAM CLARKE

School of Geography, University of Leeds

Telephone (+44) 1243 779777

e-mail (for orders and customer service enquiries): cs-books@wiley.co.uk
Visit our Home Page on www.wileyeurope.com
or www.wiley.com

Reprinted May 2004

Other Wiley Editorial Offices

John Wiley & Sons Inc., 111 River Street, Hoboken, NJ 07030, USA

Jossey-Bass, 989 Market Street, San Francisco, CA 94103-1741, USA

Wiley-VCH Verlag GmbH, Boschstr. 12, D-69469 Weinheim, Germany

John Wiley & Sons Australia Ltd, 33 Park Road, Milton, Queensland 4064, Australia

John Wiley & Sons (Asia) Pte Ltd, 2 Clementi Loop #02-01, Jin Xing Distripark, Singapore 129809

John Wiley & Sons Canada Ltd, 22 Worcester Road, Etobicoke, Ontario, Canada M9W 1L1

Wiley also publishes its books in a variety of electronic formats. Some content that appears in print may not be available in electronic books.

British Library Cataloguing in Publication Data

A catalogue record for this book is available from the British Library

ISBN 0-470-84409-4

Typeset in 10/12pt Times by SNP Best-set Typesetter Ltd., Hong Kong
Printed and bound in Great Britain by Antony Rowe Ltd, Chippenham, Wiltshire
This book is printed on acid-free paper responsibly manufactured from sustainable forestry in which at least two trees are planted for each one used for paper production.

Contents

List of Contributors

Shelley Alexander, Department of Geography, Faculty of Social Sciences, University of Calgary, 2500 University Drive NW, Calgary, Alberta T2N 1N4, Canada

Seraphim Alvanides, Department of Geography, University of Newcastle-upon-Tyne, Newcastle-upon-Tyne NE1 7RU, United Kingdom

Mark Birkin, School of Geography, University of Leeds, Leeds LS2 9JT, United Kingdom

Jan Bleeker, GGD Rotterdam e.o., PO Box 70032, 3000 LP Rotterdam, The Netherlands

Paul Boyle, School of Geography and Geosciences, University of St Andrews, St Andrews, Fife KY16 9AL, United Kingdom

António Câmara, New University of Lisbon, 2825 Monte de Caparica, Lisbon, Portugal (and YDreams, SA, Madan Park, Caparica, Lisbon, Portugal)

Tony Champion, Department of Geography, University of Newcastle-upon-Tyne, Newcastle-upon-Tyne NE1 7RU, United Kingdom

Graham Clarke, School of Geography, University of Leeds, Leeds LS2 9JT, United Kingdom

Martin Clarke, GMAP Limited, 1 Park Lane, Leeds LS3 1EP, United Kingdom

Mike Coombes, Centre for Urban and Regional Development Studies (CURDS), University of Newcastle-upon-Tyne, Newcastle NE1 7RU, United Kingdom

Massimo Craglia, Sheffield Centre for Geographic Information and Spatial Analysis, University of Sheffield, Sheffield S10 2TN, United Kingdom

Richard Culf, GMAP Limited, 1 Park Lane, Leeds LS3 1EP, United Kingdom

António Eduardo Dias, University of Evora, Evora, Portugal (and YDreams, SA, Madan Park, Caparica, Portugal)

Robin Flowerdew, School of Geography and Geosciences, University of St Andrews, St Andrews, Fife KY16 9AL, United Kingdom

A. Stewart Fotheringham, Department of Geography, University of Newcastle-upon-Tyne, Newcastle-upon-Tyne NE1 7RU, United Kingdom

Martin Frost, South East Regional Research Laboratory (SERRL), School of Geography, Birkbeck College, University of London, 7–15 Gresse Street, London W1T 1LL, United Kingdom

Stan Geertman, URU and Nexpri, Faculty of Geographical Sciences, Utrecht University, PO Box 80.115, 3508 TC Utrecht, The Netherlands

Richard Harris, School of Geography and South East Regional Research Laboratory (SERRL), Birkbeck College, University of London, 7–15 Gresse Street, London W1T 1LL, United Kingdom

Tony Hernandez, Centre for the Study of Commercial Activity (CSCA), 350 Victoria Street, Ryerson University, Toronto, Ontario M5B 2K3, Canada

Mark E.T. Horn, Commonwealth Scientific and Industrial Research Organisation (CSIRO), Mathematical and Information Sciences, GPO Box 664, Canberra A.C.T. 2001, Australia

Ken Jones, Centre for the Study of Commercial Activity (CSCA), 350 Victoria Street, Ryerson University, Toronto, Ontario M5B 2K3, Canada

Tom de Jong, URU, Faculty of Geographical Sciences, Utrecht University, PO Box 80.115, 3508 TC Utrecht, The Netherlands

Pauline Kneale, School of Geography, University of Leeds, Leeds LS2 9JT, United Kingdom

Steven Laposa, PricewaterhouseCoopers (PwC), 1670 Broadway, Suite 1000, Denver, CO 80202, USA

Paul Longley, Department of Geography and Centre for Advanced Spatial Analysis (CASA), University College London, 1–19 Torrington Place, London WC1E 6BT, United Kingdom

Paul Paquet, Faculty of Environmental Design, University of Calgary, 2500 University Drive NW, Calgary, Alberta T2N 1N4, Canada

Simon Raybould, Centre for Urban and Regional Development Studies (CURDS), University of Newcastle-upon-Tyne, Newcastle NE1 7RU, United Kingdom

Philip Rees, School of Geography, University of Leeds, Leeds LS2 9JT, United Kingdom

Peter Rogerson, Department of Geography, University of Buffalo, Buffalo NY 14261, USA

Linda See, School of Geography, University of Leeds, Leeds LS2 9JT, United Kingdom

John Shepherd, South East Regional Research Laboratory (SERRL), School of Geography, Birkbeck College, University of London, 7–15 Gresse Street, London W1T 1LL, United Kingdom

Paola Signoretta, Sheffield Centre for Geographic Information and Spatial Analysis, University of Sheffield, Sheffield S10 2TN, United Kingdom

David Simmonds, David Simmonds Consultancy, Suite 23, Miller's Yard, Mill Lane, Cambridge CB2 1RQ, United Kingdom

Andy Skinner, MVA, 26th Floor, Sunley Tower, Manchester M1 4BT, United Kingdom

John Stillwell, School of Geography, University of Leeds, Leeds LS2 9JT, United Kingdom

Grant Thrall, PricewaterhouseCoopers (PwC), 1670 Broadway, Suite 1000, Denver, CO 80202, USA (and University of Florida, Gainesville, FL 32611, USA)

Nigel Waters, Department of Geography, Faculty of Social Sciences, University of Calgary, 2500 University Drive NW, Calgary, Alberta T2N 1N4, Canada

Coen Wessels, Nexpri, Utrecht University, PO Box 80.115, 3508 TC Utrecht, The Netherlands

Leo van Wissen, Faculty of Spatial Sciences, University Groningen, PO Box 800, NL-9700 AV Groningen, The Netherlands, and Netherlands Interdisciplinary Demographic Institute (NIDI), The Hague, The Netherlands

Acknowledgements

Various people have helped in the preparation of this book, not least Alison Manson in the Graphics Unit in the School of Geography at the University of Leeds who has drawn and improved many of the figures, and Lyn Roberts at Wiley who provided support and encouragement to us as editors throughout the duration of the project. However, no collection like this would be possible without the contributions of the authors themselves and we are very grateful for the efforts that all our collaborators have made in producing their chapters in the first instance and in responding to our editorial suggestions.

1

Introduction

Graham Clarke and John Stillwell

Abstract

Applied work in geographical information systems (GIS) and spatial analysis has been a persistent component of research activity in geography and regional science for several decades. This introductory chapter of this book establishes the background to the contemporary flurry of applied activity, explains the structure of the book and introduces the contents of the chapters that constitute the various parts into which the book has been divided.

1.1 Background

The history of applied spatial modelling has been a chequered one. If we ignore the elements of simple numerical reporting in the early commercial geographies of the nineteenth century, then we can probably trace the origins of applied quantitative spatial analysis to the fields of transportation modelling, especially at the University of Pennsylvania, in the early 1960s (Herbert and Stevens, 1960; Harris, 1962). Mathematical geography became prevalent in planning circles both in the USA and in Europe from this time onwards. The early 1960s also witnessed the emergence of statistical geography: the search for patterns or similarities in spatial data sets and the testing for significance, with the aim of providing universal spatial theories and laws (or at least theories that worked well in the real world). Although this was often very empirical in nature, it seldom extended beyond the academic geography community. There are excellent summaries in Haggett (1965) and Haggett *et al.* (1977).

Foot (1981) has reviewed a number of the more well-known applications of large-scale models in the United Kingdom (UK) during the 1960s. The Portbury Dock

Applied GIS and Spatial Analysis. Edited by J. Stillwell and G. Clarke
 ISBN: 0-470-84409-4

study in 1964 was one example of an application of mathematical modelling to estimate the future flow of exports through UK ports and hence the evaluation of the need for a new major dock facility at Portbury, in South West England. The Haydock shopping study (1966) was another classic example, incorporating a spatial interaction model of retail expenditure in North West England to estimate the potential revenues for a new regional shopping centre located roughly half-way between Liverpool and Manchester (a concept ahead of its time perhaps!). Another major study was the analysis carried out to quantify the impact of a third London airport at various locations in the South East of England. Paralleling these studies, were a large number of comprehensive land-use/transportation models built as part of local authority strategic planning. Batty (1989, 1994) has reviewed these applications in detail.

The fierce critiques of urban modelling in the 1970s (Lee, 1973; Sayer, 1976) brought the first era of applied modelling to an end. Lee believed the various large-scale models to be generally too complex, mechanical and ultimately unworkable. Sayer's critique was based on the role of agents in modelling – that is, that urban dynamics were as much the result of key decisions made by urban gatekeepers as they were due to land-use/transportation changes. In 1976, models were banned from being used in many UK retail planning enquiries because of the conflicting evidence they supposedly produced. More generally in land-use modelling, May (1991) argues that the decline in planning applications using models resulted from model-builders being too keen to present blueprints rather than alternative options. Wilson (1997) queries whether it was simply easier for planners to face the very difficult political decisions associated with cities 'in a haze rather than with maximum clarity' (Wilson, 1997, p. 18). Academics, meanwhile, retreated back to theory in an attempt to address the concerns expressed by humanist and Marxist geographers. For example, the 1970s and the early 1980s witnessed a research agenda dominated by the desire to incorporate dynamics into urban models (Allen and Sanglier, 1979; Wilson, 1981; Dendrinos, 1985). As Wilson (1997) reflects: 'While it can be argued that the full breadth of the attack was misguided – Marxian economists, for example, had no hesitation in using mathematical models – the analysis does point to the need for an understanding of deeper underlying structures that might be at the core of urban evolution' (Wilson, 1997, p. 10).

For much of mainstream academic geography, however, quantitative methods had become irrelevant and indeed counterproductive as the focus switched from serving the interests of planners to looking for grand theories of social change. The cause of quantitative geography was not helped by the desertion of some of its key pioneers (especially in the UK) such as David Harvey and Doreen Massey. The UK story is mirrored in many other parts of Europe and the USA. The full story of the retreat from quantitative geography is well told by Johnston (1996).

However, some remained loyal to the cause. A few geography departments around the world remained strong quantitatively, whilst the growth of the discipline of regional science helped the surviving quantitative geographers find allies amongst regional economists and planners (Plane, 1994). Slowly, through the 1980s and into the 1990s, quantitative geography began to claw its way back into the discipline (although it would not, to date, seriously challenge the stranglehold of 'critical human geography'). The reasons for this are numerous and a fuller account is provided by Clarke and Wilson (1987). However, it is clear that there were two fundamental drivers of change. The first was the focus anew on applications. Statistical and mathematical geography became much more focused on how it could solve real location problems in fields such as health, education, retailing, transport and deprivation analysis. The pioneers here were undoubtedly the quantitative geographers at the University of Leeds who began to find commercial applications for their models in both the public and private sector, leading to the formation of GMAP in the late 1980s (see Chapter 3). The second driver of change was the developments occurring in the fields of geographical information systems (GIS) and geodemographics. Interestingly, both of these originated outside the discipline of geography. However, as they became well known in planning and business circles, the quantitative geographers were quick to realise that they provided a new platform on which to sell their wares. In the UK, a number of geography departments benefited from the Economic and Social Research Council (ESRC) initiative to create a series of research laboratories in GIS (see the final chapter in this collection by Flowerdew and Stillwell for a review of this initiative).

Alongside these two major developments came a series of other enabling factors which helped to promote applied quantitative analysis. First, the personal computer (PC) allowed the planner (both public and private sector) to have immediate desktop access to powerful software for spatial mapping and analysis. Second, spatial data was also becoming more routinely available in most application areas, including planning, as evidenced in Stillwell *et al.* (1999). Third, pressure was mounting on academics to interact more with the outside world in order to attract new income streams. Finally, we might acknowledge a genuine sea-change in attitudes of many academics – an increase in the desire to see their models usefully applied to social and economic problems – as well as a parallel change of attitude by private and public sector clients who began to recognise more seriously the benefits available from the exploitation of large data sets by new geotechnologies. The variety of planning support systems now being used in practice (Geertman and Stillwell, 2002) is one reflection of this new interest. The more generic set of issues also allowed a new emphasis on applied research in regional science (Clarke and Madden, 2001), and in non-quantitative aspects of human and physical geography (Pacione, 1999).

1.2 Aims and Contents

Against this background, the aim of this book is to illustrate the applied nature of contemporary quantitative geography through a series of case studies. It is clearly only a selection but there are still only relatively few examples of commissioned research in the literature (for further examples, see Clarke and Madden, 2001). The paucity of applied work in academic journals is perhaps not surprising since the results of many such projects are confidential, especially for private sector applications. Here, however, we have collected together a suite of case studies of funded work which has been applied to particular real-world problems confronted by either private or public sector organisations. Between them, these case studies involve the application of GIS, statistical models, location-allocation models and network or flow models. However, rather than presenting these case studies on the basis of the methodologies they adopt, the book is arranged thematically in four parts around important subject areas: business applications; social deprivation studies; transport and location problems; and national spatial planning applications.

The initial quarter of the book (Part One) contains a collection of chapters focused on applications in the private sector, predominantly drawn from retailing. The first two chapters draw upon experiences with various clients. This shows the strength of applied work in retail analysis. Both Jones and Hernandez and Birkin *et al.* draw upon their experiences with many global organisations involved in retail location analysis. Jones and Hernandez give a very useful US/Canadian overview of commercial applications of GIS and spatial analysis. Their main conclusion is that the market is moving (slowly!) from one of simple mapping to the use of more sophisticated data mining and visualisation techniques. This argument is taken up in Birkin *et al.* where the authors examine a number of difficult applied location problems that have been faced by a number of clients. They begin to show how these problems may be addressed from an analytical point of view. The study by Câmara and Dias is a very novel use of network analysis for guiding shoppers around a major shopping centre complex in Lisbon, Portugal, undertaken in cooperation with SONAE, the company responsible for running the shopping centre. As the use of mobile phones for location-based services becomes more sophisticated, it is apparent that their use for pinpointing the geographical location of various types of services will increase. There is an important role for spatial analysis in optimising the best ways to maximise access to geographical information of this sort. Finally, in this part of the book, Laposa and Thrall investigate the use of GIS and models for the study of house price variations for PricewaterhouseCoopers. Thrall has been involved with applications of GIS in the business sector for many years, specialising in the real estate sector (Thrall, 2002). The main contention of the chapter is that the use of GIS, three-dimensional modelling and visualisation of house sale prices provides greater explanatory power than typical hedonic residential price models for estimating zonal or regional average house prices.

The set of chapters that are presented in Part Two have been commissioned by various local authorities helping to expose different aspects of social problems, especially deprivation and crime. The second chapter by Boyle and Alvanides examines a problem that many urban areas face in the UK. This is the so-called 'two-speed' growth problem. On the one hand, there are areas of cities experiencing rapid growth in the number of jobs, levels of service provision, house prices and/or residents. On the other hand, other parts of cities get left behind and remain as pockets of acute deprivation. In Europe, there are European Union (EU) monies available for those cities in most need of support. However, often the pockets of deprivation are not large enough to allow credible and integrated programmes to be mounted. This was the case in point with the city of Leeds, UK. The authors of this chapter, commissioned by the City Council, demonstrate how it is possible to identify a contiguous set of enumeration wards within the Leeds area which, when aggregated together, represent a relatively large concentration of households with a high deprivation score. The initial chapter in this part of the book, by Harris and Longley, takes a more reflective look at existing deprivation indicators used in socio-economic planning. In particular, they evaluate the combination of data obtained from lifestyle databases with that available from high-resolution satellite imagery. They show how the latter can help to evaluate the degree of homogeneity in census-based deprivation scores (or income estimates) and offer a potential for representing deprivation or income clusters at more precise spatial scales. The third chapter relates deprivation in the city to the problems of child care. Sponsored by the local social service department, Craglia and Signoretta explore the use of GIS for the construction of a 'Children's Services Plan' in Sheffield, UK. In a sense, this is a children's census; data is put together on all aspects of children's health, status of care and behaviour (crime, attendance at school, etc.). Each of these variables is then ranked by small geographical area. The final product is a child deprivation score for the entire city. The final chapter in this part of the book is by Rogerson who looks at one dimension of social deprivation in more depth: namely, crime within Metropolitan Buffalo, sponsored by various city agencies. He uses GIS linked to the very latest spatial statistical analysis tools to find regions with significantly higher crime than could be expected under a random distribution of crime activity. He also uses various statistical techniques to examine the dynamics of crime patterns.

The set of chapters comprising Part Three relate to transport networks and location problems. The first chapter by Horn is a good example of using transport models to investigate the potential impact of a number of future transport plans. The case study is based on the Gold Coast, a rapidly expanding area of South East Queensland in Australia. The model is set up to help assess the viability of several road-based 'demand-responsive' transport modes (taxis, multi-taxis and so-called roving buses) designed to supplement existing bus, rail and taxi services. The analysis tool is a simulation model driven by demand simulators that replicate likely or future patterns of consumer demand. Network simulators help to service that demand by tracking the movements of individual vehicles and allocating the closest vehicle. The benefits and costs of operating such a system have been articulated and now the transport planners in Queensland have to make the ultimate decision of whether to introduce the scheme. The second chapter by Simmonds and Skinner uses a well-known land-use/transport model (DELTA) to examine the future transport plans of both West and South Yorkshire in the UK in the context of the preparation of Regional Planning Guidance for Yorkshire and the Humber (Government Office for Yorkshire and the Humber, 2001). As was noted in Section 1.1, land-use/transport models have a long history of applied success in planning. The model here helps to address two key questions: how can the authorities design an integrated and sustainable transport policy for the future, and how can the most urgent problems be specifically addressed? The following chapter by Geertman *et al.* reports the response to a classic geographical problem – where to site ambulance stations in a city context in order to minimise the time taken to access all parts of that city. The methodology adopted combines traditional shortest-path network analysis from a major proprietary GIS package with an in-house accessibility indicator package based on spatial interactions of flows. The work was commissioned by Rotterdam Municipal Health Authority. The final chapter by Alexander *et al.* reports on a very common environmental consequence of road construction, especially in areas well inhabited by wildlife. The construction of roads in such areas can cut-off access to pathways or routeways historically used by various types of wildlife. The authors present a case study of the use of GIS to identify optimal placement sites for structures that facilitate wildlife passage across, or under, highways in the Canadian Rocky Mountains.

Part Four of the book contains a set of chapters that look at large-scale national social and economic problems. These are largely funded by government departments. The first chapter by Rees *et al.* was commissioned by the UK Department of Environment, Transport and the Regions (DTLR now the Office of the Deputy Prime Minister, ODPM). A key concern of DTLR was to balance population growth across the UK. In order to achieve this, they need to equalise future regional population growth in situ, but also to reduce migration losses from northern regions (flowing to the south). The chapter describes a two-stage migration model built to enable policy makers to investigate the first-round quantitative impacts of alternative economic and policy scenarios on gross flows of population between regions. The model used is a very disaggregated spatial interaction model, calibrated using statistical regression techniques. The second chapter continues the theme of policies for regional balanced population and economic growth. Van Wissen explores the concept of regional 'carrying capacities' as a framework for establishing policies on regional economic growth in the Netherlands. The concept of 'carrying capacity' is borrowed from ecology – the maximum use of land that can be sustained over time. This idea is translated into an economic context by a model of inter-industry linkages. It is argued that the growth potential of an individual firm in a locality relates to the size and composition of the population of firms in that locality. The final model is a mixture of a spatial interaction and an input–output model.

The chapter by Coombes and Raybould reports on the process of finding sites for a potential new UK government information service (funded by the Lord Chancellor's Department). It is another classic location problem and, not surprisingly, it fits into a framework of many public sector location–allocation problems. In this case, the problem is to locate a new information service for those couples facing the prospect of filing for divorce from a possible 647 candidate sites. The chapter by Frost and Shepherd looks at another issue of increasing concern to the UK DTLR (now ODPM) – rural accessibility. As the level of service provision declines in rural areas, there is an urgent need to measure rural accessibilities. Frost and Shepherd use GIS to build a parish-based survey of access to services. They also evaluate the changing role of the local market town in rural areas, in order to be able to identify market towns with strong/weak service centre functions. The final chapter by Kneale and See addresses a key problem faced by the Environmental Agency in the UK and elsewhere – how can we improve our flood forecasting methodologies. This comes at a time when the UK has faced a number of very wet periods and the amount of flooding has been severe. Yet, despite advances in most aspects of computer technology, the ability to predict the consequences of these floods has not been very successful. The approach by Kneale and See uses neural networks to improve the forecasting process and to give more time for operators to send out alarms.

The concluding chapter of the book addresses some of the issues and concerns that confront those working in an academic environment when attempting to undertake applied research. Flowerdew and Stillwell reflect on the advantages and limitations of applied research by drawing on two different university experiences during the 1990s: the ESRC-funded Regional Research Laboratory (RRL) initiative (with comments based largely on the activities of the North West RRL at the University of Lancaster) and the Yorkshire and Humberside Regional Research Observatory established at the University of Leeds. They conclude that despite various difficulties associated with applied research (frequently done through consultancy arrangements), much valuable work has been undertaken already and the opportunities for making use of new GIS, analysis methods and modelling techniques in the future are very exciting.

We hope that the contributions assembled in this collection provide a useful representation of what has been achieved in recent years.

References

Allen, P. and Sanglier, M. (1997) A dynamic model of growth in a central plane system, *Geographical Analysis*, **11**: 256–73.

Batty, M. (1989) Urban modelling and planning: reflections, retrodictions and prescriptions, in MacMillan, B. (ed.) *Remodelling Geography*, Blackwell, Oxford.

Batty, M. (1994) A chronicle of scientific planning: the Anglo-American modelling experience, *Journal of the American Institute of Planners*, **60**: 7–16.

Clarke, G.P. and Madden, M. (2001) *Regional Science in Business*, Springer, Berlin.

Clarke, M. and Wilson, A.G. (1987) Towards an applicable human geography, *Environment and Planning A*, **19**: 1525–41.

Dendrinos, D.S. (1985) *Urban Evolution: Studies in the Mathematical Ecology of Cities*, Oxford University Press, New York.

Foot, D. (1981) *Operational Urban Models*, Methuen, London.

Geertman, S. and Stillwell, J.C.H. (eds) (2002) *Planning Support Systems in Practice*, Springer Verlag, Heidelberg.

Government Office for Yorkshire and the Humber (2001) *Regional Planning Guidance for Yorkshire and the Humber (RPG 12)*, Department of Transport, Local Government and the Regions, The Stationery Office, London.

Haggett, P. (1965) *Locational Analysis in Human Geography*, Edward Arnold, London.

Haggett, P., Cliff, A.D. and Frey, A. (1977) *Locational Analysis in Human Geography*, 2nd edition, Edward Arnold, London.

Harris, B. (1962) Linear programming and the projection of land uses, *Paper 20*, Penn-Jersey Transportation Study, Philadelphia.

Herbert, D.J. and Stevens, B.H. (1960) A model for the distribution of residential activity in an urban area, *Journal of Regional Science*, **2**: 21–36.

Johnston, R.J. (1996) *Geography and Geographers*, 5th edition, Edward Arnold, London.

Lee, D.B. (1973) Requiem for large-scale models, *Journal of the American Institute of Planners*, **39**: 163–78.

May A.D. (1991) Integrated transport strategies: a new approach to urban transport policy formulation in the UK, *Transport Reviews*, **11**: 223–47.

Pacione, M. (ed.) (1999) *Applied Geography: Principles and Practice*, Routledge, London.

Plane, D. (1994) Comment: on discipline and disciplines in regional science, *Papers in Regional Science*, **7**: 19–23.

Sayer, R.A. (1976) A critique of urban modelling, *Progress in Planning*, **6**: 187–254.

Stillwell, J.C.H., Geertman, S. and Openshaw, S. (eds) *Geographical Information and Planning*, Springer Verlag, Heidelberg.

Thrall, G. (2002) *Business Geography and New Real Estate Market Analysis*, Oxford University Press, New York and London.

Wilson, A.G. (1981) *Geography and the Environment: Systems Analytical Methods*, Wiley, New York.

Wilson, A.G. (1997) Land-use/transport interaction models: past and future, *Journal of Transport Economics and Policy*, **32**(1): 3–26.

Part One
GEOBUSINESS

2

Retail Applications of Spatial Modelling

Ken Jones and Tony Hernandez

Abstract

This chapter examines the use of geographical information systems (GIS) and associated forms of spatial analyses currently used by North American retailers, property managers and urban planners in assessing various strategic issues (see Goodchild, 2000, for a more general assessment of GIS and spatial analysis). Initially, the chapter evaluates the value and adoption of spatial analysis and associated technologies by retail sector analysts. This discussion will be followed by five recent applications. These examples, drawn from client-based and academic research, will serve to illustrate the use of spatial analysis and GIS at various scales and for a variety of users groups. These cases will examine: (i) an assessment of regional variations in retail sales performance; (ii) the impact of 'big boxes' on the competitive structure of a retail environment; (iii) the spatial variability of customer penetration associated with a destination retailer; (iv) the internal dynamics and performance of a regional shopping centre; and (v) the spatial/temporal variation of sales volumes by product category for a network of retail outlets. The chapter will conclude with an evaluation of data mining and visualisation technologies as a potential means of enhancing the value of the spatial approach to retail corporate decision making. Due to commercial sensitivity, the names of the retail organisations involved have been omitted to maintain confidentiality.

2.1 Introduction

Over the last 25 years the awareness and use of GIS technologies for retail planning in North America has increased significantly (Sheerwood, 1995; Hernandez, 1999;

Applied GIS and Spatial Analysis. Edited by J. Stillwell and G. Clarke
 ISBN: 0-470-84409-4

Hernandez and Biasiotto, 2001). Initially, GIS technology was considered by most retail organisations as having limited application, being viewed by many users as nothing more than an elegant form of digital mapping. GIS-based applications were typically housed in either the real estate or market research departments and normally were used to provide visual support for internal decision making (Reid, 1993). The early adopters, often associated with the grocery and department store chains, were inhibited by a number of constraints. These typically related to the limitations imposed by existing computer capacity, the general lack of spatially referenced databases, the elementary nature of the GIS software and the lack of trained analysts. As a consequence, early applications were associated with the automation of various traditional trade area methodologies such as customer spotting, boundary definition and market share estimates. Often retailers used simplistic trade area demographics to examine issues that related to store cannibalisation and market penetration. The next major stage in the development of spatial modelling is related to the development and widespread adoption of spatial demographic cluster systems in North America and Europe (see Flowerdew and Goldstein, 1989, for example).

In North America, two firms dominated the commercialisation of these advances – Claritas Corporation (US) and Compusearch Micromarketing (Canada). In both cases, these research firms took existing census data and developed spatially defined socioeconomic clusters for both the American and Canadian markets, respectively. Typically, these cluster profiles were developed for specific geographic areas (i.e., the enumeration area, block face or various postal geographies ZIP+4 or FSALDU). Many retail organisations subscribed to these databases and used them to assess existing store performance, evaluate future market opportunities, and/or to plan and execute marketing or direct mail campaigns. Shopping centre developers also used these systems to market various properties to prospective tenants, to examine mall performance, and to evaluate potential expansion plans. More recently, the use of GIS and associated spatial technologies has been extended into a number of new areas. This has been facilitated by the development of new databases that relate to the supply side of the retail economy, the acceptance of GIS as a mainstream technology by a growing number of retail analysts and decision makers, and the increased availability of more sophisticated spatial modelling and data visualisation software.

This chapter will illustrate the use of various spatial modelling applications in somewhat non-traditional, but burgeoning, retail applications. By adopting this approach, we hope to demonstrate the increased potential of spatial data, spatial analysis, and a spatial perspective to various levels and types of decision-making activities associated with the retail and service economy. In North America, the term *business geomatics* has been coined recently to reflect the emergence of a new interdisciplinary area that is beginning to link spatial information, spatial technologies, the spatial sciences and applications (Yeates, 2001). In the long run, this development will increase the awareness and credibility of spatial modelling and GIS applications to a much wider community in both the academic and private and public sectors.

2.2 Retail Sales Performance: A Macro Approach

The first application focuses on the analysis of the major concentrations of retail sales activity at the national or regional level using small area retail trade estimates data (SARTRE). SARTRE reports total annual retail sales for approximately 142 000 retail locations across Canada aggregated to Forward Sortation Area (FSA) level. FSAs are a unit of postal geography comprising 1452 areas covering all of Canada. SARTRE is based on a combination of data from the Retail Chain Location file of Statistics Canada (that tracks the locations of all retailers operating in Canada with four or more locations) and corporate tax returns from Revenue Canada. The increasing availability of macro data permits the retail analyst to rank major retail areas in Canada by retail sales performance. In addition, given the longitudinal nature of these data, the fastest growing retail markets can be readily identified. In this example, these spatial databases are used to link retail trade sales to store locations at a fine level of geographic detail – typically, postal areas or census areas. The example combines data from survey and taxation information for all incorporated retailers in Canada and groups these businesses according to retail category (i.e., grocery retailing, fashion, department stores, automotive). This level of supply-side information allows for a detailed assessment of retail performance by major retail category for relatively small geographical areas.

The analysis of data of this type can yield some interesting general trends at the national scale. Often retail chains seek out locations in fast-growing areas. The urban market areas identified in Table 2.1 are confined to areas with retail sales in excess of $100 million and sales growth greater than 200%. Interestingly, the two largest regional markets in Canada, Ontario and Quebec, combine for only four of the top 10 fastest growing retail concentrations (defined by FSA). A detailed analysis suggests that these high growth locations tend to exhibit a suburban character, although two of the three Edmonton locations, along with the Montreal FSA, are certainly central within those cities. The locational dynamics of retailing for the period are also uncovered when we compare the sales growth to the change in the number of

Table 2.1 *Fastest growing retail areas in Canada, 1990–95*

FSA	Market	% Retail sales growth[1]	% Retail location growth[1]
V9V	Nanaimo, BC	3283.29	133.3
L5R	Mississauga, Ont.	1817.46	94.7
V2Y	Langley, BC	1628.72	285.7
T5V	Edmonton, Alta.	1451.68	157.1
H3K	Montreal, Que.	672.28	−4.8
T5A	Edmonton, Alta.	373.92	−9.3
V3N	New Westminster, BC	276.28	2.1
G1X	Ste-Foy, Que.	246.57	8.6
M8X	Etobicoke, Ont.	223.79	−3.2
T0C	Edmonton, Alta.	213.76	−17.9

Note: [1]Percentage growth calculated as follows: (1995 value minus 1990 value) divided by the 1990 value ×100

Table 2.2 *Top 20 Canadian retail hot spots*

FSA	Major shopping node	Market	Sales score[1]	Sales/Location ($000)[2]
M5B/C	Eaton Centre/The Bay	Toronto (CBD)	8.02	1626.204
T2H	Chinook SC	MacLeod Trail, Calgary	6.76	2051.122
L3R	Markville SC	Markham, Ont.	6.02	1326.133
H3A/B	Place Ville Marie/ Eaton Ctr.	Montreal (CBD)	5.89	1469.971
V6X	Lansdowne, SC	Richmond, BC	5.86	1727.706
M6A	Yorkdale SC	Toronto	5.39	1965.889
V5H	Metrotown	Burnaby, BC	5.34	2182.542
H9R	Fairview Ctr.	Pointe Claire, Que.	5.09	2053.887
M1P	Scarborough Town Ctr.	Toronto, Ont.	5.05	1981.640
L4M	Bayfield St./ Georgian Mall	Barrie, Ont.	4.74	1860.050
V1Y	Orchard Park SC	Kelowna, BC	4.64	1452.908
T1Y	Sundridge SC	Calgary, Alta	4.59	2349.063
T5T	West Edmonton Mall	West Edmonton, Alta	4.56	1408.541
T2J	MacLeod Trail	Calgary, Alta	4.48	1547.741
S7K	Midtown SC	Saskatoon (CBD)	4.48	1198.968
N8X	Devonshire Mall	Windsor, Ont.	4.35	2074.595
L5B	Square One	Mississauga City Ctr., Ont.	4.25	1847.830
L3Y	Upper Canada Mall	Newmarket, Ont.	4.16	1594.634
N6E	White Oaks SC	S. London, Ont.	4.14	2025.862
G1V	Place Laurier	Quebec City (CBD)	4.08	1255.477

Notes:
[1] Based on the 20 FSAs with the greatest big-box square footage in 1995. The sales index values for each FSA were normalised to assure confidentiality. These values show the relationship between sales in each of these FSAs and average sales in all FSAs in the GTA. An FSA with sales equal to the average for all FSAs, would have a sales index value equal to 1.00
[2] Sales/Location calculated as retail sales in FSA divided by the number of retail locations in FSA

locations. Over the five-year period, four markets reported an increase in sales greater than 200% while the actual number of retail locations declined. This may suggest the growing importance of large, highly productive 'big box' retailers in these areas. Growth of 3000% in Nanaimo highlights the vigour of the market in western Canada.

When the retail areas in the database are ranked by their reported retail sales figures (Table 2.2), it is possible to identify major shopping nodes that exist within the top 20 markets. A number of high profile locations appear in the *Top 20 Retail Hot Spots*, along with what may be some unexpected additions. The major Toronto area regional shopping malls like the Eaton Centre, Yorkdale, Scarborough Town Centre and Square One, are joined by Markham's Markville Shopping Centre. In addition, just to the north of the Greater Toronto Area (GTA), Barrie's Bayfield St. and Georgian Mall area makes the list with the tenth highest reported sales in the country. Of note is the strong presence of Calgary's MacLeod Trail area and the Chinook Shopping Centre. It actually tops the list if the two retail areas that make up the top position are separated. Just under half of the top 20 locations are within Ontario, reaffirming the province's strong retailing position and demonstrating continued market concentration and relative affluence. The list also appears fairly compact, showing about the same range in sales score and sales per location, with the first place position doing only double that of the twentieth location.

The use of macro retail sales databases is particularly powerful for assessing general location strategies and does provide a much needed overview of the performance of the national retail system. For the trained analyst, a diversity of locations and patterns appears as the data are explored in terms of the overall strength of retail, fashion retail and general merchandising. Such a high-level database provides benchmarks for the retail organisation and provides valuable insights for retail managers and property owners into the operation of the retail spatial economy. In particular, one can assess the changing role of downtown areas, identify healthy shopping centre locations and speculate on the impact of new retail formats such as power centres or major retailers such as Wal-Mart. The data can also be input into various data visualisation programs and dynamic three-dimensional (3D) views of the retail economy can be generated (Plates 1 and 2). MineSet software produced by Silicon Graphics Inc. was used to generate the 3D maps. The software allows users to view temporal animations of spatial data.

2.3 Impact of Big-Box Retailing

The second application illustrates the influence of big-box retailing (e.g., Home Depot, Wal-Mart) on the health of street-front (strip) retailing. Increasingly, the impact of big-box/power centre development has become a major concern for urban planners, retail organisations with street locations, and financial institutions who invest heavily in retail property. One way of measuring the influence of big-box activity on the operation of the urban retail system is to examine changes in the functional composition of those retail strips that are in direct competition with these new large format retailers. This type of analysis requires an extensive spatial database of retail activity that is collected and maintained on a yearly basis. The particular case examines retail change in the City of Toronto. In this urban area, as elsewhere, strip retail areas are dominated by independent retailers, and often provide a social, cultural and economic focus for their surrounding neighbourhoods. Indeed, one of the major distinguishing features of the Toronto area from a North American perspective is the health, diversity and vibrancy of its retail streets. In total, the urban area is served by a network of over 200 retail strips that provide the citizens of the city with access to over 18 000 retail shops, restaurants and personal and business services. However, this system has been threatened by the recent arrival of big-box retailing and this change has created concerns for local planners, neighbourhood groups and local business organisations. The analysis of spatial data can help to provide empirical evidence that is needed to assess and monitor the magnitude of the impact of large format retailers on the local retail environment.

One fundamental measure of retail restructuring can be derived through an examination of changes in the economic role of retail areas and the relative growth or decline in vacancy rates. For the 1994–97 period, data on the commercial structure of Toronto's major retail streets, including vacancy rates, were collected and tabulated by the Centre for the Study of Commercial Activity (CSCA) at Ryerson University. This information is presented in Table 2.3, which provides an aggregate view of the changing structure of street-front retailing in Toronto over the 1994–97 period for selected retail categories. These categories reflect the areas of retailing that are in direct competition with the dominant big-box retailers currently in operation (for example, hardware – Home Depot; books – Chapters; office products – Office Place/Business Depot; supermarkets – Price Costco; and various other superstores – PETsMART). The data suggest that, during this period of significant big-box and power centre development in the GTA, the relative importance of retailing on the retail streets of Toronto, as measured by the proportion of all occupied stores, declined from 53.7% in 1994 to 49.5% in 1997. In addition, certain key retail categories that compete directly with a major big-box format either declined (hardware –10.4%, General Merchandise –3.9%) or remained relatively stable (food –0.8%, Pharmacy +0.4%, and furniture/housewares +1.7%). The significant decline in men's and women's fashion corresponds with the growth of discount department store chains (e.g., Wal-Mart) and the re-mixing of the traditional department stores. Complementing these changes was the trend toward a general increase in the importance of restaurants and personal and business services along Toronto's retail streets. These non-retail functions experienced a 24.1% growth in the number of stores over the study period, and all services now account for more occupied street-front units than retail activities. It is also worth observing the steady decline in the financial category within Toronto's retail strips during this period; a change that would, without question, accelerate if proposed mergers between some major Canadian banks were to be approved by the Federal Government. The inclusion of automated banking machines in many big-box outlets further emphasises the potential of new format retailing to alter street-front activities in major parts of the tertiary sector. Finally, the data suggest that the overall health of street retailing in Toronto has declined somewhat over this four-year period, as the number of vacancies increased by 10.6%. These figures, however, may well mask real changes to the overall 'quality' of the shopping experience provided along particular retail streets. Canadian business leaders have suggested that the shift from traditional retail activities, such as clothing stores and hardware shops, to cut-price stores and doughnut shops, while better than vacancies, is not always a positive change.

Table 2.4 examines the change in the relative share of the eight retail categories that are most vulnerable to big-box retailing. Over this period, the total number of stores within Toronto's retail strips grew from 13026 to 14448, an increase of 10.9%. However, for the eight selected categories (those most impacted by the big boxes), the number of stores experienced only a 2.9% increase, from 982 to 1010. Further, the total share of all stores captured by the eight categories fell from 7.3% to 6.8%. What is evident from a closer analysis of these figures is that in all eight categories studied, the percentage share of these store types declined by an average of 7%, with

Table 2.3 *Changes in the economic role of Toronto's major retail strips, 1994–97*

Activity	Number of establishments				% Growth rate 1994–97
	1994	1995	1996	1997	
Food	1694	1692	1681	1681	–0.8
Liquor	67	69	73	79	17.9
Pharmacy	238	244	245	245	0.4
Shoes	199	207	203	210	5.5
Men's clothes	184	173	167	168	–9.5
Women's clothes	565	551	539	515	–8.7
Other clothes	414	441	429	445	7.5
Furniture & housewares	720	723	712	732	1.7
Automotive	410	426	428	428	4.4
General merchandise	254	253	255	244	–3.9
Books & stationery	179	184	180	184	2.8
Florists	180	185	187	181	0.5
Hardware	222	216	209	199	–10.4
Sports	156	152	151	158	1.3
Music	264	272	275	279	5.7
Jewellery	171	178	179	195	14.0
Other retail	1076	1143	1155	1206	12.1
Total retail	*6992*	*7172*	*7095*	*7149*	*2.2*
% Retail	*53.7*	*52.3*	*51.0*	*49.5*	*–7.8*
Financial	*615*	*597*	*582*	*575*	*–6.5*
% Financial	*4.7*	*4.4*	*4.2*	*4.0*	*–14.9*
Restaurants	2531	2676	2823	2953	16.7
Recreation	179	227	247	259	44.7
Medical	405	475	502	515	27.2
Hair/Beauty	798	900	950	1002	25.6
Cleaning	367	380	382	397	8.2
Other personal	615	680	749	836	35.9
Business services	524	613	678	762	45.4
Total services	*5419*	*5951*	*6331*	*6724*	*24.1*
% Services	*41.6*	*43.4*	*45.2*	*46.5*	*11.8*
Total activity	*13026*	*13720*	*14008*	*14448*	*10.9*
Vacancies	*1259*	*1440*	*1490*	*1392*	*10.6*
%Vacancies	*9.7*	*10.5*	*10.6*	*9.6*	*–1.0*
Total stores	*14285*	*15160*	*15498*	*15940*	*11.6*

the largest declines found in office products (–23%) and hardware (–16.5%), respectively. Even categories that have only recently been affected by big-box competition experienced a decline in their street-front locations (e.g., books down 2.2% and pet stores down 2.1%) in the years in question. Given the relatively brief time period for which data are available, these changes to the retail structure of Toronto's shopping streets are striking.

Table 2.4 *Change in the composition of retail strips in the categories most impacted by big boxes: the metropolitan Toronto experience*

Category	1994		1997		Change in relative share by category 1997–94
	No. of stores	% share	No. of stores	% share	%
Supermarkets	253	1.94	261	1.81	−6.7
Sporting goods	104	0.8	111	0.77	−3.8
Books	120	0.92	130	0.9	−2.2
Office products	35	0.03	33	0.023	−23.3
Hardware	142	1.09	131	0.91	−16.5
Toys	79	0.61	83	0.57	−6.6
Electronics	188	1.44	195	1.35	−6.3
Pet stores	61	0.47	66	0.46	−2.1
All eight categories	*982*	*7.3*	*1 010*	*6.79*	*−7.0*
Total stores	*13 026*		*14 448*		

Table 2.5 *Changes in the number of independents on Toronto retail streets, 1994–97*

Category	Number of stores		Change	% Change
	1994	1997		
Supermarkets	225	220	−5	−2.3
Sporting goods	91	100	9	9.9
Books	99	118	19	19.2
Office products	32	31	−1	−3.1
Hardware	107	96	−11	−10.3
Toys	73	82	9	12.3
Electronics	171	163	−8	−4.4
Pets	39	55	16	41.0
All eight	*837*	*865*	*28*	*3.3*
All independents	*11 096*	*12 646*	*1550*	*13.9*

Table 2.5 examines the same database in order to trace the change in the number of independent retailers in the same eight categories. In this case, two categories of retailer were more affected – hardware and electronics. These categories suffered losses of 11 and eight stores, respectively, between 1994 and 1997. There was also a slight reduction in the number of independent office products retailers and supermarkets during this interval. Independent retailers in the remaining categories experienced some growth between the years in question, most notably book sellers, sporting goods, toys and pet stores, with net gains of 19, nine, nine, and 16 establishments, respectively; a reflection of growth trends in these retail sectors. Overall, independent stores on Toronto's retail strips grew by 13.9% during this period, a

Table 2.6 Major concentrations of big-box retailing in the GTA[1]

Name of node	FSA	Sales index	% increase in sales 1989–95	Big-box growth 1989–95 (000s of sq. ft.)	No. of boxes 1995	No. of boxes 1998
Orion Gate	L6W	2.46	37	353	11	14
Woodbine & Hwy. 7	L6G	n.a.	n.a.	387	10	14
Crossroads	M9N	1.82	7	32	10	11
Thickson	L1N	2.15	19	229	9	16
Pickering	L1V	3.29	16	71	9	10
Warden & Ellesmere	M1R	1.58	20	314	3	3
Heartland	L5R	1.47	3940	238	7	9
400 & Hwy. 7	L4L	3.31	214	295	6	18
Hyde Park	L6H	1.31	22	272	8	14
Burlington	L7N	0.95	–3	36	5	5
SW Etobicoke	M9B	2.89	74	200	3	3
Central Mississauga	L4Z	0.64	23	194	3	5
Oakville	L6M	0.58	343	50	8	9
Unionville/Markham	L3R	5.19	30	60	6	7
Yorkdale	M6A	4.65	10	102	6	6
Rosart Centre	L7T	0.64	79	200	1	4
Ikea	M2K	2.09	–16	0	1	1
Warden Woods	M1L	2.07	9	101	6	8
Erin Mills	L5L	2.17	13	21	5	10
Younge/ Richmond Hill	L4C	2.39	24	180	2	4
Big-box nodes		*$4702.3*[2]	*18.60*	*3335*	*119*	*171*
Total GTA		*$21454.8*[2]	*9.20*	*6237*	*251*	*320*
Residual		*$16752.5*[2]	*6.80*			

Notes:
[1] Based on the 20 FSAs with the greatest big-box square footage in 1995. The sales index values for each FSA were normalised to assure confidentiality. These values show the relationship between sales in each of these FSAs and average sales in all FSAs in the GTA. An FSA with sales equal to the average for all FSAs, would have a sales index value equal to 1.00
[2] Expressed ($millions) includes all retail sales minus automotive sales (SIC 60, 61, 62, 64, 65)

marked contrast to the trends for the selected categories that, together, experienced only a 3.3% increase. As indicated in Table 2.3, the service sector has been experiencing substantial growth over the 1994–97 period, accounting for the vast majority of new stores within Toronto's retail strips.

In order to evaluate the impact of the big-box phenomenon on the distribution of retail sales, the 20 areas in the GTA with the greatest concentration of big-box retailers were identified (Table 2.6). These areas captured 47.4% of the big-box outlets in the GTA in 1995, and collectively accounted for over 3.3 million square feet

Table 2.7 Number and probability of retail closures and distance from the nearest competing big box, 1994–97

Category	Distance to the nearest big-box competitor			Probability of closure by category
	0–2 km	2.01–4 km	>4 km	
Supermarkets	43/329 (.13)	61/375 (.16)	15/69 (.22)	.154
Electronics	60/257 (.23)	39/177 (.22)	20/136 (.15)	.208
Office products	5/40 (.13)	7/40 (.18)	2/20 (.10)	.14
Sporting goods	6/44 (.14)	9/52 (.17)	35/213 (.16)	.162
Toys	12/73 (.16)	16/93 (.17)	12/74 (.16)	.167
Hardware	8/32 (.25)	10/82 (.12)	19/170 (.11)	.13
Books	7/42 (.17)	5/39 (.13)	4/47 (.09)	.125
Pet stores	1/2 (.5)	0/2 (.00)	2/59 (.03)	.048
Probability of closure by distance band	.173	.171	.138	.161

(53.5%) of big-box expansion during the 1989–95 period. What is particularly noteworthy is that these 20 areas experienced an 18.6% increase in retail sales between 1989 and 1995. This compares with a growth of just 9.2% for the entire urban region for the same interval. In addition, some of the most impressive sales growth occurred in the region's emerging power nodes: Highways 400 and 7 (214%), Orion Gate (37%), Thickson (19%), Hyde Park (22%), and Heartland (3940%). Each of these nodes experienced a growth of at least 200 000 square feet of big-box space between 1989 and 1995 and, collectively, these power nodes accounted for 1.4 million square feet of new, big-box retail space, or 22.2% of the total growth. In comparison, the areas associated with Toronto's traditional retailing – street fronts and shopping centres experienced more limited sales growth of just 6.8%.

This section has examined the competitive pressure that big-box retailing has placed on the retail strips in Toronto over the 1994–97 period. By using the annual inventories of street retailing collected by the Centre for the Study of Commercial Activity, the *number* and *probability* of store closures for each of the retail categories in direct competition with big-box retailers were calculated for the three-year period (Table 2.7). Over the three-year period, the number of retail categories in competition with the street retailers increased from five (supermarkets, electronics, office products, sporting goods, and toys) to eight with the introduction of Home Depot (hardware), Chapters (books), and PetSmart (pet stores) into the competitive mix in the Toronto market. As Table 2.7 shows there was a 16.1% chance that retailers operating on street-front locations in Toronto in these eight categories would close. The sectors with the highest store closure rates were electronics, toys and sporting goods. Those locations within 4 kilometres of a competing big box experienced the greatest impact with a closure rate of approximately 17%. The probability of failure exhibited a slight distance decay effect declining to 14% when the nearest big-box competitor was located more than 4 kilometres away. However, what was striking was the broad areal extent of big-box retailers over space. In aggregate, the competitive effect of big-box retailers, with their large trade areas, is distance insensitive. One location can affect the entire urban system. Among the activities that did exhibit some distance sensitivity were hardware and books, while the toys category exhibited the most homogeneous spatial impact. It is of interest to note that big-box developments in the GTA occurred primarily in the outer suburbs in the early 1990s and have subsequently 'infilled' to the edge of the downtown core. During the last few years, big-box developments, and the grouping of these to form 'power centres' have significantly impacted traditional strip retailers and shopping mall tenants alike.

Over the three years for which data were available, street retailers in the eight categories in direct competition with the selected big-box retailers experienced a closure rate of 16.1%. This figure compared to a 14.8% closure rate for all other retail categories. When the data is analysed on a yearly basis there appears to be a competitive adjustment to the big-box phenomenon. In the initial years of competition, the closure rate in a category is high as the less competitive retailers are forced out of business. This adjustment phase appears to have taken place in a variety of categories – especially food, office products and toys. Furthermore, the competitive pressures in some categories (e.g., electronics) appears to be always high – a function of the dynamic nature of the product line in these retail categories. A number of forces tend to relate to the impact of big-box retailers on local retail communities – time, distance, product category and the health of the retail economy.

2.4 Market Share of a Destination Retailer

A more typical application of spatial analysis in retailing has been the assessment of market areas. Various indicators are normally used including: market share, market penetration and market opportunity. Many of these applications have been incorporated in various software packages (e.g., Huff, 2000). In this case, the market area of a major destination is examined using GIS technology (normally through 'buffer' and 'overlay' features). The ability to visualise areas or market concentration, the spatial variability of the expenditure surface and the distance decay effects provides a wealth of information to the retailer.

In Figure 2.1, the typical map of market penetration around the destination retailer is presented. Here, a highly variable market coverage is presented, with obviously the highest concentration of retail expenditures coming from areas directly north and north-west of the site. However, pockets of high expenditures are found throughout the trade area, located in both areas that are associated with high income and/or adjacent to an expressway, thus providing greater accessibility to the retail location. In assessing this map, typically one attempts to seek relationships between the levels

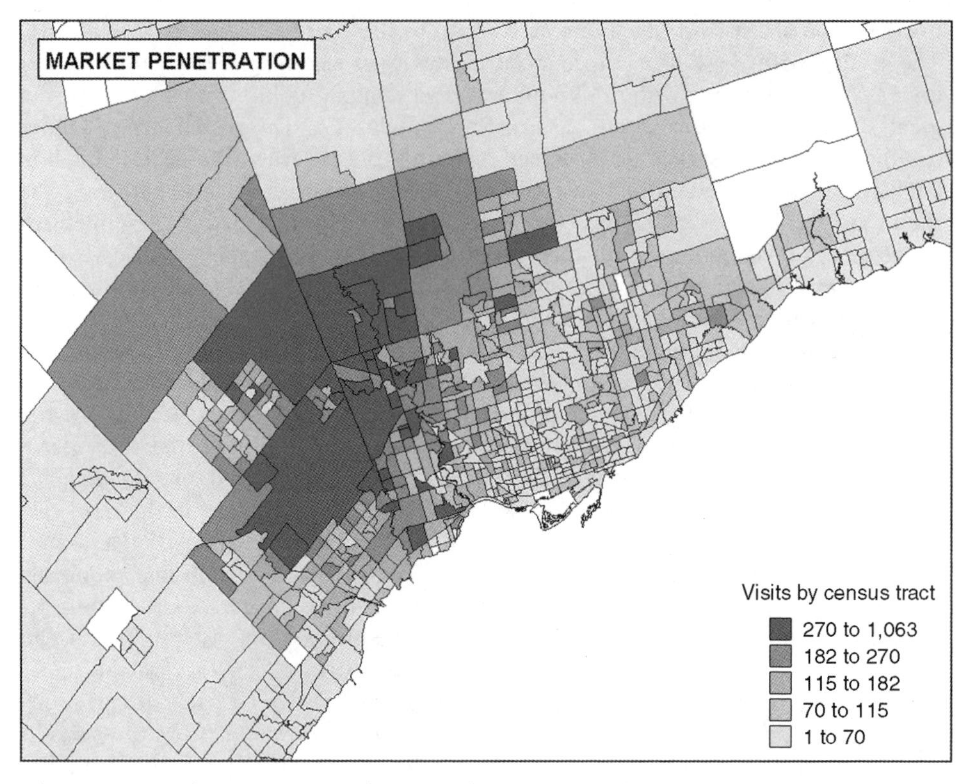

Figure 2.1 Market penetration of a major destination retailer

of market penetration, retail expenditure and the socioeconomic characteristics of the population. If strong associations are found, this information can be used in various marketing campaigns and can be used to assess the market potential of future locations.

Figure 2.2 uses the same data to develop a three-dimensional map of the trade area. This technology is being increasingly used by North American retailers to assess the spatial extent of their markets. When compared with the two-dimensional version, this representation provides much greater insights into the nature of the trade area. Clear distance decay effects become obvious and the importance of the immediate trade area becomes apparent. The value of this information is increased significantly when data are presented on a monthly or annual basis. Spatial trends become readily observed and these can be linked to a variety of corporate marketing strategies or can help assess various exogenous changes (e.g., changes in the economy or new competitive entries). The ability to incorporate simple data visualisation technologies into GIS has provided a major advance into the use of spatial mapping for many organisations and once again places more pressure on the need to collect and maintain increasingly detailed and current spatial databases (see also the case studies in Chapter 3).

Figure 2.2 *Three-dimensional trade area representation*

2.5 Evaluating Mall Dynamics

One area that has experienced relatively little attention in the literature has been the modelling of tenant performance within a shopping centre (Yeates *et al.*, 2001). This is somewhat suprising given the level of investment in these properties and the size and nature of the databases that are collected by shopping centre developers and ownership groups. Mall management is a complex process. It requires that successful shopping centre developers are able to respond to and anticipate change effectively. This reality requires the use of sophisticated management systems and the ability to mine and effectively analyse a large array of information. One means of addressing these needs is to incorporate spatial support systems into their day-to-day operations.

This application examines the development of a spatially referenced management system for a regional shopping centre. This prototype system was developed in the mid-1990s (Jones *et al.*, 1995). It linked the shopping centre developer's operational databases, market area demographics, the distribution of competing facilities and the road network. The objective of the system was to provide the shopping centre management group with the ability to examine in 'real time' the performance of the shopping centre at various spatial scales. It provided the management with a means of

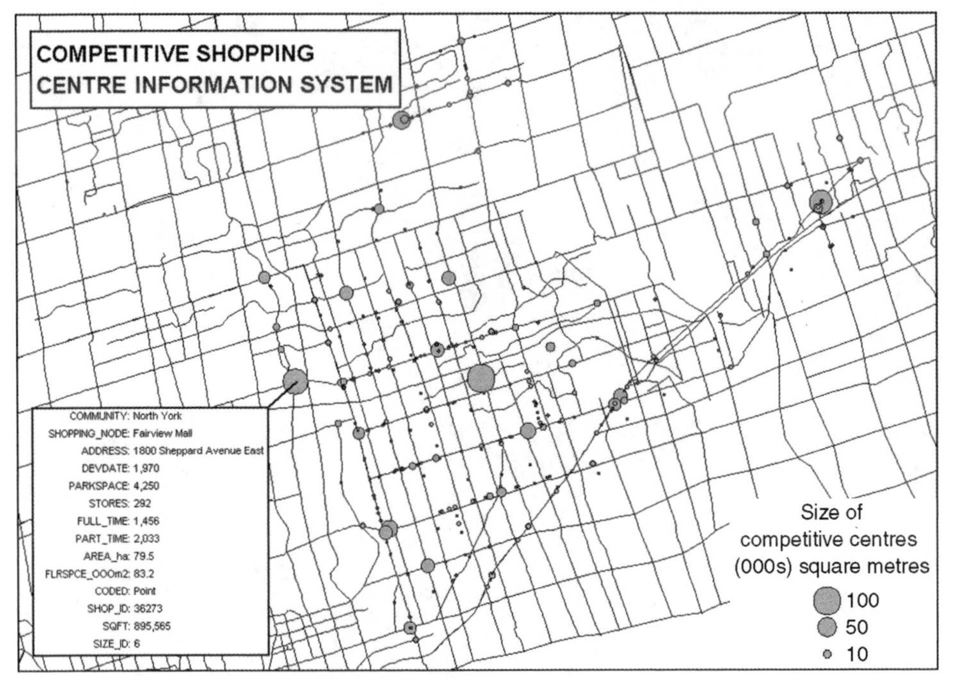

Figure 2.3 *The competitive retail system*

generating quick answers to a number of fundamental strategic management concerns. These included:

- the characteristics and changing composition of the trade area;
- areas of over and under-penetration;
- the impact of mall advertising and mall promotions;
- the location and impact of the competition;
- changes in tenant performance on a monthly and year over basis;
- the dynamics of the tenant mix;
- the effect of new tenants on mall performance;
- identification and measurement of 'mall dead spots' and vacancy rates; and
- evaluation of tenant turnover rates.

The following maps provide an illustration of the system. In Figure 2.3, the competitive retail environment is shown. With this database, the management could quickly link to supporting data that would provide a series of measures of each competitive facility in the primary trade area. These measures included such variables as date of development, number of stores, gross leasable area, number of parking spaces, employment vacancy rates, turnover rates, tenant mix and total retail sales generated in competitive retail nodes. This supply-side database was supported by a demand-side database that captured various demographic data at the enumeration

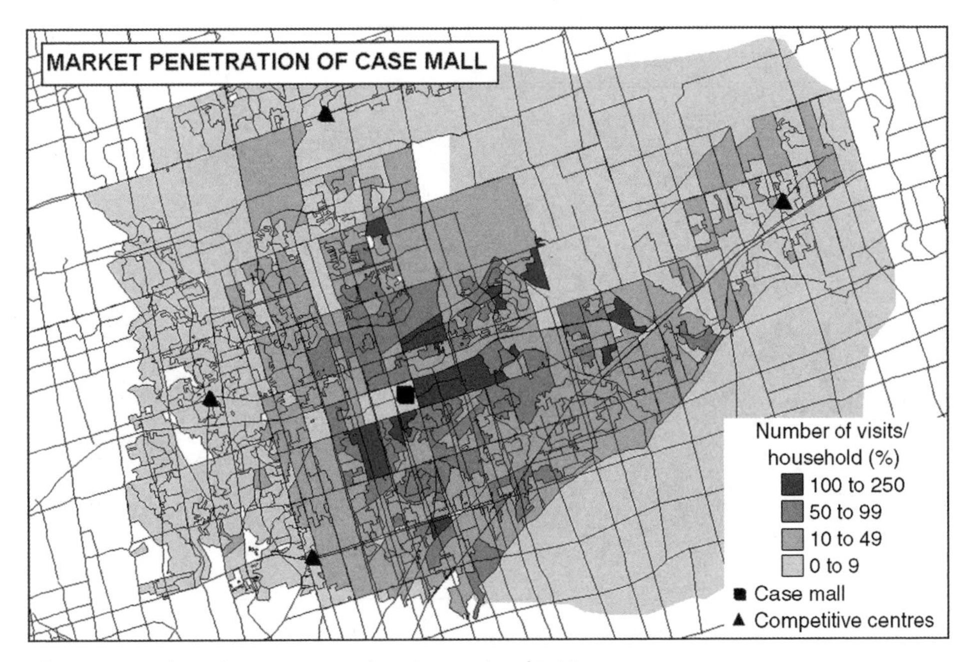

Figure 2.4 *Shopping centre market penetration by EA*

level. This data could be used when combined with customer-spotting data to (i) evaluate mall penetration by demographic segment and (ii) identify market segments that were over or under-served by the mall (Figure 2.4).

The principal advantage of the system was to develop a mechanism to monitor the performance of tenants within the mall. A GIS was developed that integrated the mall and tenant floor space with the historic performance data 'spatial record keeping' and provided mall management with a powerful, quick response analytical tool. It addressed a long-standing problem that too often shopping centre developers had no cost-effective means of systematically tracing the historical evolution of the tenant mix, structural changes or performance histories of their various properties at any spatial scale. By incorporating a GIS-based management system into their operations, shopping centre managers have a spatially based decision support system that can be applied to track and model the dynamics of tenant and mall performance. Figures 2.5, 2.6 and 2.7 illustrate the use of this system. In Figure 2.5 variations in tenant performance on an annual sales per square foot basis is depicted for a two-level mall. The map clearly shows areas of high and low performance and identifies the potential locations that should be performing at a higher level. In Figure 2.6, changes in annual sales performance are shown. Once again this system, when applied and tracked on a monthly basis, can provide mall managers with a diagnostic tool with respect to isolating those tenants that are unproductive, and more generally, identifying areas of the mall that are either gaining or losing sales. What typically emerge are well-defined spatial patterns that may reflect the importance of mall anchors, the decline or growth of certain chains, or changes in consumer

Figure 2.5 *Sales per square foot analysis*

Figure 2.6 *Tenant percentage sales change, 1994–97*

Figure 2.7 *Standardised tenant performance*

behaviours. In Figure 2.7, a more refined view of tenant performance was developed. Here, each tenant's sales performance is evaluated relative to its product category. This approach normalises the data, otherwise food court tenants would always outperform fashion tenants simply on their ability to generate much higher sales per square foot. The use of this tenant performance index provides a standardised view of mall performance and clearly identifies the best performing tenants and the hot and cold spots within the shopping centre.

2.6 Longitudinal Sales Volumes for a Retail Chain

The final application focuses on the development of data visualisation systems to analyse and track sales volume change for a retailer operating a large network of stores across Ontario. The company in this example was faced with an ever-increasing volume of transactional and operational data, in particular, weekly sales by store by product category (amounting to a substantial spatial–temporal retail performance database). The challenge for the company was in turning vast amounts of data into valuable insight and knowledge. The solution was found in the use of a geovisualisation approach using Silicon Graphics Inc.'s MineSet software and hardware.

Geovisualisation (also referred to as visual data mining) is a process of selecting, exploring and modelling large amounts of spatial data to uncover previously unknown patterns of data for competitive advantage. As MacEachren and Kraak (2001, p. 3) define, 'geovisualization integrates approaches from visualization in scientific computing, cartography, image analysis, information visualization, exploratory data analysis and geographic information systems to provide theory, methods and tools for visual exploration, analysis, synthesis and presentation of geospatial data'. There is an extensive literature within the field of data visualisation and mining (Slocum, 1999; Lloyd, 1997; Brown *et al.*, 1995; MacEachren, 1995). The traditional focus of data mining and visualisation research has been in fields such as biotechnology, engineering and medical sciences. Geovisualisation, however, is an emerging research area (MacEachren and Kraak, 2001; Slocum *et al.*, 2001).

By applying visualisations and data-mining techniques, retailers can fully exploit their data warehouses and associated large-scale relational databases, to gain a greater understanding of the markets in which they operate. By reducing complexity, encouraging model interpretation, and easily depicting multidimensional data, the visual paradigm empowers retail decision makers and potentially reduces the time and effort required to gain valuable insight from reams of data. With large amounts of data collected by retailers it is now possible to shift away from *a priori* hypothetico-deductive forms of modelling toward inductive (or querying) approaches in which models are deciphered from the data themselves.

Visualisation facilitates the development of dynamic interactive decision support tools that are a means for data exploration and provide immediate feedback to the decision maker (user). As Slocum *et al.* (2001, p. 17) note, 'developments in hardware and software have led to (and will continue to stimulate) novel methods for visualizing geospatial data'. For example, data animation can be used to depict trends and patterns by expressing how critical attributes change over key variables such as time and space. Geovisualisation tools can also depict business data using three-dimensional visualisations, enabling users to explore data interactively and discover meaningful new patterns quickly. Moreover, animated three-dimensional landscapes take advantage of a human's ability to navigate in three-dimensional space, recognise patterns, track movement and compare objects of different sizes and colours.

The use of GIS mapping is now commonplace in retail organisations in North America. GIS-derived maps are typically depicted as abstract two-dimensional plan views, such as a choropleth map of market penetration, that is viewed from directly overhead and represents data values through colour or shading, with vision the primary means of acquiring spatial knowledge (Slocum and Egbert, 1993). Three-dimensional mapping potentially assists in identifying spatial trends, intensity and variation in data. There have been a number of studies that have focused on the utilisation and potential for three-dimensional mapping (Ledbetter, 1999; Haeberling, 1999; Kraak, 1994). The market share application presented earlier in the chapter clearly illustrated the potential benefits from viewing a map in three dimensions, with spatial patterns accentuated when compared with traditional two-dimensional choropleth representation (for example, compare the visual impact of Figures 2.1 and 2.2).

Often, data sets are just too complex for representation in two or even three dimensions. By animating the display across user-defined independent variables, users can easily observe trends in extremely complex data sets. With animation, the user has the ability to discover trends, patterns and anomalies in data. The term 'dynamic representations' refers to displays that change continuously, either with or without user control. Dynamic representation has changed the way users obtain and interact with information across the full range of display technologies (Andrienko and Andrienko, 1999; Koussoulakou and Stylianidis, 1999; Bishop *et al.*, 1999; Blok *et al.*, 1999; Acevedo and Masuoka, 1997; DiBiase *et al.*, 1992; Dorling, 1992). One form of dynamic representation is the animated map, in which a display changes continuously without the user necessarily having control over that change. An argument for utilising animation is that it is natural for depicting temporal data because changes in real-world time can be reflected by changes in display time. In addition, dynamic representations also permit users to explore geospatial data by interacting with mapped displays, a process sometimes referred to as direct manipulation (Slocum, 1999).

Plate 3 provides a sequence of temporal snapshots from an animation of weekly sales data for a five-year period (1996 to 2001) that was developed by the CSCA for an Ontario-based retailer. The geovisualisation software allowed the analyst to define variables to be mapped, the time period, speed of animation, and to navigate the map area interactively whilst viewing the data animation (i.e., utilising a dynamic map representation). This provided the opportunity to search for spatial patterns and trends over a range of 'critical' operating periods during the five-year animation sequence. For example, analysing the weeks (year-over) in the run up to and after the peak December–January sales period. Plate 3 illustrates the significant effect of sales in December 2000 – the retail millennium factor – with sales dramatically increasing in December 2000 (far more than would be expected in a typical December trading period). Spatial–temporal variations in total sales provided a base level of information. The geovisualisation approach was also used to analyse spatial variation by product category over time, providing insight for merchandise-mix decisions and local area promotional campaigns. Studies to date on the merits of three-dimensional mapping and animation vary significantly, some advocate such techniques, while others highlight their limitations (Morrison *et al.*, 2000; Robertson *et al.*, 1999; Openshaw *et al.*, 1993; Slocum *et al.*, 1990). In this example, the adoption of a geovisualisation approach provided additional insight into total and category sales performance across a network of stores in Ontario, revealing patterns that would otherwise have remained hidden within the data.

2.7 Conclusions

The five examples presented in this chapter provide insights into the current use of GIS for retail applications within a North American context. They have detailed the use of spatial data and associated technologies at a variety of levels of sophistication, and reflect the broad spectrum of GIS application. These include, at the simplest level, traditional market mapping and spatial inventory reporting. Increasingly more organisations are applying a spatial approach to augment their decision support activities, for example, assessing the impact of new retail formats, evaluating retail dynamics and developing spatial management systems. More recently, the potential use of data mining and geovisualisation are being evaluated by some major retail corporations. However, what is apparent is the increasing adoption and awareness of the value of GIS by more retail organisations, but the potential benefits to be gained

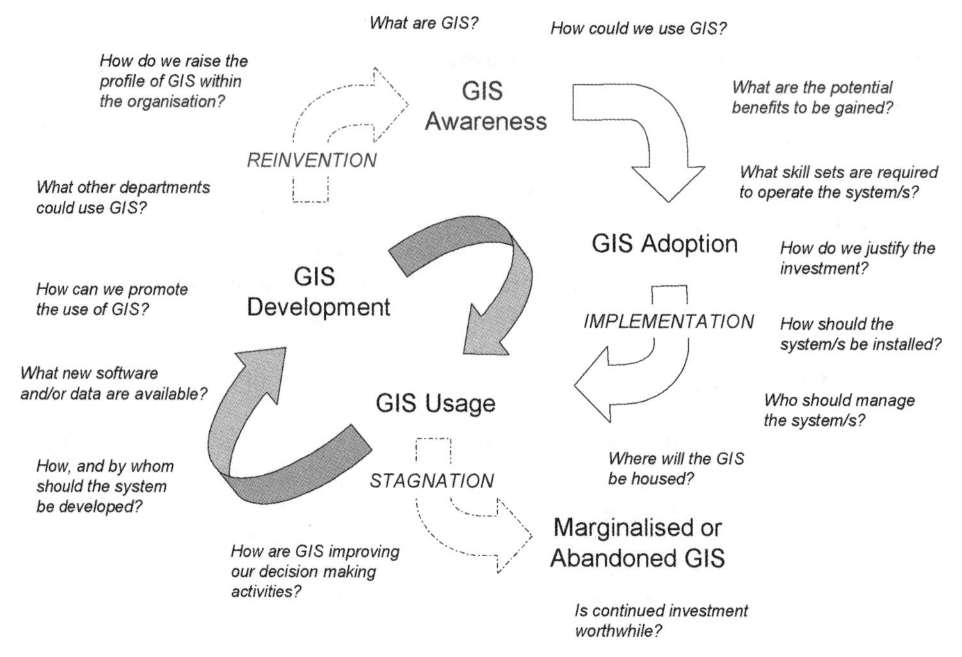

Figure 2.8 *GIS adoption history in retail organisations*

from using GIS technologies and spatial modelling are not always easily recognised. Advocates of these spatial approaches within organisations are still required if various applications of spatial models are to become common fixtures in the retail sector (Hernandez, 1999). Our assessment of the use of GIS and associated technologies is presented in Figure 2.8. In order for GIS to be successfully introduced, operated and embedded within a retailer's decision support systems, the technology requires careful planning, monitoring and management. The decision-making reality for many businesses remains, and March (1991) has summarised that they: (a) gather information but do not use it; (b) ask for more and ignore it; and, (c) gather and process a great deal of information that has little or no relevance to decisions. Traditional methods of spatial data analysis have not facilitated interactive exploration of data, and served to isolate the decision maker as the recipient of static information as opposed to an analytical actor gathering information for knowledge creation. Technological developments have enabled the development of new 'visual' spatial approaches to decision support that aim to harness the intuitive cognitive powers of decision makers and, more generally, promote the adoption and use of GIS.

If the spatial approach is to become a central element in the decision systems of major retail organisations a number of issues must be addressed. Foremost, GIS technology and spatial analysts must have access to and utilise 'real-time' data. Retail organisations must react with increasing speed to changes in the retail economy. As a consequence, the ability to integrate current data (whether it relates to sales levels, competitive effects or reliable census estimates) into the decision making of the organisation, will have a major impact on the future adoption of GIS. In addition, there must be a conscious effort to link current GIS software with more sophisticated statistical models. Without this ability the value of GIS to any organisation will be limited, as indicated further in Chapter 3 of this volume. There are obvious extensions of spatial analysis and GIS that can be developed in areas that related to data visualisation and data mining. Here, the opportunity exists for spatial analysts to distil and summarise the large databases that relate to customer sales that reside in most retail organisations but are typically not analysed in any systematic manner. Finally, the incorporation of satellite imagery and the potential of GPS-based data to improve the precision of our spatial data is another area that offers potential for the future. However, what offers the greatest source of optimism is the growing awareness of GIS in the retail industry. The term GIS is now recognised by most retail organisations. Courses in business geographics are being incorporated into many university and college business courses and business geomatics has become a recognised field of enquiry.

Glossary of Key Terms

Big box – a large-format retailer operating with depth and breadth in a well-defined merchandise category, e.g. books, sporting goods and home improvement.

Power centre – two or more big boxes sharing the same parking facilities and tenants of the same development.

Power node – a grouping in proximity to a major road/highway intersection of at least two power centres.

Market share – a measure that captures the relative share of total sales allocated to a given retailer or retail destination.

Market penetration – a measure that captures the proportion of actual against potential customers for a given area, typically based on household or population counts.

References

Acevedo, W. and Masuoka, P. (1997) Time-series animation techniques for visualizing urban growth, *Computers & Geosciences*, **23**(4): 423–35.

Andrienko, G.L. and Andrienko, N.V. (1999) Interactive maps for visual data exploration, *International Journal of Geographical Information Science*, **13**(4): 355–74.

Bishop, I.D., Ramasamy, S.M., Stephens, P. and Joyce, E.B. (1999) Visualization of 8000 years of geological history in Southern India, *International Journal of Geographical Information Science*, **13**(4): 417–27.

Blok, C., Köbben, B., Cheng, T. and Kuterema, A.A. (1999) Visualization of relationships between spatial patterns in time by cartographic animation, *Cartography and Geographic Information Science*, **26**(2): 139–51.

Brown, J., Earnshaw, R., Jern, M. and Vince, J. (1995) *Visualization: Using Computer Graphics to Explore Data and Present Information*, Wiley, New York.

Business Geographics (1993–2001) Fort Collins, Colorado: GIS World.

DiBiase, D., MacEachren, A.M., Krygier, J.B. and Reeves, C. (1992) Animation and the role of map design in scientific visualization, *Cartography and Geographic Information Systems*, **19**(4): 201–14, 265–6.

Dorling, D. (1992) Stretching space and splicing time: from cartographic animation to interactive visualization, *Cartography and Geographic Information Systems*, **19**(4): 215–27, 267–70.

Flowerdew, R. and Goldstein, W. (1989) Geodemographics in practice: developments in North America, *Environment and Planning A*, **21**: 605–16.

Goodchild, M.F. (2000) The current state of GIS and spatial analysis, *Journal of Geographical Systems*, **2**: 5–10.

Haeberling, C. (1999) Symbolization in topographic 3D maps: conceptual aspects for user-oriented design, in *Proceedings of the 19th International Cartographic Conference*, Ottawa, Canada, Section **7**: 62–9.

Hernandez, T. (1999) GIS in retailing in Canada, the UK and the Netherlands, *Research Report 1999-10*, Centre for the Study of Commercial Activity, Ryerson University, Toronto.

Hernandez, T. and Biasiotto, M. (2001) The status of corporate retail planning in Canada, *Research Report 2001-3*, Centre for the Study of Commercial Activity, Ryerson University, Toronto.

Huff, D. (2000) *Huff's Market Area Planner*, DataMetrix Inc., Tulsa, Oklahoma.

Jones, K., Pearce, M. and Biasiotto, M. (1995) The management and evaluation of shopping centre mall dynamics and competitive positioning using a GIS technology, *Journal of Shopping Center Research*, **2**: 49–82.

Koussoulakou, A. and Stylianidis, E. (1999) The use of GIS for the visual exploration of archaeological spatio-temporal data, *Cartography and Geographic Information Science*, **26**(2): 152–60.

Kraak, M. (1994) Interactive modelling environment for three-dimensional maps: functionality and interface issues, in MacEachren, A.M. and Taylor, D.R.F. (eds) *Visualization in Modern Cartography*, Pergamon, Oxford, pp. 269–85.

Ledbetter, M. (1999) 3-D visualization helps solve real-world problems, *GEOWorld*, **12**(9): 52–4, 56.

Lloyd, R. (1997) *Spatial Cognition, Geographic Environments*, Kluwer Academic Publishers, Dordrecht.

MacEachren, A.M. (1995) *How Maps Work: Representation, Visualization, and Design*, The Guildford Press. New York.

MacEachren, A. and Kraak, M. (2001) Research challenges in geovisualisation, *Cartography and Geographic Information Science*, **28**(1): 1–11.

March, J.G. (1991) How decisions happen in organisations, *Human Computer Interaction*, **6**(2): 97–117.

Morrison, J.B., Tversky, B. and Betrancourt, M. (2000) Animation: does it facilitate learning?, in *AAAI Spring Symposium on Smart Graphics*, Stanford, California, pp. 53–9.

Openshaw, S., Waugh, D. and Cross, A. (1993) Some ideas about the use of map animation as a spatial analysis tool, in Hearnshaw, H.M. and Unwin, D.J. (eds) *Visualization in Geographical Information*, Wiley, Chichester, pp. 131–8.

Reid, H.G. (1993) Retail trade, in Castle, G.H. (ed.) *Profiting from a Geographic Information System*, GIS World Inc., Fort Collins, Colorado.

Robertson, G.G., Card, S.K. and Mackinlay, J.D. (1999) Information visualization using 3D interactive animation, in Card, S.K., Mackinlay, J.D. and Shneiderman, B. (eds) *Readings in Information Visualization: Using Vision to Think*, Morgan Kaufmann, San Francisco, pp. 515–29.

Slocum, T.A. (1999) *Thematic Cartography and Visualization*, Prentice Hall, Upper Saddle River, New Jersey.

Slocum, T.A. and Egbert, S.L. (1993) Knowledge acquisition from choropleth maps, *Cartography and Geographic Information Systems*, **20**(2): 83–95.

Slocum, T.A., Blok, C., Jiang, B., Koussoulakou, A., Montello, D., Fuhrmann, S. and Hedley, N. (2001) Cognitive and usability issues in geovisualisation, *Cartography and Geographic Information Science*, **28**(1): 61–75.

Slocum, T.A., Robeson, S.H. and Egbert, S.L. (1990) Traditional versus sequenced choropleth maps: an experimental investigation, *Cartographica*, **27**(1): 67–88.

Yeates, M., Charles, A. and Jones, K. (2001) Anchors and externalities, *Canadian Journal of Regional Science*, **24**(3): 456–84.

Yeates, M. (2001) Business/retail geomatics: a developing field, *Canadian Journal of Regional Science*, **24**(3): 375–86.

3

Using Spatial Models to Solve Difficult Retail Location Problems

Mark Birkin, Graham Clarke, Martin Clarke and Richard Culf

Abstract

This chapter looks at the experience at the University of Leeds of applying spatial interaction models in a number of retail business scenarios. The argument is that retail conditions in the real world present a number of important challenges for our models. In some cases, the existing models are sufficiently robust to deal with these challenges. In other cases, we need to refine the models so that they capture more complex types of consumer behaviour. Three case study clients are presented here. The business concerns of each vary considerably. The chapter thus moves from the classic use of the models for new store openings, through to refinements associated with modelling flows and impacts of branches within shopping centres, dealing with elastic demand, incorporating shopping from the workplace and finally, through to an alternative spatial modelling framework to handle very complicated consumer markets such as petrol retailing. Hence, we conclude that applied modelling is as likely to enrich our theoretical development work as it is simply to demonstrate the usefulness of our existing methods. The link between applied and theoretical work is thus much stronger than many researchers appreciate.

3.1 Introduction and Client Needs

Spatial interaction models have a rich tradition in the geographical literature on retail market analysis. In recent years, there has been increasing commercial interest in the

Applied GIS and Spatial Analysis. Edited by J. Stillwell and G. Clarke
 ISBN: 0-470-84409-4

application of these techniques within real business environments. Such was the commercial interest in models developed by the University of Leeds that a unique university company (GMAP) was formed to professionalise the links between academia and the outside world. The range of applications and models used has been described elsewhere (Clarke and Clarke, 2001; Birkin and Culf, 2001; Birkin *et al.*, 1996, 2002a). The objective of this chapter is to reflect on the experience of developing and transferring models from an academic into a commercial environment. We will describe some of the technical developments which have been found necessary to make the models work in the real world, and to review, with examples, the degrees of success which have been achieved. It will be shown that in most cases, the models require significant extension of the basic concepts. However, in certain cases, we have concluded (reluctantly) that no amount of re-engineering can tailor the modelling approach to the requirements of difficult retail markets. In these cases, we believe that a multivariate approach to rating sites is the most sensible strategy. An alternative modelling framework of this type is also described and evaluated.

To explore these issues, we introduce three clients from the range of many that have worked with the School of Geography, University of Leeds and with GMAP (see Section 3.2). Each client had/has very different requirements in terms of store location research. Two of these are drawn from the financial services industry. The first client is the State Bank of South Australia. This is an example of the most straightforward use of spatial interaction modelling: the desire to use the models to evaluate a number of what-if proposals concerning new branch openings. This application is described in Section 3.3. The second client is a leading financial organisation based in the north of England whose spatial analysis requirements were built around two major issues. The first issue was the impact of new branch openings on its existing client base. This involved a much greater consideration of the success of the models for impact assessment. Second, was the desire to locate a network of automatic telling machines (ATMs) in good or 'optimal' locations. There was also a third concern regarding the evaluation of potential mergers and acquisitions from the perspective of the combined branch network that might be inherited: however, this is dealt with in detail elsewhere (Birkin *et al.*, 2002b). These objectives require a much more critical look at the traditional spatial interaction models typically used for such analyses. The issues these applied problems raised are addressed in Section 3.4. Finally, in Section 3.5, we review an area deemed to be extremely difficult to model – namely the petrol market. Here, the client is Exxon/Mobil. The switch from a traditional modelling approach to a multivariate approach is discussed with reference to the very difficult issues that the petrol market brings to spatial interaction modelling.

3.2 Background to Commercial Modelling

For 20 years, GMAP Limited has successfully applied geographical modelling tools within a business planning context. Clients of GMAP include retail organisations in the broadest sense, from supermarkets (Asda-Walmart, Iceland), high street multiples (Dixons, W.H. Smith) and out-of-town retailers (IKEA, MFI), to travel agents (Thomas Cook), petrol companies (BP, Exxon/Mobil), banks (Halifax, Alliance & Leicester) and automotive manufacturers and distributors (Ford, Toyota, Volvo, Mazda). The intellectual foundations of this applied research programme reside in the 1960s and 1970s, when gravity and spatial interaction models found widespread application in transportation and land-use planning, and in planning assessments of retail impact analysis (e.g., Boyce *et al.*, 1970, for examples of the former; Foot, 1981, for a review of the latter: see also Clarke and Stillwell, this volume). At this time, however, there was little or no commercial awareness or interest in these models.

Interest in commercial applications of spatial interaction models was more obvious in the 1980s and early 1990s for a number of reasons. From a retail point of view, markets were increasingly competitive, and tending towards saturation, making profitable new sites increasingly difficult to find. From a methodological perspective, the increased availability of data (about customers, markets and competitors), of PCs to process and distribute the data, and of geographical information systems (GIS) for the analysis of data, were all factors providing an impetus to market analysis. These factors were supplemented by an economic boom in the 1980s which led many UK companies to want to expand their networks. This increased interest in modelling was manifest in a number of ways:

- provision of spatial interaction modelling capabilities as standard within GIS systems (e.g., ARC/Info);
- development of specialist market analysis software (e.g., InSite, Environ, MicroVision) by business consultants; indeed, some of these business consultants were new to the scene, specialising in spatial modelling (e.g., GMAP, Geobusiness Solutions); and
- creation of dedicated site analysis and modelling departments within major retail organisations (e.g., Tesco, Marks and Spencer, J. Sainsbury, Whitbread).

There is a reasonable amount of evidence to suggest, however, that increased commercial interest in modelling has not been associated with increased technical sophistication. Despite the increasing availability of spatial interaction models, it is clear that proprietary GIS (normally standard buffer and overlay techniques) remain the most common methodology for site location (if organisations use anything over and above gut feeling – see Hernandez and Bennison, 2000). Reynolds (1991) and Elliot (1991) describe the application of such simple catchment area analysis techniques within GIS. There does not seem much evidence to support more sophisticated use of GIS since then (see Jones and Hernandez, this volume). This is somewhat surprising to the extent that our own experience demonstrates unequivocally that substantial development work has been required to transform spatial interaction models from interesting pedagogic and academic tools into valuable commercial weapons. Similarly, it seemed to be the case that, whilst the number of packages including models was increasing, the concept of the gravity model was much misunderstood and mistrusted in the commercial world. In part this was because the models were synonymous with very aggregate versions of the spatial interaction model, and in many instances simply did not work well enough. Similarly, users often did not have the experience or expertise to be able to build models from scratch and attempt the detailed disaggregation needed to make the models operational. In the next section of this chapter, we set out the classic model before discussing some of the developments and extensions required to answer the applied problems set by the clients.

3.3 Classic Model

The classic retail spatial interaction model has the following structure:

$$S_{ij} = A_i \times O_i \times W_j \times f(c_{ij}) \tag{3.1}$$

and,

S_{ij} is the flow of people or money from residential area i to shopping centre j;
O_i is a measure of demand in area i;
W_j is a measure of the attractiveness of centre j;
c_{ij} is a measure of the cost of travel or distance between i and j; and
A_i is a balancing factor which takes account of the competition and ensures that all demand is allocated to centres in the region. Formally it is written as:

$$A_i = \frac{1}{\sum_j W_j \times f(c_{ij})} \tag{3.2}$$

In summary, the model assumes that in order to forecast the flow of expenditure from an origin zone i to a destination (retail outlet or centre) j, then one begins with a fixed demand pool at i. This demand is shared between retailers in proportion to their relative size (or other measures of attractiveness), and in proportion to their geographical proximity. Individual parameters are added to these variables so that the model can be fitted (or calibrated) to real-world data (see equations below). At the very heart of what makes spatial interaction modelling so appealing as an analysis tool is that predictions are based on the replication of customer trips between individual origin and destination pairs rather than by using statistical or analogous evidence (see Birkin *et al.*, 2002a for more details on alternative methodologies for store location analysis).

For all real-world applications this model is likely to require disaggregation. This might be by demand, i.e., different parameters of the model are appropriate for different income groups, or age groups – or supply based (a new version of the attractiveness term – see below). As we shall argue below, new trends in retail geography have caused other types of refinement to be introduced.

Before we examine these new model developments, it is useful to present an application used in a classic what-if fashion for the first of our clients to be reported here – the State Bank of South Australia (SBSA). The model was of the type shown in equations (3.1) and (3.2) but disaggregated (in standard GMAP fashion) as follows:

$$T_{ij}^{lk} = \sum_{k \in j} A_i^{lk} O_i^l W_j^{lk} \exp^{-\beta_i^{lk} d_{ij}} \tag{3.3}$$

where:

$$A_i^{lk} = \frac{1}{\sum_{jk} W_j^{lk} \exp^{-\beta_i^{lk} d_{ij}}} \tag{3.4}$$

to ensure that:

$$\sum_{jk} T_{ij}^{lk} = O_i^l \tag{3.5}$$

and:

$$W_j^{lk} = R_j^{lk} Z_j^{lk} \gamma_{l(j)}^{lk(j)} \mu_j^{lk} \chi_j^{lk} \tag{3.6}$$

where:

T_{ij}^{lk} = estimated sales of product l, for organisation k in origin zone i at destination j;
O_i^{kl} = demand for product l in origin zone i;
W_j^{kl} = attractiveness of outlet j selling product l, organisation k;
β_i^{kl} = distance decay parameter disaggregated by origin i, product l, and organisation k;
d_{ij} = drive-time from zone i to outlet j;
R_j^{lk} = attractiveness of outlet j, belonging to organisation k, selling product l;
Z_j^{lk} = size of outlet j (i.e. square feet) belonging to organisation k, for product l;
μ_j^{kl} = parameter reflecting proximity of outlet j belonging to organisation k to surrounding outlets for product l;
χ_j^{kl} = parameter reflecting operating period (i.e. months open/12) for outlet j belonging to organisation k, selling product l; and
$\gamma_{l(j)}^{lk(j)}$ = brand attractiveness of organisation k, selling product l, to customers resident in i.

Figure 3.1 shows the predicted market share of SBSA across the Adelaide Metropolitan Region. It shows that generally SBSA did have a very high market share, but as with most organisations that deliver services from fixed points, that market share varied significantly across the region. The areas of low market share clearly provide an immediate snapshot of market potential. The models were then used in a systematic fashion to explore the impacts of adding a new SBSA branch in all of these market gaps. Figure 3.2 shows the market share before and after the opening of a new branch at St Agnes. It can be immediately seen that market share improved significantly. However, as important is the total revenue predicted for the new branch, and where those sales are likely to come from (including cannibalisation from the

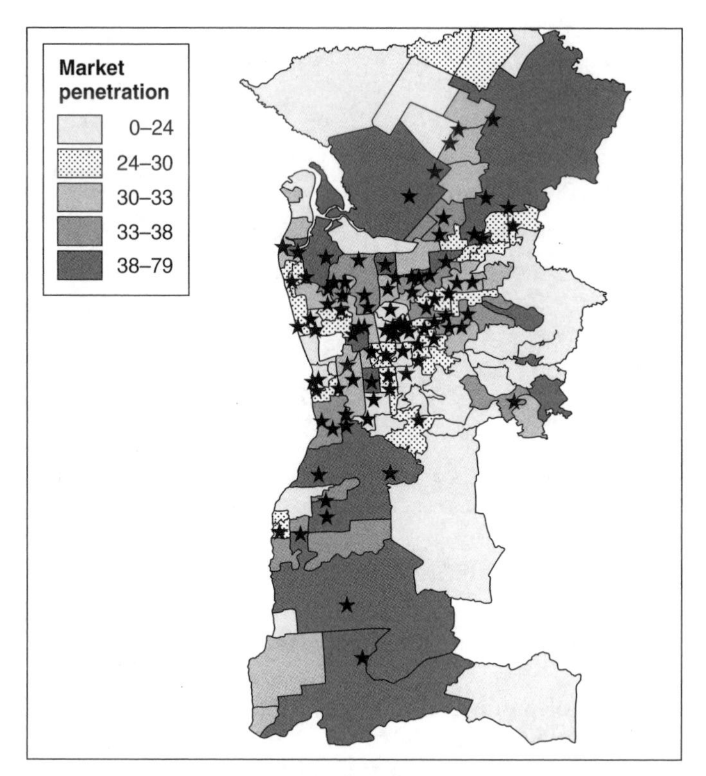

Figure 3.1 SBSA market share in Adelaide metropolitan market in mid-1990s

SBSA existing branches). Table 3.1 shows the net change in sales. It can be seen that the impacts on existing SBSA branches are not high and hence this looks like a good new location.

As we noted above, this kind of application was typical of GMAP applications in the 1980s and early 1990s. However, a more sophisticated approach to the art of store location research was required from the mid-1990s onwards. We shall explore the implications of this change in the next section.

3.4 Model Extensions

3.4.1 Introduction

The aim in this section is to explore new challenges in spatial interaction modelling, and store location research more generally, brought about by new types of application and client requirements. First, there has been an increasing concern to explore the predictive power of the models in terms of their ability to represent real deflection rates accurately. Second, there are those applications that have challenged the nature of demand: where does this come from and how can this be incorporated into

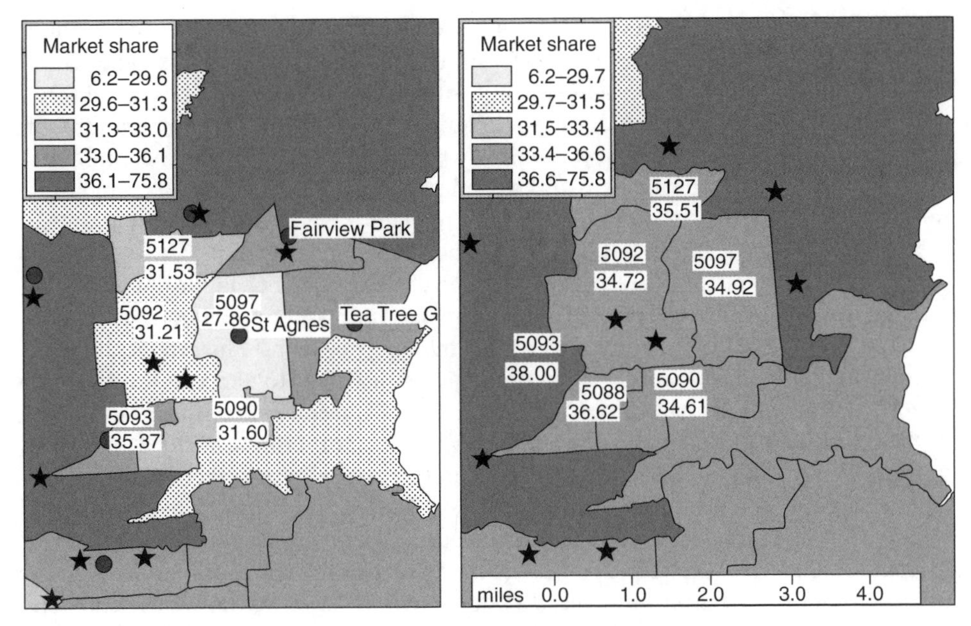

Figure 3.2 *Market shares before and after new branch opening in St Agnes*

Table 3.1 *Impacts on other branches of new branch in St Agnes*

SBSA branches	Net change in sales
Modbury Central	–13
Tea Tree Gully	–10
Clovercrest Plaza	–7
Fairview Park	–6
Golden Grove	–5
Salisbury	–5
Ingle farm	–4
Pava Hills	–3
Nailsworth	–2
Total lost to other SBSA branches	76

the spatial models? Third, there are those applications which seem to challenge the very principles of spatial interaction modelling. We shall look at each in turn.

3.4.2 Modelling Centre Deflections

One of the most interesting retail location issues over the last 20 years or so has been the decline in the number of stand-alone large stores. Although many early retail warehouses and grocery superstores occupied single-store sites the pattern since the early 1980s has been to locate within centres (warehouse parks and shopping centres often anchored by a major superstore). Similar developments are observable in non-food retail sectors: for example, the trend towards agglomerations of activity in automotive distribution – the so-called auto-mall – and the proliferation of multi-franchised outlets is transforming choosing a new car into something more akin to comparison shopping. For the retailers, whenever outlets are grouped together within city centres or retail parks, then there are potential scale economies to be obtained. In effect, these benefits may be closely related to multi-purpose trip-making, so that customers can also see benefits in visiting more than one destination within a centre or cluster. Alternatively, this may be viewed as a hierarchical choice process in which customers first choose a centre in which to shop, and only then select an outlet within that centre. This structure has been exploited and developed in the creation of the 'competing destinations' model (Fotheringham, 1983, 1986), in which a new variable is included that measures accessibility to other retail outlets. This structure is reproduced within GMAP's generic model, which is explicitly hierarchical in character, so that individual customer flows are modelled to groups of retail outlets (but the attractiveness of the group is related to the attractiveness of its component parts). Individual outlets within each group then compete for market share within the cluster. In a sense, there is also a special case of the model, in which the clusters are constituted from single outlets. This is then the traditional model, which might still be most appropriate for single-purpose trip patterns, say to stand-alone hypermarkets or schools and hospitals in the non-retail sectors.

The increasing use of centres has produced another issue of importance when dealing with classic gravity or spatial interaction models: namely, the way in which the models handle customer deflections when new stores are opened. This discovery was made through working with client B, a northern bank. For this client, financial centres were used instead of individual bank branches. That is, interactions were predicted between residential areas and financial centres (such as town centres containing four or five branches). Revenues to individual branches within the centres was then allocated according to the fascia and its store attractiveness (age of store, size, etc.). The traditional interaction model worked well for predicting revenues and catchment areas for the existing set of stores. However, for new store openings the patterns of deflections seemed intuitively wrong. Let us take the common scenario of client B wishing to open an additional branch in a large town centre where it is already present (along with a number of competitors). All things being equal, we would expect total centre sales to undergo a marginal increase in sales, but we would not expect 400 or 500 new customers suddenly to come in from surrounding towns to go to that new branch. In other words, we would expect sales for the new branch to be generated mainly from the centre's existing branches, deflecting most heavily perhaps from the existing client B branch. What we actually see from the standard model is an increase in centre sales, near incremental sales for the new branch and minimal deflections from the other client B branch in that centre.

In the 'standard' model (Table 3.2, Column 3), the introduction of a new branch has a proportionate effect on the attractiveness of the centre (via an increase in

Table 3.2 *Open a financial service branch in a large centre*

	Baseline	Standard model	GMAP model
Centre sales	18332	18782	18353
Change		450	21
Percentage change		2.4	0.1
Open branch		727	707
Existing branches	5663	5578	5288
Change		85	275
Percentage deflection		1.5	4.8

floorspace – F_j). The extent of the associated sales increase depends on the relative balance between attractiveness and accessibility. In this example the new branch generates 727 new sales, and about two-thirds of these (450) are sales which were previously made in competing centres (for higher values of beta, then we would expect to see less diversion of sales between centres as the respective catchments are more or less fixed; for lower values of beta then sales will follow floorspace, and we would expect more new centre sales). Thus, our new branch takes 450 sales away from branches in neighbouring centres, and the remainder (277) from existing branches within the same centre. The deflections within the existing centre are proportionate to the floorspace of competing financial services organisations. In the baseline, client B has a share of 5663/18 332 = 31%. Therefore client B loses 31% of 277 = 85 sales within the standard model scenario.

In the revised model (Table 3.2, Column 4), the effects are explicitly hierarchical. In the first stage, the flow to centres is modelled as a function of a basket of centre characteristics (quality of access and parking, presence of key retailers and anchor stores, pubs, restaurants and other attractors) in addition to financial services branches. Opening a new financial services branch has only a marginal effect on the character of the centre; hence the vast majority of new sales are at the expense of existing branches within the same centre. A second layer of distinction concerns the loyalty of customers to each financial services brand. Although client B can attract new customers by the addition of new branches, there are diminishing returns to scale – doubling the number of branches will not double sales, for example, because existing customer niches may become saturated. Thus, the deflections from existing branches within the estate of client B are proportionately higher than from its competitors.

In Table 3.3, we can see the same processes for a branch closure within a small centre. In this case, client B does better than expected because many of its customers choose to switch their loyalty to the remaining branches rather than defecting to an entirely new competitor.

Table 3.3 *Close a financial service branch in a small centre*

	Baseline	Standard model	GMAP model
Centre sales	4657	4157	4588
Change		500	69
Percentage change		10.7	1.5
Closed branch	558	0	0
Remaining branches	733	751	837
Change		18	104
Percentage retention		3.2	18.6

The centre effects can be expressed through a revised set of model equations:

$$T_{ij}^k = A_i^k O_i^k \theta_j^k W_j \exp^{-\beta_i^k d_{ij}} \tag{3.7}$$

where:

$$A_i^k = \frac{1}{\sum_j W_j \exp^{-\beta_i^k d_{ij}}} \tag{3.8}$$

to ensure that:

$$\sum_{jk} T_{ij}^k = O_i \tag{3.9}$$

where W_j is now a measure of the attractiveness of centre j:

$$W_j = f(X_j^1, X_j^2, \ldots, X_j^n) \tag{3.10}$$

X_j^k, $k = 1, \ldots, n$ are variables which contribute to centre attractiveness.

Flows are allocated in proportion to market share within a centre:

$$\theta_j^k = f(V_j^{k1}, V_j^{k2}, \ldots, V_j^{km}; W_j) \tag{3.11}$$

where V_j^{kl}, $l = 1, \ldots, m$ are attributes for retailer k in centre j.

Ongoing research is also examining other issues to do with impact assessment. Using a set of traditional retail models for the identification of areas poorly served by the major food retailers (so-called 'food deserts') produced a standard set of deflection rates when a new Tesco store was opened in East Leeds (Clarke *et al.*, 2002). The model predicted deflection rates of up to 60% for those stores in close proximity. However, this case study was unusual in that actual consumer deflections were collected as part of the overall food deserts project. The results of this major household survey (1000 households in the vicinity of the new store) showed more complex deflection patterns than predicted in the model (Wrigley *et al.*, 2003). Two trends were particularly noticeable. First, there was more loyalty to the discount sector than predicted by the models. It seems that some consumers felt uncomfortable in the new Tesco store – they felt more at home remaining in the discount store. Second, there was a greater switching from Tesco's other store in the region. That is, the models under-predicted the amount of switching from the rather distant centre at Crossgates (home of the other Tesco store). These issues are driving the examination of new consumer behaviour variables in the retail grocery model (Clarke *et al.*, 2003). This is another good example of the importance of applied research in terms of driving theoretical extensions of our models.

3.4.3 Incorporating Elastic Demand

A second major issue to emerge from working with client B was the concept of elastic demand. In practice, the assumption of a fixed demand pool (O_i) is not always a sufficient one. This is particularly the case when the levels of demand and supply for a product or service are interrelated. In the case of groceries, for example, it is safe to assume that customers have a reasonably fixed requirement for fruit and vegetables, meat, dairy products and so forth. These requirements are unlikely to be altered by the decision of Safeway to build a new supermarket on the doorstep. That is, although consumers may switch allegiance and frequent the new store, their average weekly budget for food is unlikely to increase significantly. On the other hand, providers of leisure services such as cinemas, sports clubs or restaurants would do well to recognise that these facilities will to some extent generate their own demand. People lacking access to a cinema, for example, will tend simply to substitute other activities, and will have a lower demand for cinema tickets than similar people who have easy access to cinemas.

Modelling frameworks for the treatment of elastic demand have been proposed by Pooler (1994) and Ottensmann (1997). The principle of these applications is that the level of demand in an area is not fixed, but is related to some measure of service provision in that area. The relationship may be regulated through the introduction of a new parameter, which can adopt some very low or zero value for activities such as the sale of retail groceries, to much larger values for a restaurant or cashpoint machine. The latter is the example emanating from client B. Figure 3.3 plots the relationship between ATM location and consumer residence in Leeds.

There are many approaches to the problem of coping with a demand base which is not fixed. One strategy is to develop a hybrid model, since the flows can no longer be strictly considered as origin-constrained. An alternative is to specify the model via a two-stage process: first, estimate the demand; second, allocate the flows. So we might start with:

$$O_i = e_i P_i w_i^{\phi} / \hat{w}_I^{\phi} \quad (3.12)$$

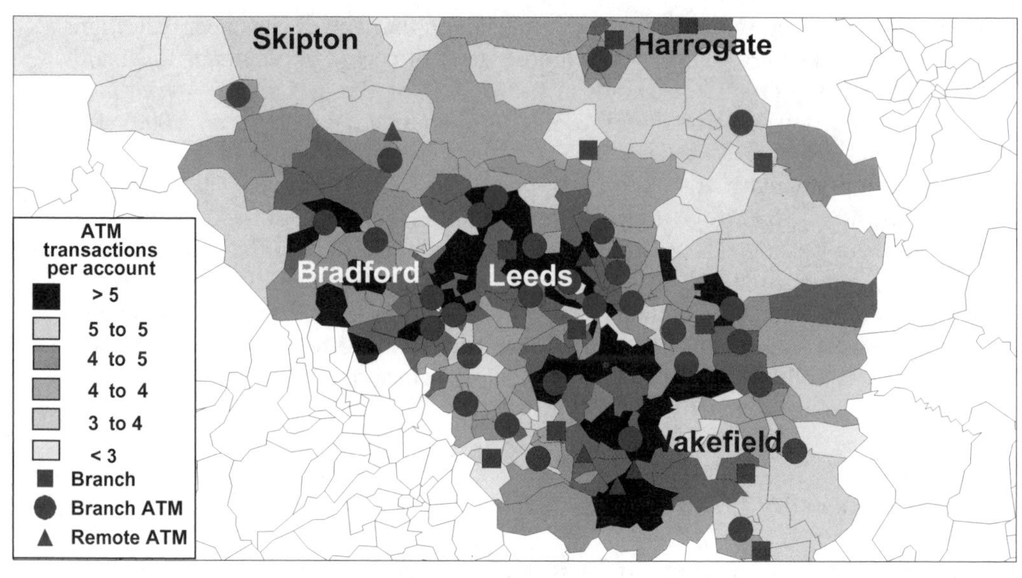

Figure 3.3 Variations in ATM usage by Leeds residents

where:

e_i = estimated sales of product l, for organisation k in origin zone i at destination j;
P_i = demand for product l in origin zone i; and
w_i = measure of access to services for residents of zone i, such as the Hansen accessibility:

$$w_i = \sum_{jk} W_j^k \exp^{-\beta_i^k c_{ij}} \tag{3.13}$$

and where:

$\hat{w}_l$ = average accessibility across the whole region; and
ϕ = is a parameter.

Having estimated the demand, we may then continue as previously with the traditional spatial model.

3.4.4 Trips from Work

Even if it is safe to assume that the demand for products and services is relatively fixed, the assumption of a simple flow from the point of consumption (residence) to the point of sale (outlet or retail centre) is an increasingly unsophisticated representation of the consumption process. To an increasing extent, retail consumption is satisfied through a complex web of multi-purpose trips, including trips to or from places of work, education or leisure.

One implication is that the demand pool needs to be segmented, at least into a portion which may be satisfied from the home, and another portion which is satisfied from the workplace. A journey to work model may be the most appropriate way of assigning customers to workplace locations since the buoyancy of retail sales within a typical city centre owes as much to the proximity of workers as to its attractiveness to suburban neighbourhoods. Our experience suggests that when analysing model outputs in the case of W.H. Smith town or city centre branches, somewhere between a quarter and a third of retail expenditure for these products was being released through the workplace rather than the residence. However, such work-based trips are also highly constrained in the geographical sense, so there will not be many trips of four or five miles undertaken within the lunch hour. There may also be an element of elasticity here between the demand components, so that householders employed within city centres may tend to spend heavily through the working week; others may be restricted to the usual evenings and weekends. This is one of the features which makes petrol forecourts so difficult to model, as we will see in the next section below. Rather than drawing from a 'captive' local market, customers will typically be intercepted at the mid-point on trips which might connect the home, workplace, shop, cinema, nightclub, holiday resort, or any other of a myriad of possible locations.

3.4.5 Conclusions

The implementation of these various model enhancements has been facilitated by the production of a combined software module incorporating a range of application building blocks. This is somewhat more complex than the basic spatial interaction model which was introduced in Section 3.2 above. Thus while a basic spatial interaction model simulation might be effected through less than a page of FORTRAN code, the 'generic model' is composed of several thousand lines of C++ code, because the model is designed to deal with many different data structures and model specifications within different markets. This latter point is illustrated in Table 3.4, which makes explicit some of the issues which we have already discussed above, i.e., that different approaches are more relevant in some circumstances than others. For example, as we saw earlier, elasticity of demand is far more important in the provision of financial services than motor vehicles.

3.5 Approaches to Modelling Difficult Markets

Some retail markets are evidently more difficult to model than others. This idea is summarised in Figure 3.4. Excellent results have been achieved in forecasting outlet revenues for retailers such as supermarkets and DIY retailers. For example, Birkin *et al.* (1996) report on an application to the DIY sector in which average forecasting errors of around 5% were consistently achieved. Of equal importance is the ability

Table 3.4 Importance of model drivers by retail market

	Supermarket	Retail	Auto	Finance	Petrol
Spatial aggregation	Fine zones	Medium zones	Coarse zones	Medium zones	Medium zones + network
Brand loyalty	Moderate/ Strong	Moderate	Moderate	Strong (transactions)	Moderate
Attraction components					
Space	5	3	4	1	4
Parking	5	1	3	1	5
Accessibility	5	4	2	3	5
Product range	4	2	4	2	1
Prices	5	5	5	5	5
Adjacencies	1	5	1	5	2
Opening hours	4	3	1	3	5
Complexity of distribution	Simple/ Moderate	Simple/ Moderate	Simple/ Moderate	Complex	Simple
Travel distance	Short	Moderate	Long	Moderate	Long
Trip type	Single purpose	Multi-purpose	Single purpose	Multi-purpose	Distress
Segmentation	Important	Fundamental	Important	Moderate	Marginal

1 = Relatively unimportant
5 = Very important

of the models to produce accurate forecasts of the impact of new store developments on an existing portfolio of stores, and to identify the geographic extent and composition of the store catchment for advertising and promotion. Certain other markets are less tractable. A notorious example is the car market. Here the issue is not so much about the ability of the models to generate accurate estimates of 'market potential', but the fact that the actual performance of dealers varies quite strongly from place to place (not least because of the persuasiveness of individual salespersons!).

In this section we draw upon experience built up from client C – Exxon/Mobil. In the petrol sector, the issues of complexity are more fundamental. Here the normal linkage between an outlet and its local market is less strong, and the range of factors which influence revenue generation is very large. That said, using standard modelling procedures had produced some impressive results in the petrol sector. Table 3.5 shows the results from a model run in Sicily. However, as can be seen, the error margins are much higher than we would expect in other sectors, and there tends to be more outliers. Those petrol stations with particularly poor model predictions turned out to be in key transient locations. To build a model which captures transient demand as well as local demand would need a full transport network model. The daily effects of price changes are also a difficult marketing activity to capture within the models. The petrol

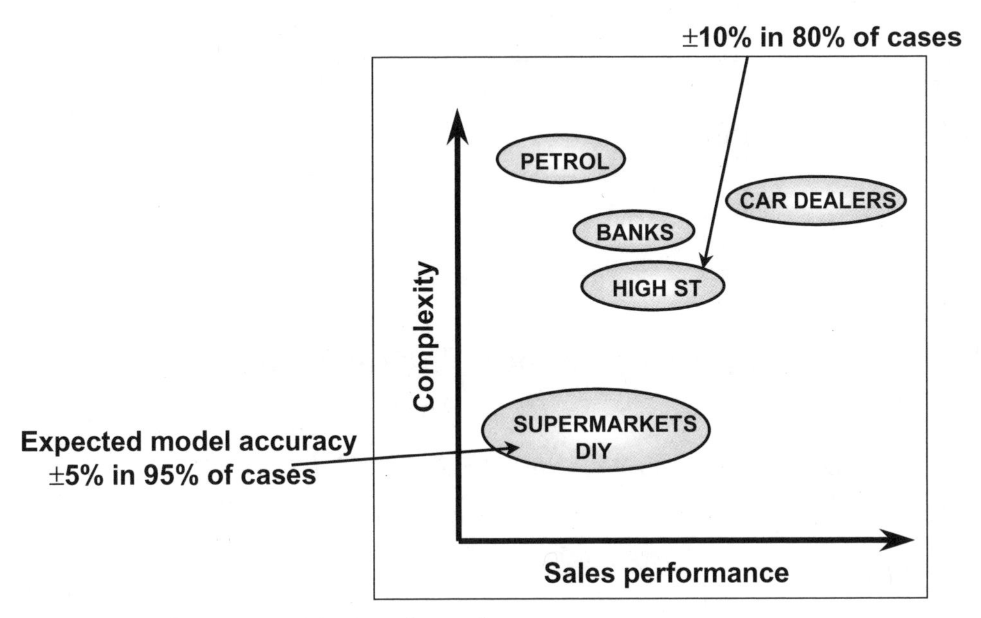

Figure 3.4 *Effectiveness of forecasts by market*

Table 3.5 *Observed versus predicted sales volumes at selected petrol stations in Sicily*

Petrol station location	Actual volume	Predicted volume	Performance index
Corso Tukory	1510	1663	0.91
Piazza Giacchery	1860	1600	1.16
Via Uditore	1002	1050	0.95
Vilae Regina Margherita	1424	1000	1.42
Corso Dei Mille	956	900	1.06
Viale Michelangelo	1089	1150	0.95
Via Aurispa	763	600	1.27
Via Sciuti	1980	2141	0.92
Niscemi	1250	1350	0.93

market is one of the most susceptible to small variations in price. Although it is relatively simple to build in a price variable to standard models (cf. Wilson, 1985), modelling the impacts of daily price changes on both local and transient demand is a more difficult issue.

In these circumstances, we have experimented with the development of multivariate statistical approaches to revenue modelling. The core of our approach is to prepare 'site ratings' for each location. There are two steps to this process: first, to identify the most important factors and to calibrate their relative importance; second, to construct individual scores for different attributes or outlet configurations. Note that while there are some similarities to straightforward regression analysis in this

Table 3.6 *Components of a site rating system*

Factor	Maximum score
Quantity	45
Quality	35
Competition	15
Concentration	5
Total	100

Summary of Site Rating System

A series of weights applied to each of the factors, reflecting relative importance.

Factor	Maximum Score	Component Weights
Quantity	45	80% Population 20% Workforce
Quality	35	50% Acorn 1-8% 10% Student% 40% Cars/Household
Competition	15	80% Weighted Comp/000 H'Holds 20% Smkts /000 Hholds
Concentration	5	100% Concentration Index
Total	100	

Score	Population	Work Force	% Acorn 1-8	% Students	Cars / H'hold (Urban)	Cars / H'hold (Rural)	Weighted Comp/000 H'Holds	S'mkts /000 H'holds	Concentration Index
1	0-3,000	0-2,000	0-10	0-2.5	1.5+	0-0.5	5.5+	0.8+	0-0.5
2	3,000-6,000	2,000-4,000	10-20	2.5-5.0	1.45-1.5	0.5-0.7	3.5-5.5	0.7-0.8	0.5-0.75
3	6,000-9,000	4,000-6,000	20-30	5.0-7.5	1.3-1.45	0.7-0.8	2.5-3.5	0.6-0.7	0.75-1.0
4	9,000-12,000	6,000-8,000	30-40	7.5-10.0	1.15-1.3	0.8-0.9	2.0-2.5	0.5-0.6	1.0-1.25
5	12,000-15,000	8,000-10,000	40-50	10.0-12.5	1.0-1.15	0.9-1.0	1.6-2.0	0.4-0.5	1.25-1.5
6	15,000-18,000	10,000-12,000	50-60	12.5-15.0	0.85-1.0	1.0-1.1	1.2-1.6	0.3-0.4	1.5-1.75
7	18,000-21,000	12,000-14,000	60-70	15.0-17.5	0.7-0.85	1.1-1.2	0.9-1.2	0.2-0.3	1.75-2.0
8	21,000-25,000	14,000-18,000	70-80	17.5-20.0	0.55-0.7	1.2-1.3	0.6-0.9	0.1-0.2	2.0-2.5
9	25,000-28,500	18,000-26,000	80-90	20.0-22.5	0.4-0.55	1.3-1.4	0.3-0.6	0.01-0.1	2.5-3.5
10	28,500+	26,000+	90+	22.5+	0-0.4	1.4+	0-0.3	0-0.01	3.5+

Scores for each factor are assigned, using each distribution, to identify appropriate cut-points

Figure 3.5 *Rating summary*

approach, and that indeed multiple regression may be used in the process of deriving the individual scores, the approach is more closely related to credit scoring (Leyshon and Thrift, 1999), and indeed to business scorecards (Ahn, 2001) in which the performance of an organisation might be evaluated against a whole raft of criteria.

Consider the example shown in Table 3.6. Here four sets of factors which drive convenience store ('C-Store') performance on petrol forecourts have been identified. Note that despite our comments above, the local trade area (or quantity of demand) is still the most important factor (generating both drive-in and walk-in traffic to forecourt shops) but accounts for less than 50% of the overall site score.

Further description of these components is provided in Figure 3.5. For example, the competitiveness of markets is assessed according to the presence of retailers of different types. Each type is in turn weighted according to its relative impact, with supermarkets having the most effect, and other forecourts the least. Summary results are shown in Figure 3.6. A typical application of this type of model would be

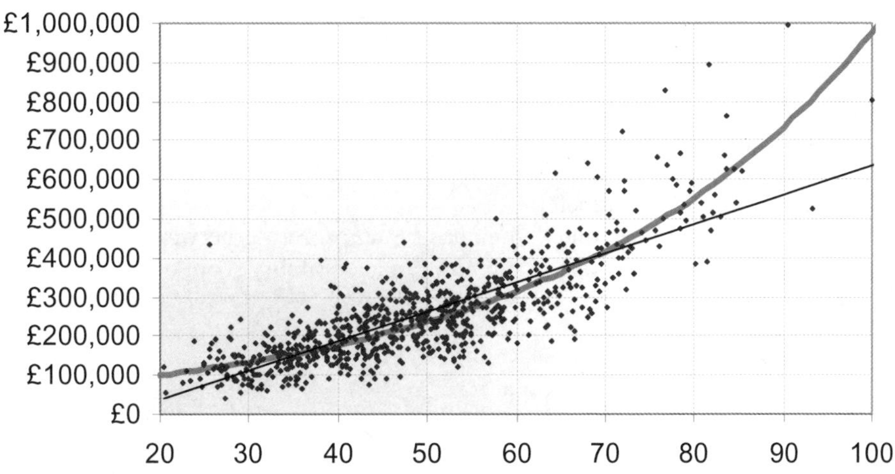

Figure 3.6 *Sample results for a site rating system*

expected to explain between 75 and 80% of the variance in C-Store performance. The tools are being extensively used to prioritise investment from a local to a global scale (i.e., not just Otley versus Ilkley, but Mexico versus Thailand!).

It may of course be possible to look for new quantitative methods to solve these difficult problems. The authors have experienced some success with genetic algorithms. These evolutionary algorithms are particularly suitable for network optimisation problems: for example, how to configure a new network of 50 car dealers or supermarkets in an Eastern European country (such problems are being faced with increasing frequency by growing retail groups in the UK and the USA). A conventional solution to problems of this type is described by Birkin, Clarke and George (1995), and a GA version by Birkin and Culf (2001). Unsupervised neural networks are also thought to be useful for problems of pattern recognition. In an abstract sense, the problem of rating sites is of this type, in which one needs to find the pattern of factors and attributes which determines outlet performance. The application of self-organising maps in the style of Kohonen (1984) (cf. Openshaw and Openshaw, 1997) has been used by GMAP to generate more effective solutions to the site rating problem described above. The weakness of this approach is that the ratings tend to be less 'clean' than those prepared through desktop analysis. In practice, a hybrid approach in which the results from some neural network-based approach are then manually tidied up has been found to be most effective (see Birkin, Boden and Williams, 2003).

Consider a simple example. We wish to rate sites according to the number of petrol pumps (two, four or six) and their accessibility (on an integer scale from one to five). The parameters to be optimised via the neural network are the importance of accessibility (relative to number of pumps), and the relative weightings between each pump size. Thus the model would be:

$$R_i = \sum_k \delta(1)_i^k a^k + \gamma \sum_m \delta(2)_i^m b^m \tag{3.14}$$

where

R_i is the 'rating' of petrol station i;
a^k is a size parameter;
b^m is an accessibility parameter;
$\delta(1)_i^k$ is a zero-one variable, one if site i has k pumps, zero otherwise;
$\delta(2)_i^m$ is a zero-one variable, one if site i has an accessibility score of m, zero otherwise; and
γ is a scalar.

An ideal solution might look like this:

$a_1 = 0.77$
$a_2 = 1.02$
$a_3 = 1.48$
$\gamma = 0.43$
$b_m = 0.48, 0.56, 0.84, 1.19, 1.21.$

Whilst this set of scores represent the best available fit to the data, it will often make sense to smooth them into a form which is slightly less accurate but intuitively pleasing and simpler to describe, for example:

Pumps – max 20 points
 2 pumps – 8 points
 4 pumps – 12 points
 6 pumps – 20 points
Accessibility – max 10 points
 1 – 3 points
 2 – 4 points
 3 – 6 points
 4 – 9 points
 5 – 10 points

In this example, if there are an equal number of sites in each category then the simple model retains 99% of the variance in the precise estimates, but involves no decimal points!

3.6 Conclusions

The application of theoretical models within an applied commercial environment has generated many extensions and enhancements. It has also posed challenging problems. To solve these problems may require imaginative extensions of an existing framework, the introduction of completely new techniques, or a recourse to old-fashioned but robust methods. A distinguishing feature of business is that novelty counts for less than performance. As the geomodelling industry continues to mature we should expect to see a greater plurality of approaches, providing demonstrable benefits to an increasingly wide portfolio of business customers.

References

Ahn, H. (2001) Applying the balanced scorecard concept: an experience report, *Long Range Planning*, **34**(4): 441–61.

Birkin, M., Boden, P. and Williams, J. (2003) Spatial decision support systems for petrol forecourts, in Geertman, S. and Stillwell, J. (eds) *Planning Support Systems in Practice*, Springer, Berlin.

Birkin, M., Clarke, G.P., Clarke, M. and Wilson, A.G. (1996) *Intelligent GIS: Location Decisions and Strategic Planning*, Geoinformation, Cambridge.

Birkin, M., Clarke, G.P. and Clarke, M. (2002a) *Retail Geography and Intelligent Network Planning*, Wiley, Chichester.

Birkin, M., Clarke, G.P. and Douglas L. (2002b) Optimising spatial mergers: commercial and regulatory perspectives, *Progress in Planning*, **58**(4): 229–318.

Birkin, M., Clarke, M. and George, F. (1995) The use of parallel computers to solve non-linear spatial optimisation problems: an application to network planning, *Environment and Planning A*, **27**: 1049–68.

Birkin, M. and Culf, R. (2001) Optimal distribution strategies: one company's experience in the application of regional science techniques to business planning problems, in Clarke, G.P. and Madden, M. (eds) *Regional Science in Business*, Springer, Berlin.

Boyce, D., Day, N. and McDonald, C. (1970) *Metropolitan Plan-making*, Monograph Series Number 4, Regional Science Research Institute, Philadelphia.

Clarke, G.P. and Clarke, M. (2001) Applied spatial interaction modelling, in Clarke, G.P. and Madden, M. (eds) *Regional Science in Business*, Springer, Berlin.

Clarke, G.P., Eyre, H. and Guy, C. (2002) Deriving indicators of access to food retail provision in British cities: studies of Cardiff, Leeds and Bradford, *Urban Studies*, **39**(11): 2041–60.

Clarke, G.P., Douglas, L. and Guy, C. (2003) The impact of new store developments on urban food deserts, *Working Paper*, School of Geography, University of Leeds (forthcoming).

Elliott, C. (1991) Store planning with GIS, in Cadeau-Hudson, J. and Heywood, D.I. (eds) *Geographic Information 1991*, Taylor and Francis, London.

Foot, D. (1981) *Operational Urban Models*, Methuen, New York.

Fotheringham, A.S. (1983) A new set of spatial interaction models: the theory of competing destinations, *Environment and Planning A*, **15**(1): 15–36.

Fotheringham, A.S. (1986) Modelling hierarchical destination choice, *Environment and Planning A*, **18**: 401–18.

Hernandez T. and Bennison D. (2000) The art and science of retail location decisions, *International Journal of Retail, Distribution and Management*, **28**(8/9), 357–67.

Kohonen, T. (1984) *Self-organisation and Associative Memory*, Springer-Verlag, Berlin.

Leyshon, A. and Thrift, N. (1999) Lists come alive: electronic systems of knowledge and the rise of credit-scoring in retail banking, *Economy and Society*, **28**(3): 434–66.

Openshaw S. and Openshaw C. (1997) *Artificial Intelligence in Geography*, Wiley, Chichester.

Ottensmann, J.R. (1997) Partially constrained gravity models for predicting spatial interactions with elastic demand, *Environment and Planning A*, **29**: 975–88.

Pooler, J. (1994) A family of relaxed spatial interaction models, *Professional Geographer*, **46**: 210–7.

Reynolds, J. (1991) GIS for competitive advantage: the UK retail sector, *Mapping Awareness*, **5**(1): 33–6.

Wilson, A.G. (1985) Structural dynamics and spatial analysis; from equilibrium balancing models to extended economic models for both perfect and imperfect markets, *Working Paper 431*, School of Geography, University of Leeds.
Wrigley, N., Warm D. and Margetts, B. (2003) Deprivation, diet and food retail access: findings from the Leeds 'food desert' study, *Environment and Planning A*, **35**(1): 151–88.

4

Location-Based Services for WAP Phone Users in a Shopping Centre

António S. Câmara and António Eduardo Dias

Abstract

Advances in mobile Internet devices together with the recent availability of location awareness for mobile phones represent new opportunities for the implementation of location-based services (LBS). In the commerce arena, LBS can have a crucial role in combining virtual and real infrastructures, thus improving the experience of visiting a commercial area like a large shopping centre. To support these synergies between real spaces and their symbolic representations, a new personal navigation technology has been developed based on mobile phones named DiWay. DiWay facilitates user navigation and bridges the gap between traditional 'bricks and mortar' commerce and electronic or mobile commerce by the use of anchor points. The most relevant of these can be memorised using the phone, a concept called Phonemarks. Furthermore it also provides security and community applications. DiWay is built upon the FluidMapping software framework that facilitates the use of heterogeneous databases and displays the results in any presentation device, including 2.5 or 3G mobile phones. The system is being deployed for one of the largest shopping centres in Europe, the Centro Colombo in Lisbon.

4.1 Introduction

Maps are symbolic representations of real spaces. The gap between reality and its symbolic representations has been shortened in recent years with digital multimedia

Applied GIS and Spatial Analysis. Edited by J. Stillwell and G. Clarke
 ISBN: 0-470-84409-4

information systems. Such systems have included maps, aerial and ground photos, video and sound. Reviews of available spatial multimedia systems may be found in Câmara and Raper (1999), Gouveia and Câmara (1999) and Câmara *et al.* (2000). Mobile devices and particularly Internet-enabled mobile phones are helping to bridge that gap further by enabling the confrontation of reality with their digital representations in real time.

Such devices are creating opportunities for new LBS that rely on having a spatial background usually in map form with geo-referenced points of interest and other relevant information. LBS also depend on knowing the position of the user in space. LBS applications for the common mobile phone user include personal and car navigation, point-of-interest search, e-commerce, emergency services and entertainment. LBS are supported by context-aware systems relying on the dissemination of contextual information (Dias *et al.*, 1998; Ryan *et al.*, 1998; Ryan, 1998; Abowd *et al.*, 1997a). Their main focus regarding context is on location, thus falling into the field of location-aware computing systems. These systems have been thoroughly discussed in Dias *et al.* (1996, 1997), Davies *et al.* (2001), Beadle *et al.* (1997, 1998), Abowd *et al.* (1997), Frisk (1998) and Pradhan *et al.* (2001). A good review of location systems for location-aware applications can be found in Hightower and Borriello (2001).

The Colombo Centre (http://www.colombo.pt) is the largest shopping centre located in the Iberian Peninsula. It has over 40 million visitors a year, 420 shops, 65 restaurants, 10 movie theatres and many other facilities. The parking lot has capacity for 6000 cars. In such a large shopping centre, guidance is needed to locate shops and products. A system that facilitates car finding in the parking lot may be also helpful. The need for systems, minimising those problems, is well understood by the companies that manage shopping centres. SONAE, the company that runs Colombo Centre, has developed paper-based diagrammatic maps and signs available in several locations to help locate stores. It has also created a system based on the parking cards and fixed machines that stamp the position of a car in the parking lot.

Both systems have shown limitations as follows:

- paper-based representations cannot be queried and are usually thrown away after use. Moreover, any change in the Colombo Centre's structure requires new representations. These two issues bring up both financial and environmental problems;
- diagrammatic maps also represent a level of aggregation not suitable to detailed information needs from the customers, which are often centred on products;
- information structure on diagrammatic maps is static which limits its use by those with specific search needs; and
- the car location system is expensive to maintain and forces the visitor to find the nearest stamping machine.

Our group has developed a navigation technology, called DiWay, based on mobile phones that overcomes these limitations and, at the same time, presents additional advantages. These include the facilitation of mobile commerce, social and entertainment activities and emergency services. DiWay is based on the use of wireless application protocol (WAP) using images and text, and the short message service (SMS) relying upon on text messages. It will also accommodate multimedia message service (MMS) sending images, sounds and text upon request or permission of the user. DiWay architecture is supported by a view of geographical information systems (GIS) as being composed by a data tier, a logic tier and a presentation tier. The data tier facilitates the access of geo-referenced data. The logic tier incorporates spatial analysis tools to determine shortest paths, near points of interest and shortest circuits. Finally, the presentation tier enables the visualisation of maps, images and text on the Web, mobile Internet, and SMS and MMS systems. DiWay may be seen thus as a gateway for mobile access to a customised GIS of a shopping centre. Early tests at the Colombo Centre, carried out in cooperation with SONAE, have shown that the system may be adapted to other shopping centres of the SONAE group. The success of that adaptation depends on minimising the burden placed on the user in future developments of DiWay.

This chapter discusses the problem underlying the need for DiWay, and the methodological issues behind DiWay such as micro-detailed geo-referencing, database development, algorithms behind navigation and user interface design. The DiWay application to the Colombo Centre is reported for illustrative purposes. Future developments of DiWay are also listed.

4.2 Problem Definition

The geo-referencing of points of interest in urban areas is becoming increasingly detailed to satisfy user needs. Maps of major cities available on the World Wide Web illustrate this trend. Shopping centres provide the setting to go one step further. Geo-referencing points of interest such as shops is not enough; one may be interested in locating shelves or even products in a shop, and certainly one is interested in having the parking places geo-referenced. The definition of spatial location, at this refined level, facilitates navigation, commerce and social activities.

Exploration of a place is usually associated with the use of symbolic representations such as maps that are then matched against the reality they try to synthesise. This matching is facilitated by the use of 'anchors': a discrete number of points of interest displayed in maps that are landmark shops or other easily recognisable locations. In a shopping centre, there are certainly shops more easily noticed than others. From the perspective of different visitors, one may conclude that every existing shop may be an anchor for a certain visitor. By the same token, every product in a shop may be a point of interest and, clearly, the same is true of every parking place in the centre.

The Colombo project has shown that the definition of anchors has been limited by the lack of the availability of back-office data to support the potential uses of the system. That data may be only location (such as in the case of the parking places), but it may also include other data such as the Web address, list of products and prices. In the end, anchors have to be defined by the shopping centre operators. Traditional maps and wayfinding signs of shopping centres are thus limited to aid generic navigation:

- they do not help a visitor to find a particular product on a shelf or remind him or her of the place he or she left the car;
- they do not help him or her in obtaining more information on a product or even where and how to buy it; and
- they are of little use in helping the visitor in meeting other visitors at a certain time in a certain place.

Mobile phones, and particularly Internet-enabled mobile phones, provide an alternative platform for navigation in shopping centres: they can display maps showing the visitor's location, offer directions and present product information. They may be used to buy products, locate friends and in emergency situations. However, the development of navigation systems for mobile phones to be used in shopping centres presents key research problems such as:

- indoor positioning, as traditional mobile phone positioning systems (based on the Global Positioning System or other radio-based triangulation systems) are not successful in indoor spaces;
- the selection of anchor point technology to provide information to the mobile phone user;
- the definition of the most appropriate spatial analysis tools to aid the users in navigation;
- the software framework to support heterogeneous data sources and present information in different platforms (Web, WAP, SMS, MMS, UMTS, VoiceXML); and
- user interface issues in the design of the DiWay system for mobile phones.

4.3 DiWay Methodology

4.3.1 Introduction

The DiWay methodology comprises proposals to solve the research problems mentioned above. It is based on the representation of a shopping centre as a set of graphs, each graph representing a floor. In the parking lot floors, each node represents one parking spot and the arcs are the paths to elevators and stairs. In the shopping floors, each node is a shop, restaurant or other facility. The arcs are the shopping centre streets. Each shop's node may have associated secondary nodes representing shelves or products. Links between floors are represented by special arcs, which have specific properties indicating the type of connection (stair, elevator, escalator, ramp) and the floors they link to. Using these properties the system is able to generate paths according to specific user needs. If, for example, a handicapped person wants to visit the centre using a wheelchair, DiWay will only use floor-linking arcs with an elevator or ramp connection thus providing the user with a personalised path.

Each node is represented by an alphanumerical code defined hierarchically (Figure 4.1). Country, city and shopping centre sub-codes are automatically provided by the operator's location technology, so that the user only has to enter between three (shop) and seven characters (shop + object). In the DiWay methodology, these codes are

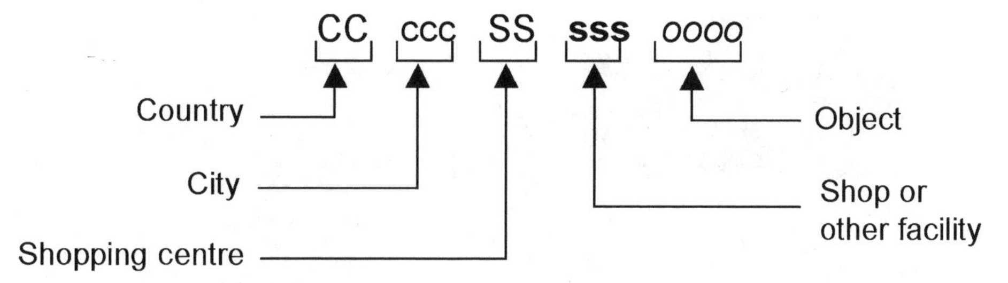

Figure 4.1 *The hierarchical structure of the spatial bar code (SBC)*

named spatial bar codes (SBCs) and identify uniquely an object anywhere in the world. Each node is considered to be an anchor point. It is possible to geo-reference them *a priori*, thus establishing a correspondence between their coordinates and the SBCs, by digitising the plans of the shopping centre.

4.3.2 Indoor Positioning

Anchors may be tags or emitters with associated SBCs. The SBCs may be presented together with the corresponding WAP address or SMS number (Figure 4.2). This concept solves the positioning problems in indoor spaces where both satellite and radio frequency triangulation systems are often inadequate.

4.3.3 Anchor Point Technology

For the time being the best option is to use plastic tags as in Figure 4.2. The user can then use this code to locate him or herself to the system, writing it on a WML form or in the text of an SMS. For the third-generation mobile phones, and even for some 2.5G smart phones, a Bluetooth-based emitter will probably be the best solution (Held, 2000). Using Bluetooth emitters as anchors will allow users to pinpoint their location automatically without having to read or write any SBC. Whenever they ask for some information, DiWay will tell the phone to read the nearest SBC and display the adequate contextual information. Tests with Bluetooth-based anchors are already being conducted using a PDA to simulate a 3G phone.

The anchors may be used to support navigation, as a reference for meetings and emergency services and to facilitate mobile commerce as discussed for the Colombo Centre case study.

4.3.4 Spatial Analysis Tools

The spatial analysis problems involved in shopping centre navigation are essentially the definition of the shortest path between an origin and a destination. Typically, the

Figure 4.2 *Use of a tag to anchor a shop in the Colombo Centre*

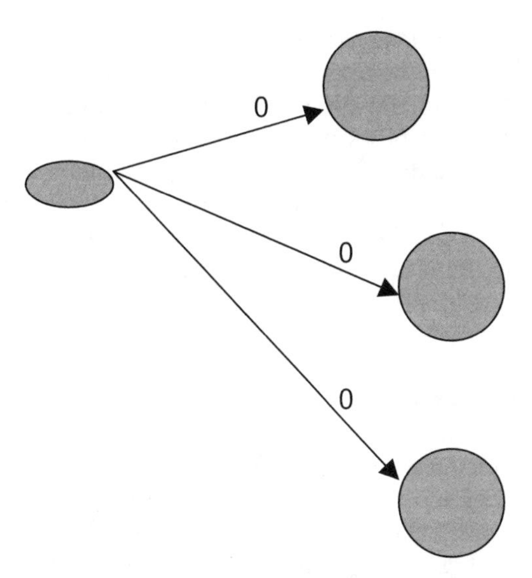

Figure 4.3 *Adaptation of Dijkstra algorithm to solve the problem of having alternative origin points: the distance between the virtual origin point and the potential departing points is 0*

visitor wants to find the route between an entrance of the shopping centre and a shop or a product. Conversely, he or she may be interested in finding the way between the exit and the car in the parking lot. For both problems, DiWay applies the Dijsktra algorithm (see the description in Cormen *et al.*, 1990). The Dijsktra algorithm requires that weights are assigned to each arc (or shopping street). Those weights represent distances. The algorithm then proceeds to determine the shortest distances from the origin to all other points in the network. It also provides the sequences of arcs that are associated with those distances.

DiWay applies a modification of the Dijkstra algorithm to solve two other problems: finding the shortest path between a destination and several possible origin points; and determining the nearest point of interest. The former problem is of interest when one wants to select which entrance to use to be closer to a certain shop. The latter occurs when someone wants to look, for example, for the nearest bookshop to his or her current location.

To solve both problems, artificial nodes and arcs with null weights are considered. When there are several origin points and one destination, an artificial node is inserted as a virtual origin point. The distances between the virtual and the real origin points are zero (Figure 4.3). The Dijsktra algorithm is then applied with the virtual origin point as the starting node. The origin point that is part of the shortest path is the entrance that the user should select.

In the nearest-point problem, several destination points may be considered (i.e., all the bookshops in the centre). In this case, the arcs from the real destination points (the available bookshops) to a virtual destination point have weights equal to zero. The Dijsktra algorithm is applied considering the virtual destination point. The bookshop that is part of the shortest path is the nearest bookshop to the user.

Other analytic problems of interest are node-covering algorithms (see Princeton's Department of Mathematics website on the travelling salesperson problem). One may be interested in defining shortest circuits or paths covering a predefined number of shops. The solution to these problems is not included yet in the DiWay methodology. Examples may include the planning of a visit to a number of predefined shops (i.e., to a bookstore, sports store, furniture shop and clothes store), trying to spend the minimum amount of time departing and arriving at the parking place. Inside a supermarket, one may want to define the minimum path from the entrance to the exit booth while picking up predefined items, provided that these are geo-referenced.

4.3.5 Software Framework

The software platform developed by YDreams to support LBS is called FluidMapping. It was developed on Microsoft.NET, taking advantage of the flexibility, security and performance of this platform. FluidMapping is structured into three tiers: the data tier, the logic tier and the presentation tier (Figure 4.4). These tiers have been designed to:

- handle heterogeneous sources of information therefore allowing the integration with already deployed database services. This is important for a system like DiWay where data are provided by several different entities using different information systems;

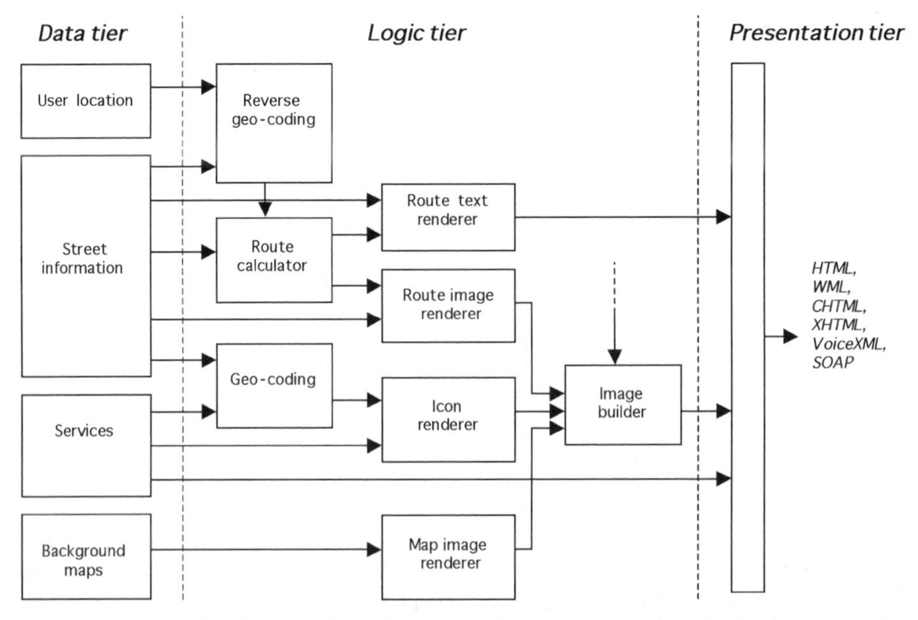

Figure 4.4 *Example of a typical application architecture using the FluidMapping platform*

- be incorporated into locally executed applications (classic architecture), client/server environments (three-tier architecture), or distributed environments (distributed three-tier, or *n*-tier). This is therefore a flexible solution allowing the same service to run locally (e.g., on a kiosk or a PDA) or remotely (e.g., on a mobile phone, Internet-enabled PDA or interactive television) or even as a web-service;
- allow the stratification of products using different combinations of modules. In this way, the same framework can be used to deploy different products or services. Using FluidMapping, DiWay can easily be converted to a museum navigation service; and
- allow the presentation of the services in all wired and wireless devices (SMS, MMS, HTML, WML, CHTML, XHTML, VoiceXML and SOAP).

4.3.6 Data Tier

There are four types of information that are essential for location-based services: user location, street information, background maps and the services (or contents) themselves.

User Location

The user location is what differentiates the LBS from the traditional wired services. The location can be found by automatic technologies such as enhanced observed time difference (E-OTD), time of arrival (TOA), cell ID and assisted-GPS (A-GPS) or simply supplied by the end user. The existing automatic methods can only give the area where the user is located; they still cannot provide the store or object location. If the service needs a more accurate position, the user still has to type it or say it, depending on the device he or she is using. Future outdoors solutions will minimise this problem. DiWay type propositions, based on Bluetooth code emitters, will solve the indoor positioning problem.

Street Information

The street information database includes all the information for the streets. It is used for the calculation of the route between two points, the determination of the nearest point and to find the geographic position of a point given its address (geocoding). This information is created using the FluidMapping Graph Editor.

Background Maps

The background maps are used for the graphical visualisation of the route and position of a point. These maps are related to the street information but are handled separately. For example, WAP uses black and white maps, while the Web may use coloured maps. However, the road information can be exactly the same. The FluidMapping platform has editors for raster and vector maps. When using raster maps, the adequate maps are automatically tiled to form the background map. Vector maps can either be constructed using the appropriate editor or imported from other applications like MapInfo.

Services

The services databases include the contents that will be geocoded and made available to the end user. The FluidMapping framework only needs the address of the points of interest to handle the contents automatically. The operator usually controls the user location, and the shopping centre's administration, the street information and background maps databases. Third-party companies usually supply the services databases through syndication. This means that the operator may not have control over how the data is supplied. As a result, there can be a great variety of transportation protocols, query methods and data structures involved. The conversion and integration problems are usually significant.

To minimise these problems, FluidMapping adopted the Microsoft BizTalk server solution (http://www.microsoft.com/biztalk/). BizTalk provides powerful and easy-to-use tools to control the flow of information and its conversion into formats that the YDreams platform can interpret. The use of this product considerably reduces the time necessary to integrate a new third-party data provider.

The data tier framework of FluidMapping is developed on top of the latest generation of Microsoft database-access components called ADO.NET. These are very generic components allowing a common data retrieval method for all types of databases, static files or data generated dynamically by other components, simplifying the code of each component.

4.3.7 Logic Tier

The logic tier framework consists of several modules. Key modules were shown in Figure 4.4 and are discussed below. These components use XML to communicate between them thus providing total control over the data flow. This data can also be redirected, at any place, to third-party components.

Geocoding: this module uses the street information database to convert an address into geographical coordinates.

Reverse geocoding: uses the street information database to convert geographical coordinates into addresses.

Map image renderer: generates the background map image given the geographic coordinates, image size and source.

Route calculator: calculates the route between two geographical points. The different methods of calculating routes (shortest, fastest, cheapest) are implemented in different modules so that they can be used separately, depending on the product.

These modules share the same interfaces so their interchange is straightforward.

Route text renderer: processes the route calculation result so that the name of the streets, shops, direction changes and other useful information are added to it.

As for all the other components the output of this module is in XML. For example, the direction changes are coded as tags, instead of text, so that they can later be translated into any language using a simple XSLT rule.

Route image renderer: draws the route so that it can be superimposed on the background map. It provides total control over the line colour, style and width.

Icon renderer: calculates the relative position of the icons to be superimposed on the background map image. It gives total control over the icon design.

Image builder: allows the combination of several layers of images. It gives total control over the layer order, position, size and opacity. This module also enables the addition of other elements to the image such as legend, copyright notice and watermark. The image can be saved in any of the Web file formats (PNG, JPG, WBMP and GIF) and other graphical formats.

4.3.8 Integration

The logic tier framework is currently implemented on the Microsoft.NET framework (http://msdn.microsoft.com/net/) and can take advantage of all the related technologies on the Windows 2000, Windows XT and Windows CE platforms. On client/server architectures, the end user does not need to have any of these platforms to be able to access the services. The framework is located on the server and only the output is sent to the client, meaning that it is independent of the client's platform.

In case the framework is run locally on devices with Windows CE platform, like pocket PCs and auto PCs (http://msdn.microsoft.com/library/default.asp?URL=/library/wcedoc/apcintro/apcintro.htm), it can take advantage of the CE edition of the Microsoft SQL Server database (http://msdn.microsoft.com/library/default.asp?URL=/library/wcedoc/sqlce/portal_6ol0.htm) to store the data on a local database. To keep memory usage down, this database can store just the data related to the services close to the user location. This database can be automatically updated through wired or wireless synchronisation. The framework can also be wrapped in a SOAP compatible Web-service that allows its access to third parties through a custom API.

4.3.9 Presentation Tier

The use of XML on the framework allows the customisation of its output to virtually any device, from regular HTML for web browsers, passing by several wireless types such as WML, CHTML and XHTML, to VoiceXML, that allows the use of voice on a regular cellular phone.

The DiWay interface consists of the display of maps and associated geo-referenced content by request on WAP enabled mobile phones. Typical queries include point-of-interest search (which may comprise nearest point-of-interest search), information on anchor points, directions for personal navigation and emergency services.

The display of maps is limited by:

- the size of most WAP-enabled phone screens; as a result, display clutter usually occurs; and
- the monochromatic nature of those screens; colour cannot be use to differentiate floors, shops or products.

Interaction with WAP phones is further hampered by the multidimensionality of the keyboard (Matilla and Ruuska, 2000): most keys stand for one number and three letters. To minimise these problems, DiWay has incorporated the following principles.

- Minimise display clutter. Two levels are provided: the overall view of the floor and the local area where the visitor is at a certain moment. Both are represented diagrammatically. Text labelling is done for major anchors at the global level. At the local level, text is only used to label the anchor of interest; numerical labelling, complemented by text, is used for adjacent anchors and streets.
- Minimise text input. Menu options are offered for point-of-interest search. Users only need to type three letters to obtain a list of matching names for streets and points of interest.

An additional feature of the interface is the Phonemarks (Figure 4.5). By using the Phonemarks feature the user stores anchor points for later exploration in the fixed or mobile Internet platforms. If those anchor points reflect commercial items this

Figure 4.5 Phonemarks – whilst visiting, users can send codes of preferred objects or shops to a descriptive file. This will allow the user to visit, at a later date, Web sites or WML decks that are associated to these codes

feature may be used for mobile commerce. Phonemarks are also used to store the location of cars in the parking lot.

The DiWay system may also use text messages (SMS) and, in the recently released phones, text, image and sound messages based on MMS. At this point, it is not yet clear if navigation systems based on MMS, that place less of a burden on the user, may overcome mobile Internet alternatives. If flat rates are adopted as in Japan, mobile Internet will become less expensive, and allow for a thorough use of mobile applications.

4.4 Application

DiWay is being implemented for the Colombo Centre in a demonstration project that includes the following features.

- Store and product finders, to find stores that sell a particular product. These are displayed on diagrammatic maps of the shopping centres. Paths from origin to destination are provided on those maps and by textual directions using key anchor points.
- Parking place finder. Using the Phonemarks feature, the visitor stores the spatial code of the place where he or she parked the car. The parking place finder will later help him or her in finding the way back to the car (Figure 4.6).
- Product interaction, to enable the visualisation of detailed product information and facilitate its mobile purchase (Figure 4.7). The user types in the code for the product (or receives its code via a Bluetooth anchor) and accesses all the corresponding stored information. Furthermore, if the store is closed the user can buy the product through DiWay or use the Phonemarks feature to store the product's code for a later review.

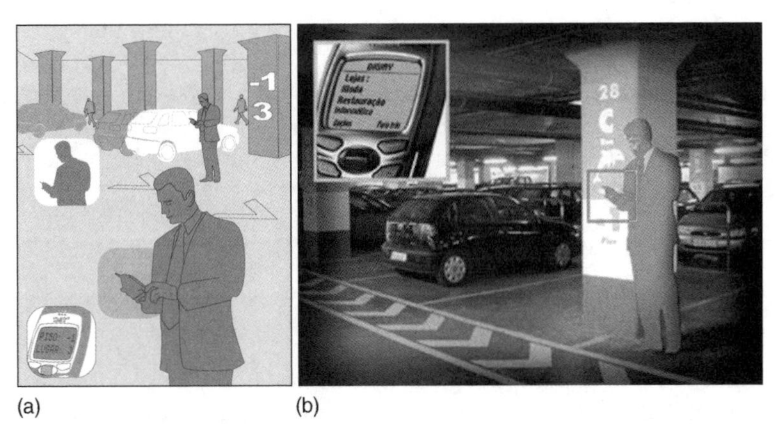

(a) (b)

Figure 4.6 Using Phonemarks to (a) store the position of the car and (b) locate the car in Colombo Centre's parking lot

Figure 4.7 Interaction with a product using the mobile phone

- Casual encounters, to facilitate the meeting of people with identical profiles. This is a community-building feature that uses the centre as a link from the cyberspace to the real world, allowing people to establish virtual communication and eventually meet using the directions provided by DiWay. This works like a treasure-hunting game where the treasures are real people. Profiles are filled on the Web prior to the visit, or through DiWay upon entrance to the shopping centre. In order to stimulate sponsorship of the system, directions will be given using shops as points of interest and the meeting points will be established at main sponsor stores.
- Emergency alerts are possible by pressing e after entering the nearest spatial code (or in the future receiving it by activating its emission from the nearest Bluetooth anchor).
- Information on transportation to and from the centre. This feature includes an automated service to request taxis, without forcing the user to make any phone call. DiWay will both inform the user and the taxi of the most adequate meeting point.

The experience so far has shown that frequent WAP users find the system useful. However, such users total less than 10% of the Colombo Centre visitors that have mobile phones. A much higher percentage regularly use SMS and is expected to adopt MMS. Thus, it is clear that unless third-generation devices enable an easier use of the mobile Internet, DiWay and other navigation systems will be based on MMS systems that will enable the display of maps, sounds, images and even 360 degree views.

4.5 Conclusions and Further Developments

The emergence of mobile computing has opened new opportunities for LBS. Indoor applications of LBS in shopping centres are most relevant to facilitate navigation, mobile commerce and community activities. This chapter has described a methodology developed with those goals in mind for mobile phones that are mobile Internet,

SMS and MMS enabled. The approach, named DiWay, includes detailed georeferencing of anchor points and other relevant objects, spatial analysis tools and an appropriate interface for mobile phones. DiWay is based on a software platform called FluidMapping. It relies on a view of GIS as composed of modular components. FluidMapping is designed to handle heterogeneous databases and multiple presentation devices such as personal computers, personal digital assistants and mobile phones. It also includes modular spatial analysis components. In this chapter, FluidMappping was applied to a shopping centre represented as a graph where nodes were shops and parking places and the arcs were the shopping streets. The spatial analysis component used in DiWay was based on the Dijkstra algorithm. This algorithm was applied to determine the shortest path and the nearest point of interest. The potential application of node-covering algorithms to determine optimal circuits was also discussed.

The DiWay approach was illustrated with the development of a system for the Colombo Centre in Lisbon. This system has been developed in cooperation with SONAE, the operator of the centre. It was designed to replace the existing methods of assisting visitors in their shopping experience. The experience so far has shown that the system is only beneficial for seasoned WAP users. These make up less than 10% of the total number of visitors to the Colombo Centre. MMS, which enables the visualisation of maps, sounds, images and 360 degree views, may be an alternative for future implementations. DiWay will benefit also from the development of the third-generation mobile phones and the increasing use of Bluetooth chips. Third-generation phones will offer a larger bandwidth and colour. Bluetooth chips will provide an infrastructure to facilitate indoor positioning and avoid dialling codes which is the current positioning solution.

References

Abowd, G., Atkeson, C., Hong, J., Long, S., Kooper, R. and Pinkerton, M. (1997) Cyberguide: a mobile context-aware tour guide, *Wireless Networks*, **3**: 421–33.

Abowd, G., Dey, A., Orr, R. and Brotherton, J. (1997a) Context-awareness in wearable and ubiquitous computing, in *Proceedings IEEE 1st International Symposium on Wearable Computers*, 13–14 October, Cambridge, MA. Available as an html document in http://www.cs.gatech.edu/fce/pubs/iswc97/wear-poster.html.

Beadle, H., Harper, B., Maguire Jr., G. and Judge, J. (1997) Location aware mobile computing, in *Proceedings IEEE/IEE International Conference on Telecommunications (ICT'97)*, Melbourne, April, pp. 1319–24. Available as an html document at http://www.elec.uow.edu.au/people/staff/beadle/badge/ ICT97/ict97.htm.

Beadle, H., Maguire Jr., G. and Smith, M. (1998) Location based personal mobile computing and communication, in *Proceedings 9th IEEE Workshop on Local and Metropolitan Area Networks*, Banf, May. Available as an html document at http://www.elec.uow.edu.au/people/staff/beadle/badge/ieee_lan_98/.

Câmara, A.S. and Raper, J. (1999) *Spatial Multimedia and Virtual Reality*, Taylor & Francis, London.

Câmara, A.S., Scholten, H. and Remedio, J.M. (2000) Spatial multimedia, in Openshaw, S. and Abrahart, A. (eds) *GeoComputation*, Taylor & Francis, London.

Cormen, T., Leiserson, C. and Rivest, R. (1990) *Introduction to Algorithms*, McGraw-Hill, New York.

Davies, N., Cheverst, K., Mitchel, K. and Efrat, A. (2001) Using and determining location in a context-sensitive tour guide, *IEEE Computer*, **34**(8): 35–41.

Department of Mathematics, Princeton U. (2001) Travelling Salesman Web site at http://www.math.princeton.edu/tsp.

Dias, A.E., Romão, T., Pimentão, P. and Câmara, A. (1998) Keeping contextual awareness: the TOI system, in *Proceedings CD-ROM GIS PlaNet'98*, Lisbon, Portugal, 7–11 September.

Dias, A.E., Romão, T. and Pimentão, P. (1997) Traveller's on-line information system – TOI: information system and location modules, in *Proceedings JEC-GI'97 Joint European Conference and Exhibition on Geographical Information*, Vienna, Austria, 16–18 April, IOS Press, pp. 238–47.

Dias, A.E., Santos, E., Pimentão, P., Pedrosa, P. and Romão, T. (1996) Bringing ubiquitous computing into geo-referenced information systems, in Rumor, M., McMillan, R. and Ottens, H. (eds) *Geographical Information – From Research to Application Through Co-operation*, IOS Press, Amsterdam, The Netherlands, pp. 100–3.

Frisk, C. (1998) Location and context awareness in adaptive personal communication services. Available as an html document in http://www.pcs.ellemtel.net/Claes/ pcc/context.html.

Gouveia, C. and Câmara, A.S. (1999) Multimedia and urban planning, in Stillwell, J., Geertman, S. and Openshaw, S. (eds) *Geographical Information and Planning*, Springer Verlag, Heidelberg.

Held, G. (2000) *Data Over Wireless Networks: Bluetooth, WAP and Wireless LANs*, McGraw-Hill, New York.

Hightower, J. and Borriello, G. (2001) Location systems for ubiquitous computing, *IEEE Computer*, **34**(8): 57–66.

Mattila, K. and Ruuska, S. (2000) Designing mobile phones and communicators at NOKIA, in Bergman, E. (ed.) *Information Appliances and Beyond*, Morgan Kauffman, San Francisco.

Pradhan, S., Brignone, C., Cui, J., McReynolds, A. and Smith, M. (2001) Websigns: hyper-linking physical locations on the web, *IEEE Computer*, **34**(8): 42–8.

Ryan, N. (1998) FieldNote Desktop: an experimental spatio-temporal information system, 4th International Colloquium on Computing and Archaeology, Bilbao, May. Available as an html document at
http://www.cs.ukc.ac.uk/research/ fieldwork/papers/Bilbao/FieldNote Desktop.html.

Ryan, N., Pascoe, J. and Morse, D. (1998) Enhanced reality fieldwork: the context-aware archaeological assistant, in Gaffney, V., van Leusen, M. and Exxon, S. (eds) *Computer Applications in Archaeology*. Available as an HTML document at
http://www.cs.ukc.ac.uk/research/fieldwork/papers/CAA97/ERFldwk.html.

5

Mass Appraisal and Noise: the use of Lifestyle Segmentation Profiles to Define Neighbourhoods for Hedonic Housing Price Mass Appraisal Models

Steven Laposa and Grant Thrall

Abstract

This chapter presents a case study on the application of geographic information systems (GIS) with the mass appraisal of 1400 residential properties located in an urban area. The objective of the case study is to estimate the total value of the residential portfolio using 850 sales from the multiple listing service. The residential properties are in an urban area subject to extensive vandalism, crime and vacant homes. The authors contend that the use of GIS, and special three-dimensional (3D) modelling and visualisation of sale prices, provides a higher explanatory power than typical hedonic residential price models. The study shows how the application of lifestyle segmentation profiles (LSPs) such as ESRI Bis (formerly CACIs Coder/Plus) ACORN geodemographic database can support valuation modelling.

5.1 Introduction

The practice and implementation of mass appraisal models are widespread in the real estate industry. Local tax assessors use a form of mass appraisal modelling to

Applied GIS and Spatial Analysis. Edited by J. Stillwell and G. Clarke
 ISBN: 0-470-84409-4

estimate residential property values for property tax assessments (for a discussion and overview, see International Association of Assessment Officers (IAAO), 1999). Implementations of computer-assisted mass appraisal (CAMA) applications are widespread throughout the assessment industry. Mass appraisal models have been used in litigation cases and portfolio investment analysis for similar properties, such as fast food or convenience stores. Mass appraisal is a statistical procedure of defining a representative sample of a larger population with the purpose of estimating the aggregate value of the overall population. According to the Appraisal Institute Foundation, a mass appraisal includes:

- identifying properties to be appraised; defining market area of consistent behaviour that applies to properties;
- identifying characteristics (supply and demand) that affect the creation of value in that market area;
- developing a model structure that reflects the relationship among the characteristics affecting value in the market area;
- calibrating the model structure to determine the contribution of the individual characteristics affecting value;
- applying the conclusion reflected in the model to the characteristics of the property(ies) being appraised; and
- reviewing the mass appraisal results.

Lead author, Steve Laposa, is global Director of Real Estate Research for PricewaterhouseCoopers (PwC). In that capacity, he was responsible for supporting a team assigned to appraise approximately 1400 residential properties in a major midwestern city. Grant Thrall, was invited to contribute his GIS and business geography expertise to the team. The properties were located in an older, downtown neighbourhood. The properties were owned by a client of PwC; the client requested a total value of those properties in order to establish a market value which would be used to contribute those properties for the creation of a new public–private joint venture. Mass appraisal therefore derives a reasonable estimate of the aggregate value of the properties. Individual property appraisal would instead estimate the value of each individual property, which could be aggregated for an estimate of total value. However, the cost and time involved in executing 1400 individual appraisals would be too great and would not therefore meet the clients objective. Mass appraisal generates comparable results at a much lower cost and in a much shorter time horizon. Hedonic mass appraisal models assume there to be a high level of homogeneity in the population. Hedonic mass appraisal models are a special case of the more general hedonic model. Hedonic models rely upon various forms of regression analysis, where a dependent variable is hypothesised to be explained or described by a set of independent variables (for discussion, see Thrall, 2002, Chapter 4).

Residential home values within small local neighbourhoods are generally homogeneous in terms of size, number of bedrooms, lot size, age and other commonly used descriptive characteristics. Large variances between adjoining homes are usually not found. However, large variances between adjoining homes can be found in older urban neighbourhoods. In the United States older urban neighbourhoods often exhibit great structural and value differences from street to street, and even within a block. So, the assumption of homogeneity in the population breaks down. Older urban, downtown residential neighbourhoods, often contain deteriorating homes due to abandonment, vandalism, fire, crime and lack of standard maintenance. Thus, the average price per square foot over a large sub-market does not necessarily reflect the value of an individual home. This provides a problem for the usual procedures for mass appraisal.

A key ingredient to mass appraisal models is the assumption that variance is small between residential home values within a neighbourhood; the smaller the geographic delineation of the neighbourhood, the smaller would be the expected housing price variance. The assumption of small variance excludes the existence of large price per square foot variance between nearby homes. In other words, the land price surface is expected to be smoothly unfolding across space, and not discontinuous (see Thrall, 1998a, for a discussion on the error of this assumption).

Geographic surface interpolation procedures assume that space is smooth and continuous. These procedures assume that values between locations do not change abruptly, allowing local values to be accurately represented by a larger spatial trend. In many instances, these assumptions accurately represent physical phenomena. However, real estate is more problematic. Ownership and build characteristics change from parcel to parcel. So, spatial trend models may be used to represent general trends, but not individual parcel values. When the variation between parcels is very large, as in our study, large area spatial trends have considerable error.

In this chapter, we deal with two issues of mass appraisal: first, how can one specify an appropriately small geographic neighbourhood in a manner that is both defensible and repeatable between neighbourhoods, and even between cities? Within the resulting small geographic neighbourhood, land prices would be expected to have a small (minimum) variance between dwellings; second, when large per square foot variances between nearby homes do occur, what can explain the variance and how can that explanation be incorporated into a mass appraisal model? This chapter addresses the use of mass appraisal modelling incorporated with GIS and databases. Our contribution to the literature is the use of lifestyle segmentation and 3D GIS applications in addition to standard housing hedonic models. First, we provide a discussion of the explanatory variables we use in our mass appraisal, including lifestyle segmentation profiles which are particularly important for neighbourhood delineation. We then present a discussion of our methodology and our conclusions.

5.2 Explanatory Variables and Lifestyle Segmentation Profiles

It has long been recognised that house prices are highly dependent upon neighbourhood. Homer Hoyt in his renowned study demonstrated that like-kind neighbourhoods were so spatially interdependent that they formed geographic clusters giving rise to Hoyt's classic hypothesis of the sectoral nature of cities (Hoyt, 1933, 1939, 1966). Within a sector, a variety of measurements are correlated with housing prices; the most commonly used in hedonic housing price models are listed in Table 5.1. The first column of Table 5.1, AM, are measurements of access to place specific

Table 5.1 Independent variables by geographic scale

Accessibility measures (AM)	House features (HF)	Neighbourhood attributes (NA)
Highway	Living area	Average house size
Central business district	Bathrooms	Size of neighbourhood/subdivision
Major retail	Pool	Floodplain
Minor retail	Fireplace	Census tract land characteristics
Major employment centres	Foundation type	% land area in single-family dwellings
University/colleges	Age	% land area in multi-family dwellings
Hospitals	Heating type	% land area in offices
	Air-conditioning	% land area in industry
	Roof type	% land area in parks
	Wall type	Census tract population characteristics
	Garage type	% non-white
	Lot size	Median income
		School district

phenomena that are often cited as being important in households' decisions to locate which translates into a housing price effect. Features of a house (HF) affect the market value of a dwelling; the larger the square footage, and the more bathrooms, the greater will be the market value of the house. Fireplaces and pools are expensive to build (often, dwellings with greater construction costs will have greater market value) and neighbourhood attributes (NA) generally affect housing prices. Neighbourhoods dominated with large houses are generally expected to have greater housing values than neighbourhoods with smaller houses. Hedonic housing models will have house price or square footage price as the dependent variable, and sets of variables from the three columns of Table 5.1 as independent variables (see Thrall, 2002, Chapter 5 for further discussion, and Thrall, 1998b, 2002, Chapter 3, on how land prices are related to the leading independent variables).

A typical hedonic housing price model (see also Case *et al.*, 1991; Clapp *et al.*, 1995) might be expressed, using terms from Table 5.1, as

$$P = P(AM, HF, NA) \tag{5.1}$$

where AM are a representative set of accessibility measures, HF is a representative set of housing features, and NA is a representative set of variables that characterise a neighbourhood. Regression coefficients are estimated for a hedonic expression based upon equation (5.1). Dummy variables might be added to the hedonic expression of functional (5.1) to capture the difference in how the independent variables affect the dependent variable between the various neighbourhoods. The mechanics of appropriate estimation of functional expression (5.1) are well known and successfully applied in CAMA (computer assisted mass appraisal) for the determination of estimates of market value of individual dwellings. However, two problems can arise that can make for unreliable estimates of a hedonic expression based upon functional form (5.1): spatial autocorrelation and biased data sets, and appropriate geographic delineation of the neighbourhoods.

Spatial autocorrelation is a measurement of how closely correlated near neighbours are. If near neighbours are highly correlated with one another, then there is said to be a high level of spatial autocorrelation. From the above discussion, housing prices are generally expected to be highly correlated with prices of neighbouring dwellings. After all, it is said that real estate is 'location, location, location', and indeed if that is true, then real estate prices will be dependent upon the spatial association with neighbouring properties. Spatial autocorrelation in hedonic housing models arises because an observation's price highly depends upon the price of a near neighbour, and that dependence declines the further apart are the observations in geographic space. The degree of spatial autocorrelation may be measured. If there are high levels of spatial autocorrelation, then regression coefficients will be biased in an unknown direction and thereby make for unreliable housing forecasts. An accepted and most frequently used method for correcting for spatial autocorrelation is by way of appropriate independent variable choice, where the independent variables capture the spatial trend. The problem has been, however, the identification of independent variables that can capture the uniqueness of the neighbourhood. A neighbourhood can be a geographically contiguous area, or noncontiguous, where similar physical characteristics of the properties may or may not be important. However, usually, similarity among the inhabitants is important when defining a neighbourhood.

Data observations that form the basis of the calibration of a hedonic housing price model are assumed to be unbiased representations of the population that the model is intended to represent. Constraints on experimental design might not allow the data observations to represent the population adequately. Constraints on the experimental design might give rise to small variations among the observations in the traditional independent variables listed in Table 5.1 – namely the experimental design might have constrained the observations to have been selected from a region where the properties were physically similar. In addition, house prices within the region might have substantial sub-market variation.

In this chapter we propose a solution for when housing observations are taken from a housing stock that is highly homogeneous in physical characteristics, taken from the same region of the city, while at the same time whose prices are observed to be highly variable though spatially autocorrelated. The experiment reported on here is to estimate housing prices of 1400 properties that are overall of the same age and physical characteristics, and taken from what colloquially would generally be accepted as the same sector of the city. Yet, the sales of the subject properties revealed that there was a wide variation in actual prices paid for these properties. The traditional hedonic housing price approach, following accepted practices and including those physical variables listed in Table 5.1, did not yield acceptable results. The researchers then were confronted with discovering the underlying cause of housing price differentiation, where those housing prices are not a function of traditionally measured physical characteristics.

Since the observations were taken from a small geographic area, and since the experiments revealed that whatever physical variation there was that existed between the dwellings were not determinants to revealed house prices, the researchers then hypothesised that some underlying social forces were the cause of variation in housing prices. The problem was how to measure those small geographic-scale neighbourhood-specific social forces in a repeatable unbiased manner? In other words, we preferred to use some automated procedure to define a neighbourhood versus relying upon our own prejudices, biases and limited information to define a neighbourhood. Therefore, we turned to methods and procedures of social area analysis.

While lifestyle segmentation profiles (LSPs) have their roots in academia, they recently have received more attention by practitioners with requirements to use consistent criteria to define neighbourhoods on the basis of the type of people that live there. LSPs provide a multidimensional view of people, in contrast to conventional univariate census variables such as census tract median income. LSPs allow for a differentiation between neighbourhoods on the basis of the social makeup of the inhabitants of those neighbourhoods. If neighbourhoods are defined on the basis of dominant LSPs, then within a geographic region of a city, a neighbourhood need not be spatially contiguous as has been the convention in hedonic modelling with dummy variables representing neighbourhoods. LSPs are used to define a social neighbourhood. It is our contention that LSPs can identify geographically small sub-markets, and thereby capture the small-scale component that gives rise to near neighbour spatial autocorrelation for some properties, and discontinuities for other properties. In our study, we used ESRI Bis (formerly CACI) ACORN LSPs because they have a geographic resolution of a ZIP + 4 code that we considered a necessary and sufficient level of geographic resolution. Other LSPs such as Mosaic (Thrall *et al.*, 2001) have the geographic resolution of the census block or block group, and are appropriate for other forms of analysis. For a discussion of LSPs from Claritas Corporation, see Hartshorn (1992, p. 255) and Weiss (1988).

Social area analysis was developed in the post World War II era, by sociologists interested in the social structure of urban sub-markets (see Shevky and Williams, 1949; Shevky and Bell, 1955; as quoted in Hartshorn, 1992, p. 248). Geographers, adopted the social area analysis work of sociologists (see for example Berry, 1964, 1969, 1970) and used factor analytic methods to classify neighbourhoods. CACI Marketing Systems commercialised the concept through the sale of neighbourhood classification data, which CACI referred to as ACORNs, an acronym for 'A Classification of Residential Neighbourhoods'. CACI Marketing Systems became well known to the business community for its ACORN LSPs. CACI Coder/Plus, the product used in this research, integrates the assignment of geographic coordinates, ACORN LSPs, and ZIP + 4 codes to each record observation based upon its street address. Recently, ESRI (www.esri.com) purchased CACI Marketing Systems and markets the same product under the name ESRI Bis. ACORN categories are accurate and differentiated at the geographic scale of the ZIP + 4 code (for further discussion on the derivation and use of LSPs, see Thrall *et al.*, 2001; and Thrall, 2002).

5.3 Data and Methodology

In order to estimate the values of 1400 homes, the authors collected 850 actual home sales from a local Multiple Listing Service (MLS). They selected actual

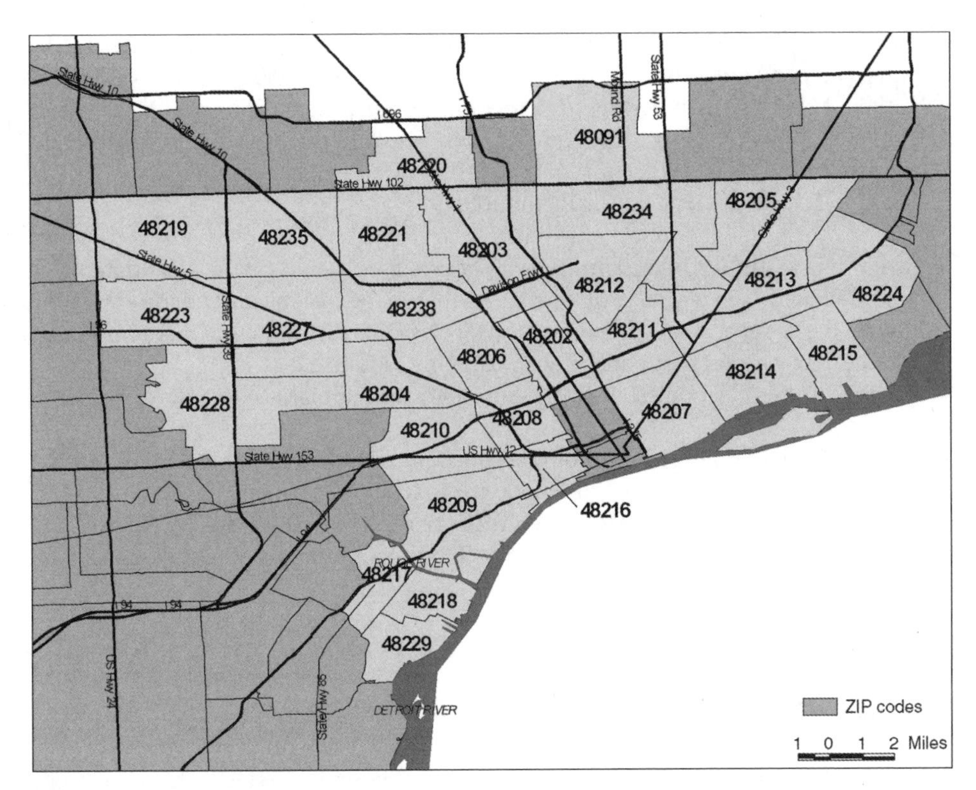

Figure 5.1 Targeted geographic area by zipcode

home sales for the prior year and limited their sample to (1) homes sold under \$80 000 based on discussions with field inspectors, and (2) geographic proximity to the 1400 homes. Figure 5.1 illustrates the zipcodes selected for the MLS home sale database.

Plate 4 is a geographic visualisation of the MLS properties selected for use as comparables. Addresses of the properties were used to geocode. ESRI Bis (Thrall, 1998b) was used to assign geographic coordinates. ESRI Bis assigns geographic coordinates on the basis of the most accurate coordinate that can be derived between each data record's street address, ZIP + 4, or ZIP code; ESRI Bis is an 'intelligent' geocoder in that addresses will be corrected and standardised using current US postal records. Among the corrections are street address format, name spelling and the addition of ZIP and ZIP + 4 codes. Standardised FIPS codes are also added for documenting the census polygon (block, block group, tract) in which the address is located. Because the neighbourhood is 'older urban inner core', ESRI Bis intelligent use of the US Census TIGER/Line files results in street address assignment of the geographic coordinates (Thrall and Thrall, 1993). ESRI Bis also assigned its ACORN LifeStyle Segmentation Profiles (LSPs) to the data observations; all properties within the same ZIP + 4 are assigned the same ACORN LSP classification. ArcView (www.esri.com) software was used to visualise the geographic data. ArcView was commanded to display the MLS parcels as a point data file, and to shade those points thematically by three categories of range: \$0 to \$20 K, \$20 K to \$60 K, >\$60 K. Qualitative interpretation of the map includes that there are clusters of observations within each price category; however, observations from all price categories can be observed throughout the map.

There were several steps in our methodology:

- review client-supplied data on the 1400 homes and ensure that critical data fields are populated, i.e., age, size of home, number of bedrooms and condition of the home. Field inspectors coded each property for fire damage, vandalism and whether the property was abandoned (dummy variables);
- geocode the 1400 homes, identify zipcodes, and integrate ACORNs for each property;
- select MLS actual home sales from the zipcodes identified in step 2. Obtain similar data fields as found in step 1 and code with ACORNs;
- run standard pricing models on MLS database, including transformations and stratified models based on price ranges. Test for differences across zipcodes and LSPs; and
- use 3D GIS applications on MLS data.

5.4 Analysis

Figure 5.2 shows the distribution of MLS home prices and correlation between home price and size in square footage. Table 5.2 confirms the lack of correlations between home price to number of baths, number of bedrooms and square footage.

Figure 5.3 is a boxplot of house prices by ACORN categories and the number of MLS home sales by each ACORN (see Table 5.3 for a complete listing of ACORN LSPs). Only eight LSPs were represented in our data set. ANOVA tests confirm significant differences between ACORN categories.

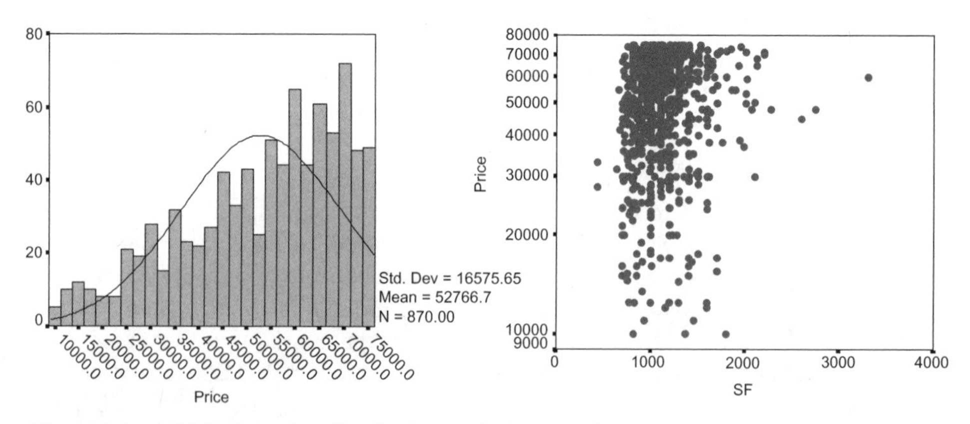

Figure 5.2 *MLS home price distribution and size correlation*

Table 5.2 *Correlation matrix*

		Price	Baths	Bedrooms	SF
Price	Pearson Correlation	1.000	.023	.042	.109**
	Sig. (2-tailed)	.	.495	.214	.001
	N	870	870	870	870
Baths	Pearson Correlation	.023	1.000	.279**	.319**
	Sig. (2-tailed)	.495	.	.000	.000
	N	870	870	870	870
Bedrooms	Pearson Correlation	.042	.279**	1.000	.629**
	Sig. (2-tailed)	.214	.000	.	.000
	N	870	870	870	870
SF	Pearson Correlation	.109**	.319**	.629**	1.000
	Sig. (2-tailed)	.001	.000	.000	.
	N	870	870	870	870

** Correlation is significant at the 0.01 level (2-tailed)

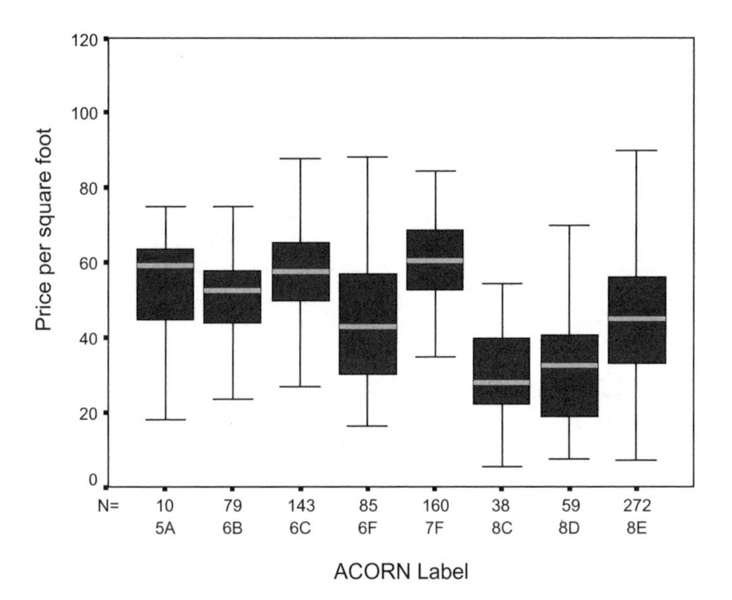

Figure 5.3 *Sale price per square foot by ACORN category*

Table 5.4 presents results from a multiple regression model with price as the dependant variable, and square footage, number of baths and number of bedrooms as independent variables. The regression model's explanatory power is around 1%.[1]

So what is the implication of the model of Table 5.4 having only an R^2 of 1%? First, the traditional hedonic mass appraisal model as described above is inappropriate for neighbourhoods such as that represented in our data set. The traditional hedonic

[1] The authors attempted several other regression models with similar results.

Table 5.3 ACORN lifestyle segmentation profiles

ACORN	Description
1A	1A Top One Percent
1B	1B Wealthy Seaboard Suburbs
1C	1C Upper Income Empty Nesters
1D	1D Successful Suburbanites
1E	1E Prosperous Baby Boomers
1F	1F Semirural Lifestyle
2A	2A Urban Professional Couples
2B	2B Baby Boomers With Children
2C	2C Thriving Immigrants
2D	2D Pacific Heights
2E	2E Older Settled Married Couples
3A	3A High Rise Renters
3B	3B Enterprising Young Singles
4A	4A Retirement Communities
4B	4B Active Senior Singles
4C	4C Prosperous Older Couples
4D	4D Wealthiest Seniors
4E	4E Rural Resort Dwellers
4F	4F Senior Sun Seekers
5A	5A Twentysomethings
5B	5B College Campuses
5C	5C Military Proximity
6A	6A East Coast Immigrants
6B	6B Working Class Families
6C	6C Newly Formed Households
6D	6D Southwestern Families
6E	6E West Coast Immigrants
6F	6F Low Income: Young & Old
7A	7A Middle America
7B	7B Young Frequent Movers
7C	7C Rural Industrial Workers
7D	7D Prairie Farmers
7E	7E Small Town Working Families
7F	7F Rustbelt Neighbourhoods
7G	7G Heartland Communities
8A	8A Young Immigrant Families
8B	8B Social Security Dependents
8C	8C Distressed Neighbourhoods
8D	8D Hard Times
8E	8E Urban Working Families
9A	9A Business Districts
9B	9B Institutional Populations
9C	9C Unpopulated Areas

approach does not work because the data do not conform to required assumptions of the modelling procedure, namely that the distribution of both dependent and independent variables are not normal in our data set. The distribution of these variables confirms the lack of homogeneity within the sub-market of the study.

Table 5.4 *Multiple regression model output*

Variables entered/removed[b]

Model	Variables entered	Variables removed	Method
1	SF Baths, Bedrooms[a]	.	Enter

[a] All requested variables entered
[b] Dependent variable: Price

Model summary[b]

Model	R	R square	Adjusted R square	Std. error of the estimate	Durbin-Watson
1	.115[a]	.013	.010	16494.82	.814

[a] Predictors: (Constant), SF, Baths, Bedrooms
[b] Dependent variable: Price

Coefficients[a]

Model	Unstandardised coefficients		Standardised coefficients	t	Sig.	95% confidence interval for B		Collinearity statistics	
	B	Std. Error	Beta			Lower bound	Upper bound	Tolerance	VIF
1 (Constant)	47400.153	2849.657		16.634	.000	41807.110	52993.195		
Bedrooms	−1090.939	1118.223	−.043	−.976	.330	−3285.683	1103.804	.597	1.674
Baths	−450.813	1735.245	−.009	−.260	.795	−3856.591	2954.965	.888	1.126
SF	8.057	2.567	.139	3.139	.002	3.018	13.095	.582	1.719

[a] Dependent variable: Price

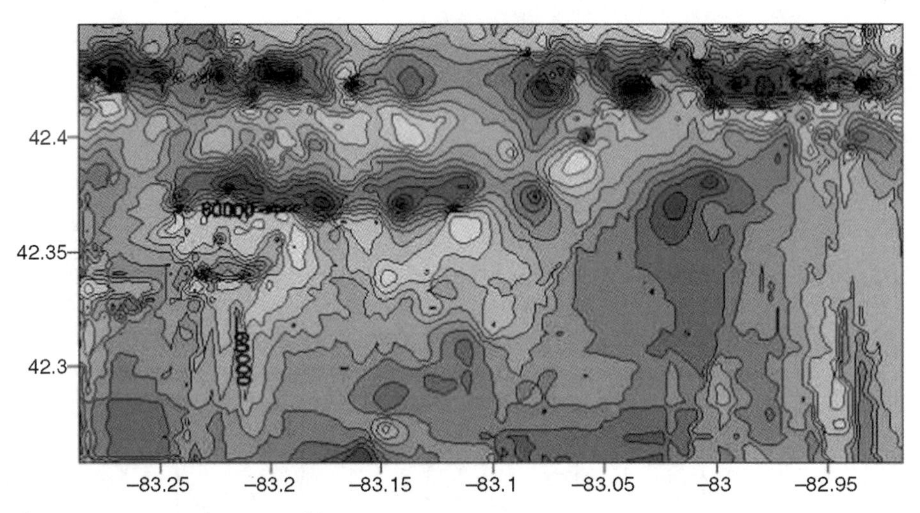

Figure 5.4 *Map of estimated home prices*

Figure 5.4 is a geographic visualisation of the total price surface. The darker shades are isoline peaks, while the lighter shades are isoline valleys. Hence, the higher the price of the MLS comparable property, the darker the area; the lower the price, the lighter the area. Thrall (1998) argued that isoline maps cannot be used to derive estimates of properties within a contour; however, he reasoned that their use was appropriate for qualitative analysis to generate hypotheses about the underlying geographic nature and regularity of the phenomena being studied (in this instance, neighbourhood variation prices of homogeneous properties). Visual analysis of Figure 5.4 reveals a clear east–west street trend, with adjacent streets having abrupt and discontinuous (versus a gradual smooth) change in housing prices. Figure 5.4 was constructed using the same data set as elsewhere in the paper; SURFER (www.goldensoftware.com) was used to derive the isolines. Qualitative analysis of Figure 5.4 gave rise to the hypothesis that prices were highly dependent upon neighbourhood, where neighbourhood was defined as a street face.

The US Postal Service defines unique ZIP + 4 codes along a street face. A second hypothesis arose from the qualitative analysis as well: that since various and nonspatially contiguous ZIP + 4 codes shared the same price isoline, and since the properties were highly homogeneous between neighbourhoods, then the underlying geodemographic characteristics might explain the variation in prices between ZIP + 4 neighbourhoods, and commonality of prices between various ZIP + 4 neighbourhoods.

Figure 5.5 corroborates the qualitative analysis of Figure 5.4. The spatial trend is strongly east–west, with abrupt changes along parallel lines (streets) from north to south,[2] particularly in the northern study area. So, even though the physical

[2] Field inspectors confirmed that houses on main neighbourhood arteries were generally in acceptable quality condition whereas similar sized homes on adjacent streets were of lower quality.

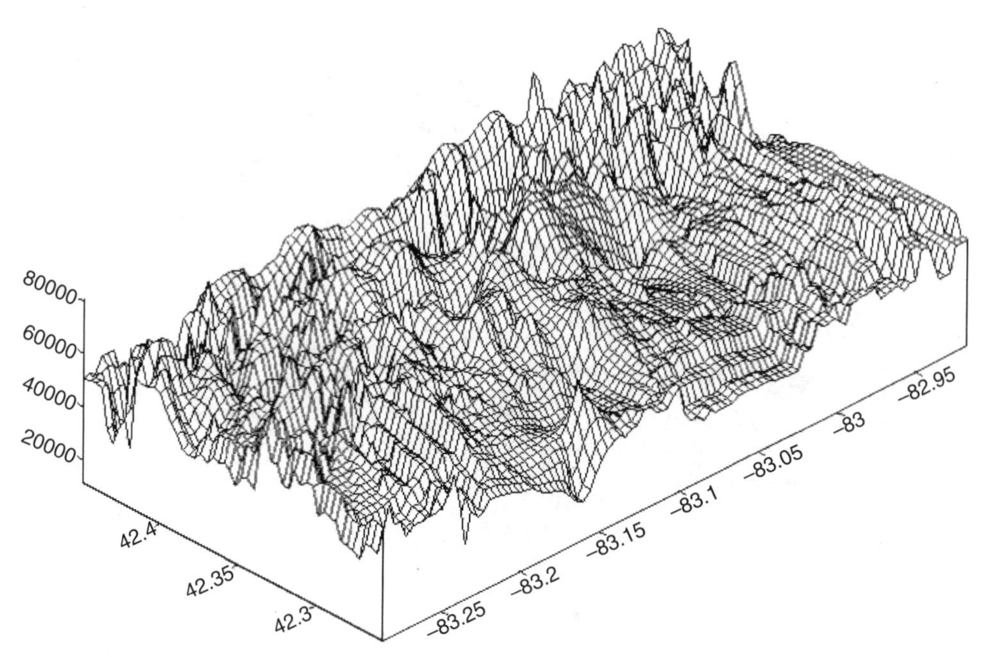

Figure 5.5 *3D map of home prices in study area*

properties were highly homogeneous, there is a strong geographic trend among the properties within what would be considered a sub-market. Therefore, any global averaging of housing values following conventional appraisal techniques would result in an underestimate of properties on some streets in the north, an overestimate of some properties on other streets in the north, and a highly mixed quality of assessment in the southern area of the sub-market.

Plate 5 provides the necessary geographic documentation (Thrall, 1998a) showing the location of the point observations along with the neighbourhoods generated by the isolines. Spatial smoothing can create surfaces even when no observations are present, as evident in the southeastern quadrant of the study area. In other words, the estimate of the surface in the southeast is not reliable. However, because of the apparent geographic regularity and the high density of the observations that give rise to the east–west trend and the abrupt changes from north to south, qualitative analysis here may be considered as reliable. This corroborates our hypothesis that LSPs at the ZIP + 4 scale might capture the underlying cause of spatial variation in prices of the dwellings. In other words, we hypothesise that the variation in prices within LSP categories is less than the variation in prices between LSP categories. Consumers are responding to very small-scale neighbourhood specific geodemographic characteristics, and that response creates a price effect. If so, then neighbourhoods within the study sub-market can be geographically discontinuous, while at the same time sharing LSP characteristic. Plate 5 was created in the same way as the other surface maps, but with optional parameters set differently in SURFER software.

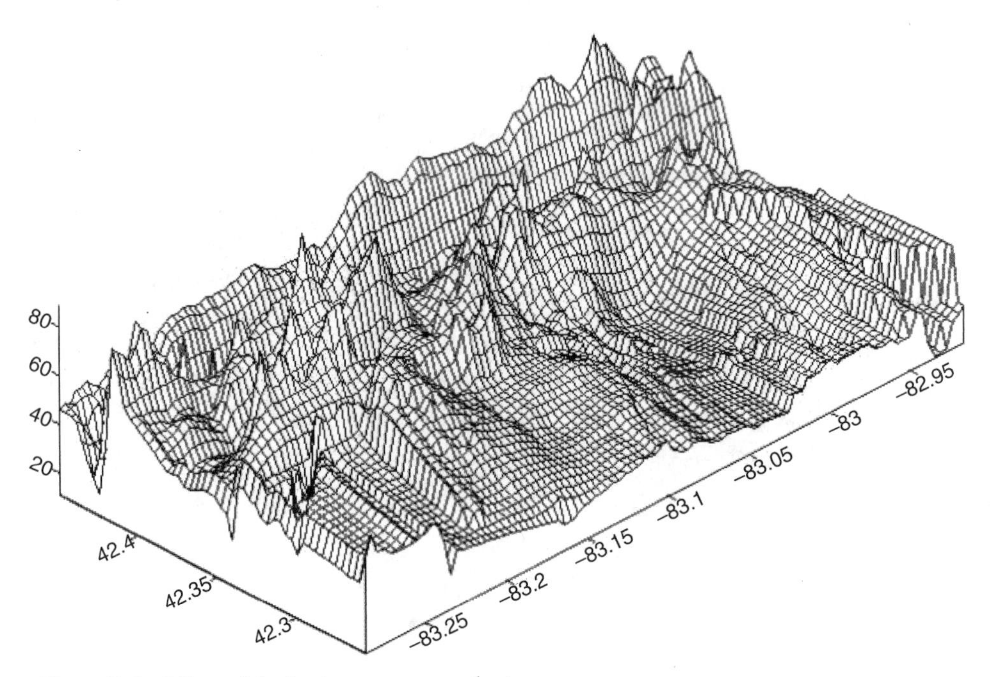

Figure 5.6 *3D model of price per square foot*

Figure 5.6 increases the vertical scale of price thereby allowing the analyst to observe and confirm the qualitative analysis presented in the other figures.

5.5 Conclusion

Practitioner applications often challenge the envelope of tidy academic solutions. Certainly as academics we would not have designed an experiment with an inherently problemsome design as this. However, practitioners are often called upon to provide acceptable solutions to real-world problems that are not as tidy as that derived from an ideal laboratory situation.

The authors support the inclusion of LSPs, such as ESRI's ACORN, in hedonic mass appraisal models. Visualisation with 3D maps assists in improving one's intuition about the geographic problems inherent in hedonic mass appraisal of urban residential neighbourhoods. A long-standing issue in mass appraisal has been how to define appropriately a neighbourhood and sub-market. Here we have demonstrated that LSPs can effectively define neighbourhoods, including neighbourhoods within a larger sub-market, that are not spatially contiguous. That is, within a sub-market, ZIP + 4s with the same or similar LSP can be grouped together to define a neighbourhood. Average square footage prices of houses can be effectively derived from those dwellings within a neighbourhood so defined. Future applications of this procedure might improve suburban residential hedonic mass appraisal that have proved problemsome because of their heterogeneity. The client accepted the value of our mass appraisal, and used that as a basis to improve their judgement and complete negotiations for the successful transaction.

References

Berry, B.J.L. (1964) Cities as systems, *Papers of the Regional Science Association*, **13**: 147–64.

Berry, B.J.L. (1969) The factorial ecology of Calcutta, *American Journal of Sociology*, **74**: 445–91.

Berry, B.J.L. (1970) *Geographic Perspectives on Urban Systems*, Prentice Hall, New Jersey.

Case, B., Pollakowski, H.O. and Wachter, S.M. (1991) On choosing among house price index methodologies, *AREURA Journal,* **19**: 286–307.

Clapp, J.M., Dolde, W. and Tirtiroglu, D. (1995) Imperfect information and investor inferences from housing price dynamics, *Real Estate Economics,* **23**: 239–69.

Hartshorn, T. (1992) *Interpreting The City: An Urban Geography*, Wiley, New York.

Hoyt, H. (1933) *One Hundred Years of Land Values in Chicago*, University of Chicago Press, Chicago.

Hoyt, H. (1939) *The Structure and Growth of Residential Neighbourhoods in American Cities*, Federal Housing Administration, Washington, DC.

Hoyt, H. (1966) *According to Hoyt – 50 Years of Homer Hoyt (1916–1966),* The Homer Hoyt Institute, Washington DC.

IAAO (1999) *Mass Appraisal of Real Property*, International Association of Assessing Officers, Chicago.

Shevky, E. and Bell, W. (1955) *Social Area Analysis: Theory, Illustrative Application and Computational Procedures*, Stanford University Press, Stanford.

Shevky, E. and Williams, M. (1949) *The Social Areas of Los Angeles: Analysis and Typology*, University of California Press, Los Angeles.

Thrall, G. I. (1998a) Common geographic errors of real estate analysis, *Journal of Real Estate Literature*, **6**(1): 45–54.

Thrall, G.I. (1998b) CACI Coder/plus version 1, Review, *Geo Info Systems*, **8**(6): 43–6.

Thrall, G.I. (2002) *Business Geography and New Real Estate Market Analysis,* Oxford University Press, New York and London.

Thrall, G. I., Casey, J. and Quintana, A. (2001) Geopsychographic LSPs, *GeoSpatial Solutions*, **11**(4): 40–3.

Thrall, G.I. and Thrall, S. (1993) Business GIS data, part two: highend Tiger/Line, *Geo Info Systems*, **3**(9): 66–70.

Weiss, M.S. (1988) *The Clustering of America*, Harper & Row, New York, pp. 4–5; as quoted in Hartshorn, 1992, p. 255.

Part Two
SOCIAL DEPRIVATION

6

Targeting Clusters of Deprivation within Cities

Richard Harris and Paul Longley

Abstract

This chapter considers the relative merits of census-based and lifestyles approaches to identify and represent spatially concentrated areas of urban deprivation (and affluence). Using census-based geodemographic indicators and lifestyles data supplied by a UK-based data warehouse company, we appraise three broad approaches: conventional deprivation indicators based on census information; hybrid approaches that link census data to more specific social survey information; and, finally, approaches that make use of private sector lifestyles surveys that can be combined with high resolution imagery of the built environment to more precisely determine social and built conditions. We argue that appraisal of these different methodologies and of the representations of socioeconomic geography they create enhances our understanding of uncertainty in GIS and other geographic modelling. It also casts light on the role and substance of generalisation in urban analysis. We explore these issues in relation to the practice of marketeers working within the geodemographic and lifestyles industries.

6.1 Introduction

Urban geography seeks to identify and understand the geographical patterning of urban landscapes and urban populations in relation to the processes that give rise to those patterns. By forming general understandings of such processes and of the inter-relationships between processes, people and places there is a possibility both to compare the socioeconomic, demographic or environmental structures of different places at particular points in time, and to predict and simulate how settlements might

Applied GIS and Spatial Analysis. Edited by J. Stillwell and G. Clarke
 ISBN: 0-470-84409-4

evolve in the future. These understandings are essential to the governance and management of urban environments and systems. However, the detailed richness of any (specific) representation is always depleted by the process of generalisation when hard decisions are made regarding what to deem as irrelevant detail in order to establish baselines of comparability. Rarely is all of what is discarded wholly irrelevant to the process of building representations (Longley *et al.*, 2001, Chapter 3).

Donnay *et al.* (2001) make this clear in the context of urban remote sensing and image classification. They see advantages of a generic model in the power of constructing a relatively simple (and understandable) representation of the properties of a large number of objects and in the portability of that general model from one location to another, permitting comparison between places. Yet, the match between a generalised model and the specific patterns observed in any scene normally will not be exact; there is inevitable uncertainty associated with the application of a generic model to a particular setting. Such uncertainty is likely to be greater than that associated with more specific, finely focused case studies. But if the merits of generalisation are accepted as worthwhile, then attention focuses upon how to minimise the uncertainty associated with the practice of abstraction and generalisation; how to ensure general models are 'fit for purpose' and usable.

In the past, outdated, imprecise, coarsely aggregated and indirect data sources have done urban geographers few favours in their search for empirical regularities and general understandings of urban growth, structure and morphology. Given those circumstances it is unsurprising that many geographers have become disenchanted with working with inert spatial mosaics and sometimes absurdly surrogate data, instead channelling their efforts to detailed, qualitative and ideographic approaches. At its best, urban geography continues to generate valuable insights into urban lifestyles and landscapes. However, there is also a gathering sense that the subject has become overly preoccupied with introspective and individualistic case studies and anecdote at the expense of more system-wide understandings of urban environments and systems. It is ironic (or perverse) that the relative abandonment of systematic, quantitative approaches in much of urban geography has come at a time when the data infrastructure for representation building has improved enormously, offering more direct and meaningful sources of information and analysis. In the UK, for example, Ordnance Survey (OS) recently has launched MasterMap, an ambitious project to provide a consistent, national framework for referencing geographic information. The foundation of MasterMap is a database containing information about polygons that together represent 400 million man-made and natural features (for example, individual fields, lakes or pillar boxes) and includes National Grid, GPS and detailed topographic information for all features (see www.ordsvy.gov.uk).

We are not suggesting that only that (large, and increasing) subset of phenomena that can be incorporated in quantitative representation should be deemed worthy of academic study; nor are we implying that better data and better data-handling technologies are a panacea to the limits of GIS-based representation. However, it is increasingly apparent that the worlds of business and commerce, with their pragmatic tactical and strategic concerns, better understand the importance and power of generalisation than much of academia. We view this as an odd, undesirable and somewhat self-defeating state of affairs when policy-makers at many levels of governance (in both the private and public sectors) are increasingly becoming minded to ask geographical questions, and look to geographical data and analysis for the answers. Consider, for example, the UK Government's Policy Action Team (PAT) 18 report on better information which states that 'comprehensive and up-to-date [geographical] information about deprived neighbourhoods is crucial to the National Strategy for Neighbourhood Renewal [. . .] Far too often information is seen as an additional and peripheral issue in regeneration activities, rather than as central to them. In reality, information is one of the key tools in the drive towards a more inclusive society' (PAT 18, 2000, p. 5).

One outcome of the Action Team's report was the setting up of a Neighbourhood Statistics Service (NSS) for the UK (see www.statistics.gov.uk). Yet, despite this important initiative we are actually little closer to answering one of the main questions that acted as a catalyst to creating PAT 18 and the NSS in the first instance. Specifically, at what scale does deprivation and social exclusion exist and persist? That is a question that the information currently provided by the NSS cannot answer because it is constrained to a particular, possibly inappropriate, zonal geography prior to dissemination (namely, to the 1991 UK Census ward geography). Use of the information presumes the census geography is suited to the task of targeting deprivation and measuring variation in socioeconomic conditions – a presumption that we have challenged (Harris and Longley, 2002).

Elsewhere we have followed the lead of business and commerce and looked to the datasets they are using to try and produce more finely attuned and targeted measures of deprivation (Harris and Frost, 2003; Harris and Longley, 2002). We have sought to navigate a path between the polarities of 'scientific but outdated and dull' academic research, and 'unscientific but rich and timely' analysis of private sector databases (Longley and Harris, 1999). In this chapter we continue along that path in the context of studies of deprivation in urban areas. We argue that clearer thinking about the conception, measurement and analysis of deprivation has far-reaching implications for anyone with interests in social exclusion. We focus on establishing the scale at which private sector sources of income information can be utilised to provide robust and comparable estimates of geographical concentrations of deprived (or, conversely affluent) income groups.

6.2 Geographies of Deprivation

The conception of deprivation is very much a contested domain: different conceptions lead to different geographical representations of physical and social conditions. Any representation can come to reflect the priorities of those who do the representing. The nomenclature of deprivation is often linked to notions of poverty, inequality and social exclusion. It can also have an absolute or relative meaning. Few could argue that the estimated average 24000 people who die each day worldwide from hunger – 75% of whom are children aged under five – are not deprived (The Hunger Site: www.thehungersite.com). However, it is a moot point whether those who (are said to) wear new clothes, own a microwave and also a VCR can honestly be described as living in poverty (*Daily Telegraph*, 2000 responding critically to a report by the Joseph Rowntree Foundation: Gordon *et al.*, 2000).

In generally affluent societies such as the UK's, few people are deprived of access to food, clean water, health, shelter and education (the UN's definition of absolute poverty). However, that does not mean that levels of wealth, well-being and social opportunity are harmonious across society – they are not. Hall and Pfeiffer (2000, p. 81) write that 'the optimistic high-growth period after World War II saw poverty in retreat; even experts could not imagine that inequality would increase again – yet in many countries, and cities, it has' (they cite London, New York and Los Angeles as examples). The emphasis on inequality is indicative of much deprivation-related research in the UK. For example, Shaw *et al.* (1999) and Dorling *et al.* (2000) focus upon inequalities in health outcomes in Britain. Measures of deprivation assume relative connotations, often defined against mean or median household. But this raises the question of what constitutes an inequality? The answer is necessarily subjective. Ultimately, the way in which deprivation is conceived conditions the way that it is measured (Yapa, 1998).

Even if there was consensus about the physical and social measures and manifestations of deprivation, to identify, measure and target geographical clusters of the deprived 'on the ground' requires consistent and task-relevant geographical information to be available to the analyst or policy maker. The usual areal 'containers' of deprivation in the UK – that is the zones to which social and physical characteristics are ascribed – are (1991) Census wards, containing on average a household population of 2081 but varying from a minimum of 153 households to a maximum of 9591 (Openshaw, 1995b). Are these geographical zones really the best suited for measuring and monitoring deprivation? The decision is not neutral in its effects: the choice of geography can alter the apparent intensity and extent of deprived conditions in various locations (Openshaw, 1984). Scale and aggregation effects have profound impacts upon the representation of measured attributes.

Differences of conception compounded with vagaries of measurement have far-reaching consequences for the identification and representation of deprived areas. Harris and Longley (2002) contrast the outcome of Lee's (1999, p. 178) study of local conditions with those of Gordon and Forrest (1995). The former, calculated using a central government measure (Department of the Environment, 1981), suggested that 40% of the most deprived wards in England and Wales are located in Greater London and South East England. The latter (the 'Breadline Index', described below) suggests that the same two regions actually account for 24% of the most deprived wards.

Harris and Longley (2002) also consider Lee *et al.*'s (1995) review of areal measures of deprivation, including the (then) Department of the Environment's 1981 and 1991 Indices (Department of the Environment, 1981, 1994), Jarman Scores (Jarman, 1983), the Townsend Index (Townsend *et al.*, 1988) and an index developed from the Breadline Britain poverty study (Gordon and Forrest, 1995; Gordon and Pantazis, 1995). Lee *et al.*'s study emphasises how apparent similarities or differences between various indices are a function of the variables that make up a particular index and also the scale at which the comparison is made. Lee *et al.* conclude that the various indices for measuring and mapping deprivation differ in terms of how they are constructed, what their constituent variables purportedly represent and how they are weighted. Despite this, at the coarse Census district scale of analysis, the different indices produce similar deprivation profiles and identify similar lists of the most deprived areas. However, at the ward level and below, the opposite is true – different indices produce quite different results. Multivariate indices produce better representation of the multifaceted nature of deprivation than univariate measures and can identify different populations and areas as being deprived. Yet, a large proportion of the households identified using univariate measures such as unemployment or overcrowding will also be flagged in multivariate indices.

6.3 Census-based and Hybrid Measures of Deprivation

The indices that Lee *et al.* (1995) review are typical of those that are reliant (either directly or indirectly) upon data collected in the UK Census of Population – the only comprehensive data source for mapping the widely recognised indicators of deprivation or poverty within small areas (Lee, 1999, p. 172). The 'preferred status' of the Census arises from the high proportion of the population that it enumerates and the high public sector quality control standards that govern its collection. Nevertheless, its reputation was tarnished in 1991 because of the 'missing millions' omitted (following the debacle of the Poll Tax: see Openshaw, 1995a, p. 14) and the confusions surrounding whether students should be enumerated at home or at their term-time address (Mitchell *et al.*, 2002). The overall response rate for the 1991 Census was about 97% of Britain's 55.94 million residents (Champion, 1995), but with increased under-enumeration of particular sub-groups of the population, notably city dwelling, young males. The response rate for the 2001 UK Census is estimated at about 98% but with an increased non-response rate for some questions. Questions that in 1991 met with a non-response of less than 1% tended to have, in 2001, a non-response of 1–5%. Questions with a non-response rate of 1–5% in 1991 had a rate of 5–10% in 2001. Almost one in five individuals did not complete the 2001 census question enquiring about their professional qualifications (http://www.statistics.gov.uk/census200/pdfs/dataquality.pdf). The trend to increased under-enumeration is worrying, though in an international context it is worth mentioning that the overall response rate achieved by the 2000 US Census was only 67%. Nevertheless, it is clear that modern lifestyles and attitudes make it increasingly difficult for even compulsory public sector surveys to achieve complete enumeration of the population. Ironically, the rather narrow range of questions which the UK Census asks sheds little light on those modern lifestyles and attitudes (Longley and Harris, 1999).

A particular problem in the context of deprivation analysis is that UK Census data contain no direct indicators of income, which most would agree to be a direct and relevant measure of deprivation. Analysis is also constrained by the 10-yearly interval of Census data collection and the resulting problem of data ageing. In the UK there is still no strategy for either enriching the Census using a wider range of pertinent indicators or improving temporal frequency at the neighbourhood scale. In the US and UK alike there is, however, an open discussion of replacing census collection with a fusion of other public or private sector databases to obtain the same sorts of information collected by the Census (and more). At their best, the data arising out of such approaches would be more wide-ranging and more frequently updated than conventional censuses. However, the vagaries of data fusion raise important issues of measurement error in the source data and of tracking and quantifying error propagation as the different datasets are merged.

The limitations of the Census for socioeconomic research apply equally to business and service planning applications using geodemographic classifications, such as ACORN, Super Profiles or other census-based typologies (Longley and Clarke, 1995; Birkin *et al.*, 2002). Sleight (1997) describes geodemographics as the 'analysis of people by where they live' – an apt description for area-driven deprivation research also. Indeed, Voas and Williamson (2001) describe geodemographic classifications as no more than 'dressed-up measures of affluence' (conversely, measures of deprivation). Further similarity is found in the efforts of practitioners from both research fields to 'freshen-up' census data by linking them to more up-to-date and relevant but less geographically replete samples of survey data. Batey and Brown (1995), for example, describe how the Census core of the Super Profiles classification is 'cocooned' by an updateable layer of non-census data.

In the context of deprivation measures, the 'Breadline Britain' index is an interesting hybrid measure that has been described by Lee *et al.* (1999) as the best index for identifying the most deprived wards (in Britain) at a national level. Full details of the design and implementation of this index are given by Gordon and Forrest (1995), Lee *et al.* (1995), Gordon and Pantazis (1997a, 1997b), Gordon *et al.* (2000) and also Harris and Longley (2002). In brief, it is based on a quota sample of 1319 individuals aged 16 or above supplemented by an additional 512 quota interviews carried out in low income areas: areas identified as Group G ('Council Estates') of the 1981 ACORN geodemographic system (Gordon and Pantazis, 1997a). Interviewees were asked to identify from a list of 44 items (that covered a range of activities and possessions) those that were felt necessary and important to everyday living (Lee *et al.*, 1995). Items attracting a total of 50% or more support were deemed as socially perceived necessities by 'a democratic majority' (Gordon *et al.*, 2000, p. 13). A poverty line was then set at households lacking three or more necessities. On this basis the sample was classified according to an 'in poverty' (1) versus 'not in poverty' (0) dichotomy. It was assumed that few differences exist between different sections of the population as to what is regarded as a necessary item.

The results of the Breadline Britain sample survey were then generalised to the population at large. The survey data were limited to 1831 individuals. Because no similar information existed for the rest of the population it was not known (at a national level) who else in the population owned fewer than three of the necessities. To resolve this problem a link was created between the sample data and the more restricted range of variables available in the 1991 Census small area statistics by identifying variables common to both. A logit analysis was undertaken at the household scale, with deprived/not deprived as the response variable. The explanatory variables were the socioeconomic indicators of the survey that could be paired with like variables in the census (Gordon and Forrest, 1995). The regression parameters arising from the analysis were subsequently transferred from the household scale to weight corresponding census variables at both the ward and census district scales (Lee *et al.*, 1995). The resulting Breadline Britain Index is used to estimate the percentage of 'poor households' using the following regression relation:

$$y' = 0.2025x_1 + 0.2174x_2 + 0.1597x_3 + 0.1585x_4 + 0.0943x_5 + 0.1079x_6 \qquad (6.1)$$

where y' is the estimated percentage of poor households (per census ward or district) and x_1 to x_6 refer to the following 1991 census variables respectively: % households not owner-occupied; % households with no car; % lone parent households; % households where household head is of partly-skilled or unskilled occupation; % economically active population unemployed; and % households containing a person with a limiting long-term illness. Each percentage figure pertains to a census ward or a census district, depending on the chosen scale of analysis – the parameter estimates are (wrongly) taken to be independent of the scale of analysis and hence remain constant.

Although the method of the Breadline Index is an innovative means to break out of the straightjacket of conventional, census-based deprivation measurement, the user remains constrained by two main shortcomings. First, the Index is based on a small sample and generalised to the whole UK in a geographically invariant way. A single, national set of parameter estimates is created by the linkage mechanism although it is likely that the weight given to each census variable will be more appropriate in some locations and some scales of analysis than others. Second, the Index is used (and is relatively robust) only at the census ward and district scales. There is increasing evidence that even small areas conceal diversity in population characteristics and social conditions (e.g., Longley and Harris, 1999).

6.4 'Small' Areas and the Problem of Diversity

'And cities? They are diverse' (Amin *et al.*, 2000, p. 22).

A significant weakness with the use of census data for deprivation and other area-based analysis is the forced assumption that populations within 'small' areas – meaning census enumeration districts (EDs) or wards – are homogeneous in terms of the socioeconomic conditions in which they live; or, at least, that suppression of within-area diversity is a sacrifice worth making in order to forward the cause of generalisation. Much of our research has contested that assumption (or suppression), focusing on diversity *within* administrative areas. That is not to say that geographical trends and patterns do not exist *between* administrative areas; on the contrary, our argument has been that they do (Harris, 2001; see also Voas and Williamson, 2001). But what of trends and patterns that transcend administrative areas, or are hidden within them? The principal issue is not whether there is a geographical patterning imprinted on the socioeconomic landscape around us – undoubtedly there is – but, instead, the extent to which such trends and patterns are obscured or clarified by the nature of areal aggregations, including those of the census, or by sources of geographical information that are hardly fit for purpose.

If imperfections and mismatches between spatial distributions are likely to exist, a role for GIS is in quantifying the sources and operation of uncertainties through internal and external validation. Internal validation might, for example, entail experimentation with the scale and aggregation effects arising from different combinations of the smallest available areal units of analysis (e.g., Openshaw, 1984; Alvanides *et al.*, 2002). Such experimentation can generate a large range of outcomes but, at the end of the day, there is no incontestable 'optimal' solution. Worse, the results must still take the within-area distributions of the elemental units of analysis as uniform. External validation entails use of ancillary sources that are not constrained by the same (or, ideally, any) zonal geography to reveal more about small area diversity.

A simple example is given by Figures 6.1a and 6.1b, showing the southern half of a map of unemployment in Bristol, for which the unemployment attribute is divided into quartiles. Note that the spatial distribution appears different whether a count of unemployment (Figure 6.1a) or a rate of unemployment (Figure 6.1b) is derived. Consequently it is unclear whether the zone labelled A belongs in a functional sense to the zones to its east or to its west. There is little way of resolving this issue using

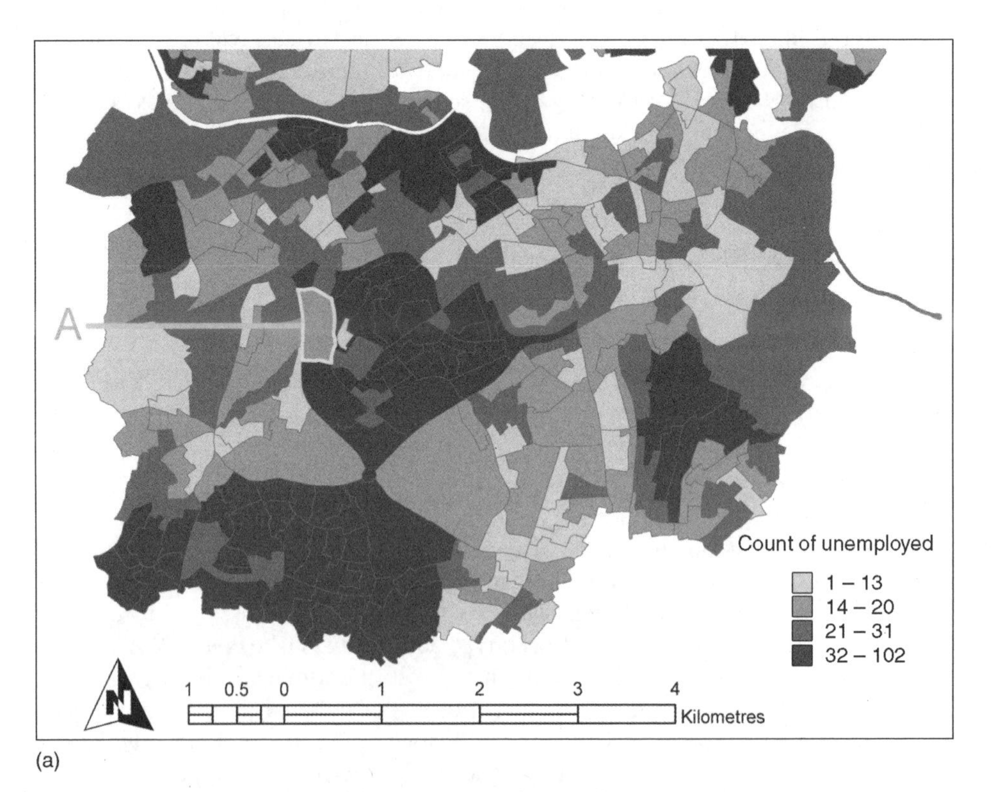

Figure 6.1 *Measures of unemployment in South Bristol, UK*

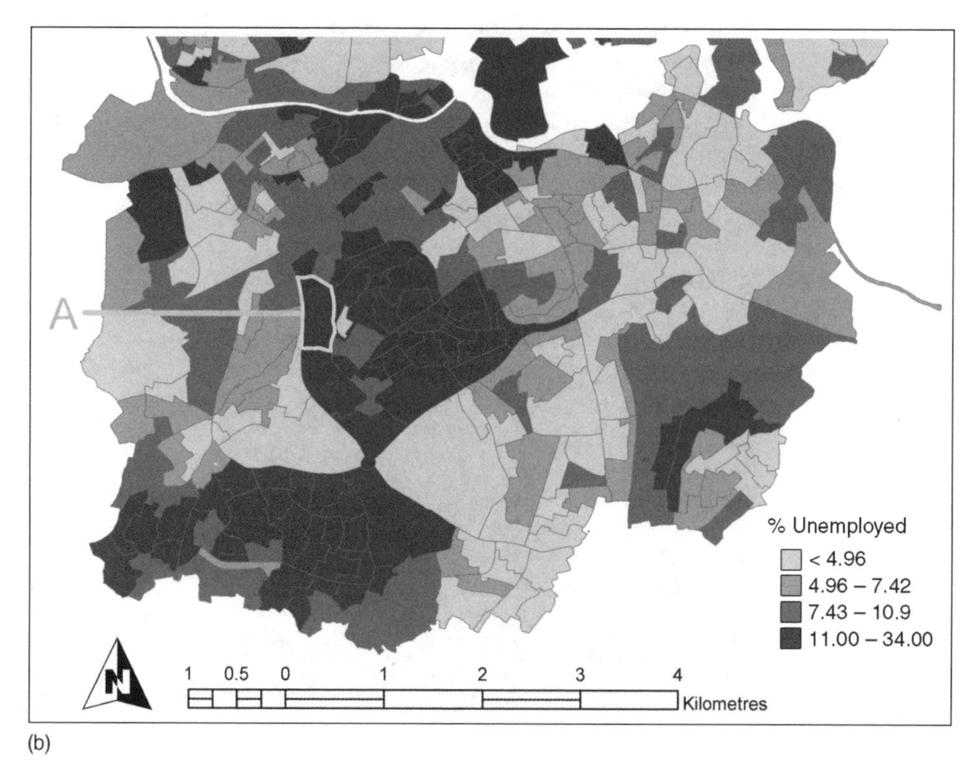

(b)

Figure 6.1 *Continued*

census data alone. Yet if the area is viewed using high resolution aerial photography, it becomes clear that the area is of mixed land-use: part residential; part industrial; part open space (Figure 6.2). The (small) population of the zone is concentrated to the north-west. Within GIS, a procedure known as dasymetric mapping (Donnay and Unwin, 2001) could be used to mask out unpopulated areas partially in order that census enumeration district (zone) counts might be averaged only over the built-up portion of the zone. In the same spirit, Longley and Mesev (2002) integrate remotely sensed and census data to produce a hybrid representation of the spatial distribution of occupied built structures, and thence create fine-scale density profiles of a city (see also Harris, 2003).

In each of the examples, the ancillary source (aerial photography or satellite imagery) does not identify unique geographical individuals: the element of analysis is determined by the granularity of the image; each individual, pixelated element might enumerate more than one, zero or indeed fractions of human individuals. A different approach is to use ancillary sources that contain individual population records. In this we have taken our cue from the marketing industry, where the era of information commerce and 'one-to-one' marketing has given rise to detailed, disaggregate datasets, commonly about people, their 'lifestyle preferences' and their shopping habits (Longley and Harris, 1999).

Figure 6.2 High resolution aerial photograph showing land use in zone A. Source: Cities Revealed (1996)

With the variety of analytical tools now commonplace in GIS software, there exists potential to adopt more flexible and geographically sensitive approaches to classification and modelling of socioeconomic conditions. This provides some defence against the charge that today's planners and policy makers have much better information about the attributes of the built environment from the greatly increased resolutions offered by remotely sensed (RS) imagery (Donnay *et al.*, 2001) than they do about the most important part of an urban settlement: the people who live and work there (Harris, 2003).

6.5 Refocusing on Income

To summarise, the lack of conventional survey alternatives has forced reliance on the census as the only viable source of fine-scale socioeconomic data (Green, 1997). Where alternative public sector data do exist they tend to be *ad hoc* and geographically limited, generated in response to the specific needs of individual councils, planning departments and research projects (e.g., Longley *et al.*, 1989) and thus not well suited for more general model building or comparative analysis. However, the data blockages to less ambiguous measures of poverty are not insurmountable. Official (governmental) datasets do not provide the only useful sources of usable information in today's information commerce, although academics and marketeers do appear to have differing perspectives on what is attainable. In the academic domain, for example, Green (1997, p. 181) laments the '. . . dearth of spatially disaggregated [income] data at the local (particularly sub-county) scale'. In a similar vein, Lee (1999, p. 175) notes that the use of unemployment rate as a principal indicator of poverty is largely a response to the lack of household income data. By contrast, private sector companies such as EuroDirect offer commercial products which allow unit postcodes to be selected on the basis of the household income levels found within them (EuroDirect, 1999). A recent study by the market research group CACI also focused upon household income and claimed to give a truer picture of wealth differentials at disaggregate scales than measures provided by the unemployment register (Baldwin, 1999).

There are a number of advantages to the use of private sector 'lifestyles' datasets that can be defined as composite databases created and updated from sources of individual and household data including shopping surveys, loyalty card information, credit ratings and the guarantee returns from consumer products. The advantages include the fact that lifestyles data can be made available to describe households at the scale of the unit postcode. This frees analysis of heterogeneity in social conditions from the out-dated 'mosaic metaphor' which underpins much census-based urban geography (Johnston, 1999; Raper, 2000). The data are also more up to date (shopping surveys of this nature typically are conducted annually) and thus present a more relevant picture of conditions in fast-changing urban systems. Most importantly of all, lifestyles datasets often do have an income question, the nature of which we consider below. This obviates the need to fall back on the indirect surrogate measures that characterise census-based deprivation studies. It is disingenuous to overlook these advantages, even if the quality of the data is at times considered uncertain or unknown (Goodchild and Longley, 1999; Harris and Longley, 2002).

Harris and Frost (2003) argue that despite the uncertain quality of lifestyles data, they can still be used to identify, at the unit postcode level, the location of every survey respondent (in the Borough of Brent, London, in their case study) who indicated their annual family income to be less than £10 000. The respondents to lifestyles surveys are self-selecting from the mailing lists and other sample frames used by lifestyles companies (there is no compunction upon households to complete and return the survey). Nevertheless, if low income respondents are shown to be clustered spatially, then there is cause to believe that there is a clear areal basis to income deprivation, particularly if a reasonable number of the clusters are shown to be within, or in proximity to wards identified as deprived according to more orthodox measures. Lifestyles data and more conventional sources can be used to cross-validate each other, or be combined to produce more data-rich models of the urban fabric (Harris and Longley, 2001).

To map or to create spatial statistics describing any of the data stored in a data warehouse necessarily requires a way of assigning location to each individual or household record. One way of doing this is to use (GB) Ordnance Survey's (OS) Address-Point product to ascribe a UK National Grid reference of up to 0.1 metre precision to the full property address. An alternative is to use OS Code-Point data which are cheaper for commercial organisations to purchase and are less voluminous to process. Whereas Address-Point assigns (in principle) a georeference to each residential and commercial address in Britain, Code-Point is intended for analysis at the slightly coarser, unit postcode scale (with an average of 12–16 residential delivery points per postcode). The Code-Point file lists, for each postcode in Britain, the grid reference of a point that is located within the footprint of the property closest to the population-weighted centre of the postcode. A Code-Point is defined by first calculating the average Easting and Northing location for all Address-Points within a postcode, and then by identifying the Address-Point closest to that location. The closest Address-Point defines the Code-Point for the postcode (although the grid reference is degraded to a maximum of 1 metre precision). Further comparison between the Address-Point and Code-Point products is made by Harris and Longley (2000).

Code-Point, Address-Point and the MasterMap products each offer ways of ascribing point locations to new souces of attribute data at ever finer levels of granularity. A problem, however, is that points are, by definition, dimensionless and, in a strict sense, occupy zero area on a map. This is clearly a problem if the intention is to use such data to explore and represent socioeconomic distributions! Commonly the problem is 'resolved' by representing a point using a circular symbol, the radius, circumference, area or shading of which can be arbitrary, tailored to enhance the aesthetics of the map or proportional to the value of the phenomenon at the point. Alternative, space-filling methodologies are less cosmetic and may cause permanent changes either to the source data structure or to the attribute table. A variety of kernel, neighbourhood, density and other estimation techniques exist that convert vector points to a near-continuous surface of raster cells, or append a measure of the local intensity of a given phenomenon to the dataset, on a point-by-point basis.

In the context of modelling lifestyles data to produce local indicators of socioeconomic conditions, the advantages of kernel or neighbourhood estimators have been documented by Harris (2003). By analysing each record – each individual or household – as part of a larger and geographically clustered group the sample size is increased and the error component is (in principle) reduced. This approach helps counter the problems of sample and response bias that affect lifestyles datasets to varying degrees (for example, there is evidence that postal shopping surveys tend to miss young, male adults: Longley and Harris, 1999). However, it also raises questions concerning the most appropriate scale at which the geographical clusters should be defined; what is 'the' correct level of aggregation? What neighbourhood, kernel or search window size should be used? This is an important consideration, since the choice has a considerable effect on the amount of smoothing applied to the data. Local estimation techniques are inherently spatial statistics in that the results (the output) are by no means independent of the geographical scale of the analysis (the size of the search window).

In their study of small area income differences within Brent, London, Harris and Frost (2003) choose a circular search window of radius 500 m. But their choice was essentially arbitrary, based only on experience of 'what seems to work'. Even assuming that the chosen scale was indeed appropriate to the analysis, a second question is immediately raised: is the search window size equally appropriate across the entire study region or should it vary geographically in order to adapt to local conditions?

An example of the latter, adaptive approach is Bracken and Martin's population surface algorithm (see Martin, 2002) designed for remodelling census population data into more realistic portrayals of population distributions. The algorithm incorporates a kernel, the size of which is sensitive to local population densities: the kernel is smallest where densities are highest; widest where densities are lowest.

An analogy to fixed versus adaptive search windows is provided by the histogram. A histogram can be plotted using equal class intervals (fixed bin widths) but unless the observations are uniformly distributed between the classes, differing numbers of observations in each class will result. The alternative is to adopt a quantile classification – as in Figure 6.1 – for which the numbers in each class are harmonised by using unequal bin widths (class intervals). The relatively simple procedure outlined by Harris (2003) uses the 'quantile approach', finding the n_i points from a total set of N that are calculated as being at least distance from a single point i (which is also drawn from the set of N). These n_i points, together with point i itself, are then taken to form a 'cluster' or cloud of $n_i + 1$ points and a total population size of m:

$$m = p_{0i} + p_{1i} + p_{2i} + \ldots + p_{ni} \tag{6.2}$$

where p_{0i} is the population at point i, p_{1i} is the population at the (first) nearest neighbourhood point to i, p_{2i} is the population at the second nearest neighbour, and so forth to the nth neighbour. Each of the points is a residential (unit postcode) CodePoint, and the straight-line distance measure does not attempt to accommodate explicitly the effects of built form or spatial structure (e.g., where actual travelled routes differ significantly from the straight-line crow's distance between points or where, in reality, a physical barrier exists that clearly separates two or more points, all be they in seeming close proximity).

The procedure is then repeated for each of the remaining ($N - 1$) points in turn. The method therefore creates as many clusters as there are points (N). Most households will be allocated to more than one cluster because any individual point can be a nearest neighbour of more than one other. The value of n_i, the number of CodePoints in each group, will vary from cluster to cluster as it is m (the total population size) that is held constant. The value of n_i is governed by both the variable size of postcode populations and the local sample/response rates for the lifestyles survey data that Harris (2003) analysed. Having grouped the households, Harris was able to identify the most frequently occurring (modal) property type within each cluster and also the proportion of households residing in that type. If there are three property classes and the number of households belonging to each class is z_1, z_2 and z_3 respectively, then, assuming z_1 is the modal type:

$$z_1 > z_2 \quad \text{and} \quad z_1 > z_3 \tag{6.3}$$

and

$$p_{z1} = z_1 / m \tag{6.4}$$

where

$$m = z_1 + z_2 + z_3 \tag{6.5}$$

or

$$m = p_{0i} + p_{1i} + p_{2i} + \ldots + p_{ni} \tag{6.6}$$

where p_{z1} is the proportion of all households belonging to the modal type.

The proportion – a simple measure of within-cluster homogeneity – was assigned back to the point (i) at the centre of each cluster and this information was used to guide the selection of training samples to classify a remotely sensed image into classes of residential building densities. Training samples were drawn from a Bristol study area where it was most certain that a particular property type existed: areas with the greatest local homogeneity of a dominant property type (Figure 6.3).

If the procedure described by Harris (2003) circumnavigates the issue of what is the correct search window size to aggregate the lifestyles data, it immediately encounters a related problem instead: what is the correct class membership size, m? The Bristol study took a value of $m = 50$ but this was largely chosen to illustrate the analytical methodology rather than for any more considered reason. For the rest of this chapter we therefore consider the effects of changing the value of m on small area measures of modal income and of income homogeneity in London.

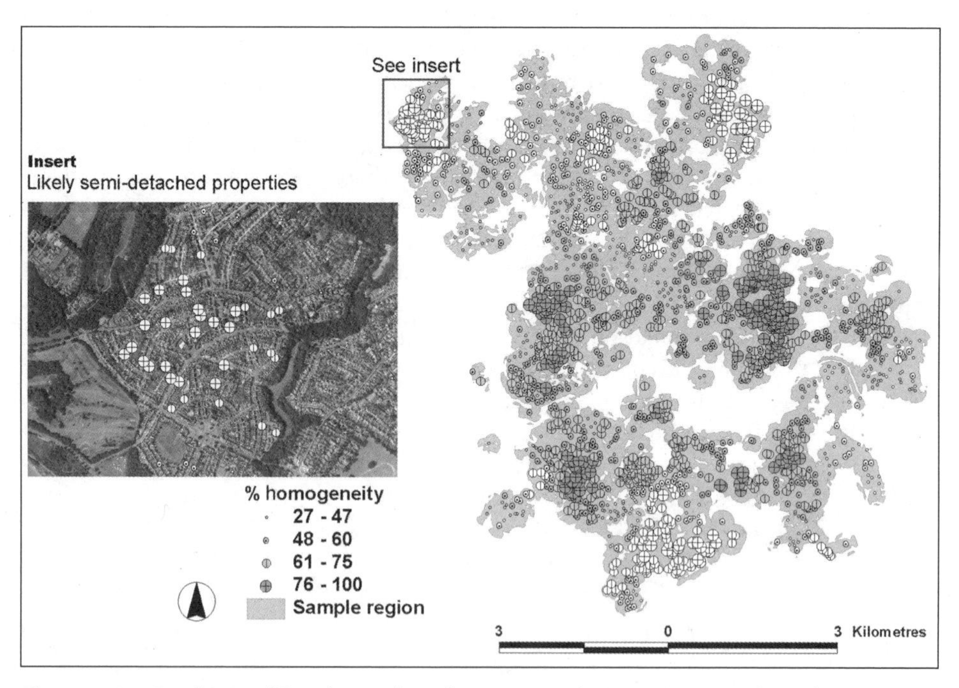

Figure 6.3 Combining lifestyles and CodePoint population maps to select, from a Cities Revealed image, training samples with high likelihood of being semi-detached properties. Source: Harris (2001)

6.6 Modelling Income Concentrations in London

Figure 6.4 shows a map of low income clusters in Inner London derived using the methodology described by Harris (2003) but with a value of $m = 75$ and actually undertaken for the whole Greater London region (only Inner London is shown to retain visual clarity of the income distributions). The source data is a commercial lifestyles database compiled from replies to a national consumer survey. That survey was postal, undertaken in England, Wales and Scotland during the spring of 1999, and sent out to a high proportion of households for whom at least one member was listed on the UK electoral register for the three countries. The full lifestyles database contains a wide range of socioeconomic, behavioural and consumer information that describe approximately one million households. We have selected from that dataset those households that gave an indication of their gross household income and have then georeferenced those records, by residential postcode, to an OS CodePoint. This leaves us with a sample of 65 685 households in 42 883 unit postcodes, all of whom

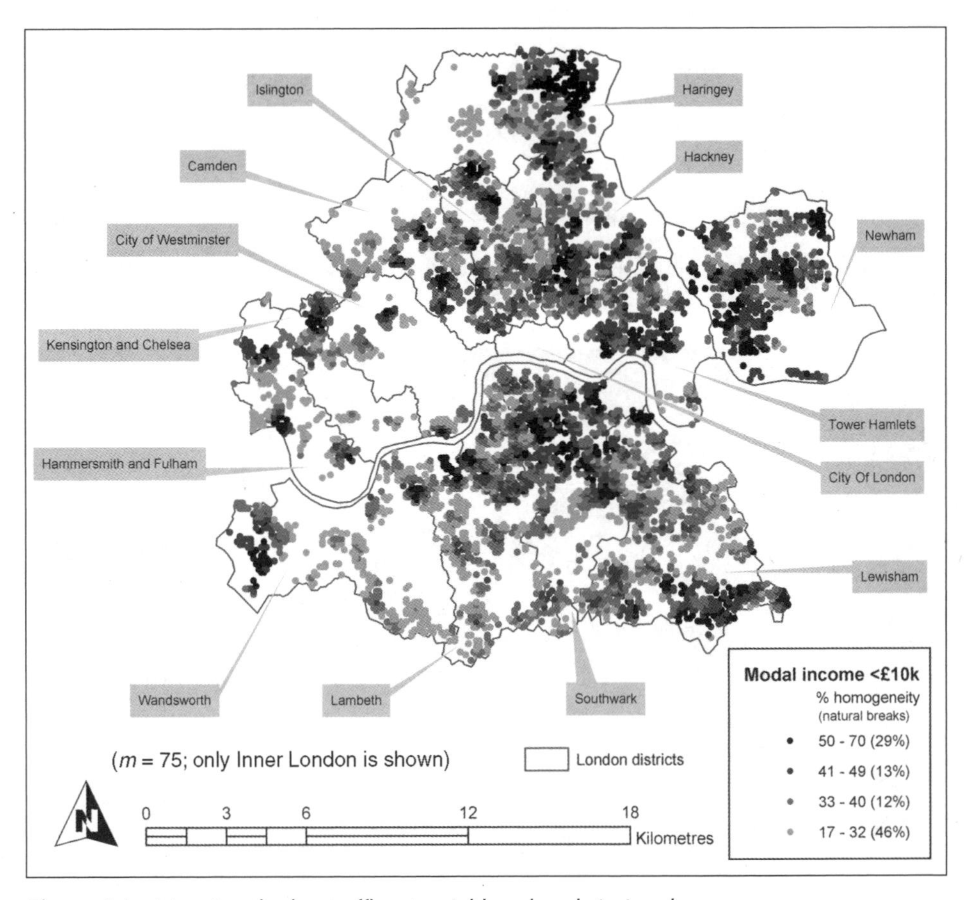

Figure 6.4 *Mapping the least affluent neighbourhoods in London*

have a household income estimated to be in one of six income bands: <£10000; £10000–£19999; £20000–£29999; £30000–£39999; £40000–£49999; £50000 and above. According to Department of Transport, Local Government and the Regions' projections from 1996 household numbers, there are 3.13 million households living in Greater London (http://www.statistics.gov.uk/statbase/Expodata/Spreadsheets/D4524.xls). Hence, the lifestyles income sample is of a size equivalent to 2.1% of households (consistent with Harris and Frost (2003) who estimated the sample size to be just under 2% for the Borough of Brent within London).

Applying the modelling procedure described in Section 6.5 permits the determination of which one of the six income types is dominant in each small area cluster (formed around each of the 42 883 CodePoints) and also what proportion of sampled households are of the modal type per cluster. In Figure 6.4 we have grouped the classes to emphasise areas (in Inner London only) where the lowest incomes (<£10000) dominate, and in Figure 6.5 the highest (£50000 or above). The average circular search window required to find the n_i neighbours that give a total population of $m = 75$

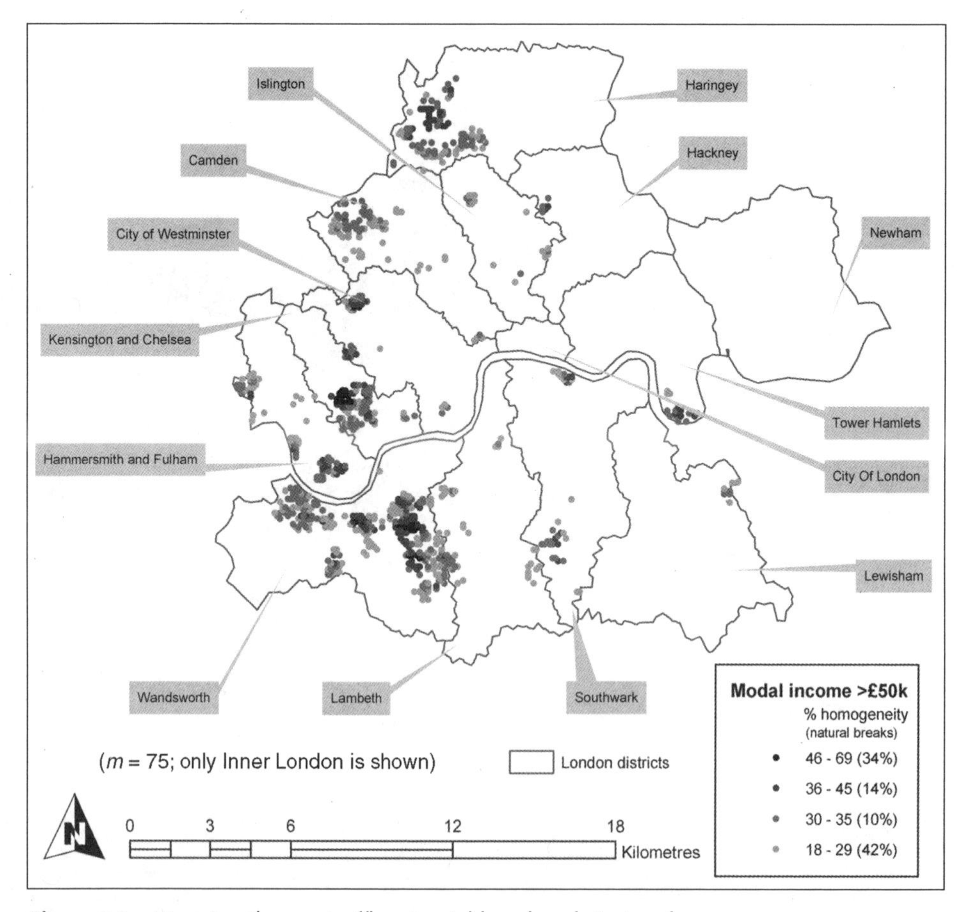

Figure 6.5 *Mapping the most affluent neighbourhoods in London*

per cluster has a radius of 452 metres. We have already noted that there is no reason to suppose that the distribution of lifestyles respondents is evenly distributed across the study region and, accordingly, the radius varies from a minimum of 141 metres to 3999 metres for the set of 42883 points. The interquartile range is much smaller, however: from 361 metres to 510 metres. Both Figure 6.4 and Figure 6.5 show the proportion of households within the neighbourhood that belong to the modal class: higher proportions suggest spatially concentrated areas of deprivation or affluence.

With respect to our earlier discussion (Section 6.5), why did we choose a value of $m = 75$? In fact, a number of values for m were considered ($m = 25, 50, 75, 100, 200$). The results of the analyses with each of these values are summarised in Table 6.1. The top seven rows of the table (excluding the headers) give summary measures of the search window radii (in metres) required to cluster a total population of m by summation of the n_i records nearest to each of the 42883 target points. For example, when m is set at 200, the mean average search window size required to gather the n_i points together to meet that population threshold is 768 metres but ranges from 361 metres to 4000 metres. The interquartile range is much smaller: from 608 metres to 854 metres. The general trend, to be expected, is that as the total population size required m decreases, so a smaller search window is required to find the n_i points that will meet the population threshold. Taking the mean, the third quartile (Q3) or the first quartile (Q1) as a measure of the level of aggregation required to meet each

Table 6.1 *Summary of analysis for various values of m*

m	200	100	75	50	25
Mean radius (m)	768	528	452	364	252
Maximum radius (m)	4000	3999	3999	3966	3820
Minimum radius (m)	361	200	141	100	0
Range	3639	3799	3858	3866	3820
Q3	854	600	510	412	300
Q1	608	412	361	283	200
IQ range	246	212	149	129	100
R ($m = 200$)		0.82	0.74	0.63	0.45
R ($m = 100$)	0.82		0.91	0.77	0.55
R ($m = 75$)	0.74	0.91		0.85	0.61
R ($m = 50$)	0.63	0.77	0.85		0.71
R ($m = 25$)	0.45	0.55	0.61	0.71	
Mode: < £10k (%)	64	57	55	52	47
Mode: £10k–£19.9k	20	23	23	24	25
Mode: £20k–£29.9k (%)	9	12	13	14	16
Mode: £30k–£39.9k (%)	0	1	2	3	5
Mode: £40k–£49.9k (%)	0	0	0	1	1
Mode: £50k+ (%)	6	6	6	6	6
No change from $m = 200$		74	69	62	54
No change from $m = 100$	74		82	70	58
No change from $m = 75$	69	82		77	61
No change from $m = 50$	62	70	77		68
No change from $m = 25$	54	58	61	68	

value of m, then it can be shown that the increase in resolution is in the order of 9:1 as m is decreased from 200 to 25 (the radius is one-third the length, hence the area is one-ninth), and of the order of 4:1 as m is decreased from 100 to 25.

Notwithstanding the apparent increase in precision, the next five rows of Table 6.1 warn that a low value of m is not necessarily to be preferred. These rows record the Pearson correlation coefficients for each set of 42883 cluster homogeneity values at the five values of m. The homogeneity values give the proportion of the cloud or cluster of $n_i + 1$ points around each of the N points that are of the modal income type for that group of $n_i + 1$ points (i.e., the value p_{z1} in equation 6.4). The set of N proportions is calculated for each of the five values of m. This gives $(m - 1)! = 10$ pairs of cluster homogeneity values, for which the highest correlation, at $R = 0.91$, is obtained between the sets for $m = 75$ and $m = 100$ (Table 6.1). The lowest correlation is (unsurprisingly) between $m = 25$ and $m = 200$. From the table it appears that the homogeneity values for $m = 75$ and $m = 100$ exhibit greatest correspondence with the sets of values calculated for the remaining values of m.

The next six rows of Table 6.1 show the proportion of the N cluster centroids assigned to each of the six income classes at each value of m. Looking across the <£10k row it can be seen that the results are not spatially invariant: as the search windows reduce in size so the proportion of clusters with a modal income in this low income class decreases from 64% ($m = 200$) to 47% ($m = 25$): a 27% decrease. That the results should change with the geography of the problem is not surprising since what we are undertaking is explicitly spatial analysis in the sense defined by Longley *et al.* (2001, p. 278): 'spatial analysis is a set of methods whose results change when the locations of the objects being analysed change'. Here, whilst the location of the points is not actually changing, the boundary of the object class (or cluster) into which they are placed, is. It is this property of geographical contingency that makes it difficult to prescribe the correct value of m. However, bearing in mind the quest for generalisation discussed in our opening section, this is not to say that some values of m are not better suited to particular applications than others. Our criteria remain to construct small area measures of income that do not mask diversity at finer scales but that do aggregate the lifestyles data into representative clusters. Looking again at Table 6.1, and at the distribution of cluster centroids across the income classes, it is seen that a degree of consistency in the results emerge when $m = 75$ or $m = 100$ (we also saw this with the Pearson correlations). Interestingly, the proportion for the highest income class (£50k or above) is constant at 6% for all the values of m tested.

The final five rows of Table 6.1 indicate the movement across the income classes as m is changed. They record, for example, that of the 42883 points, 74% remain ascribed to the same income class regardless of whether $m = 200$ or $m = 100$. As m is further reduced from 200 to 75, 69% remain in the same income class, and 54% as m is reduced from 200 to 25. The empirical case study confirms the intuitive expectation: that the results of the analytical method depend on the chosen value of m (and hence the search window size). However, the results are by no means unstable. The greatest invariance to scale appears to be when $m = 75$ or $m = 100$. To recall, we chose the former value for the results displayed by Figures 6.4 and 6.5. With $m = 75$, the average search window size has a radius, r, of 452 m (Table 6.1), giving an average area of 0.64 km^2 over which the lifestyles data are aggregated (= πr^2). This is certainly greater apparent spatial precision than is offered by census ward estimates and also by most census classifications at ED scale. Most importantly, the method offers a direct way of estimating the concentrations of specific income groups using a 'bottom-up' approach that is not constrained to fit within an externally imposed, administrative geography such as that of the census.

6.7 Conclusion

The scale and pace of change in developed countries today is such that we cannot rely upon census data alone. In the UK, the terms of the 1922 Census Act are not interpreted today as providing a mandate for collecting income data. Yet income data, together with richer information on consumption in its broad senses, are central to creation of indicators that can facilitate sensitive spatial policy and rational resource allocation. Such information is, as the report of PAT 18 suggested, essential if the UK Neighbourhood Renewal Unit is to meet its challenge to ensure that 'within 10–20 years no-one should be seriously disadvantaged by where they live' (http://www.dtlr.gov.uk/neighbourhood/).

Developments in remote sensing now allow deprivation arising out of attributes of the built environment to be made consistent with the scale at which housing is in poor condition, or to the scale at which positive or negative externality effects of infrastructure or services are experienced. Yet much remains to be done in creating relevant, timely small area measures of social conditions. Hybrid approaches such as the Breadline Britain Index seek to lever maximum value from the census, yet ultimately only partly compensate for the absence of income measures and sensitivity to local neighbourhood effects. Lifestyles data include an income question, yet the validity of income measures at fine spatial scales can be hampered by the small and geographically variable number of responses received. In our view, it is necessary to undertake some level of aggregation when analysing the lifestyles data, yet to do so in such a way as to be sensitive to local patterns within the data. Aggregating the income information into artificial census units does not meet this challenge so here we have outlined a 'bottom-up' approach to representing income clusters at more precise scales than those offered by more orthodox measures.

The apparent 'Hobson's choice' either of using overly aggregated, surrogate and outdated data on social conditions or abandoning generality and quantitative analysis altogether seems to us a peculiarly academic construct that could be dismantled by an increased, and mutually beneficial, dialogue with business users of geographical information, even withstanding concerns about privacy and ethical issues of data usage (Curry, 1998; Longley *et al.*, 2001, Chapter 19).

Acknowledgements

The research was funded under the (UK) National Environment Research Council, Urban Regeneration and the Environment Programme (http://urgent.nerc.ac.uk/), grant GST/02/2241. The authors are grateful to Claritas UK for supplying the lifestyles data analysed in Section 6, GeoInformation Group for the Cities Revealed data and Ordnance Survey for the CodePoint data. Any errors arising from the interpretation and analysis of these data are our own.

References

Alvanides, S., Openshaw, S. and Rees, P. (2002) Designing your own geographies, in Rees, P., Martin, D.J. and Williamson, P. (eds) *The Census Data System*, Wiley, Chichester, pp. 47–65.

Amin, A., Massey, D. and Thrift, N. (2000) *Cities for the Many Not the Few*, The Policy Press, Bristol.

Baldwin, P. (1999) Postcodes chart growing income divide, *The Guardian*, October 25.

Batey, P.W.J. and Brown, P.J.B. (1995) From human ecology to customer targeting: the evolution of geodemographics, in Longley, P. and Clarke, G (eds) *GIS for Business and Service Planning*, GeoInformation International, Cambridge, pp. 77–103.

Birkin, M., Clarke, G. and Clarke, M. (2002) *Retail Geography and Intelligent Network Planning*, Wiley, Chichester.

Champion, A.G. (1995) Analysis of change through time, in Openshaw, S. (ed.) *Census Users' Handbook*, GeoInformation International, Cambridge, pp. 307–35.

Cities Revealed (1996) Orthorectified, aerial photography, http://www.crworld.co.uk.

Curry, M. (1998) *Digital Places: Living with Geographic Information Technologies*, Taylor and Francis, London.

Daily Telegraph (2000) The 'absolute poverty' of language, 11 September.

Department of the Environment (1981) *Information Note No. 2*, DoE, London.

Department of the Environment (1994) *Information Note on Index of Local Conditions*, DoE, London.

Donnay, J-P. and Unwin, D.J. (2001) Modelling geographical distributions in urban areas, in Donnay, J-P., Barnsley, M.J. and Longley, P.A. (eds) *Remote Sensing and Urban Analysis*, Taylor and Francis, London, pp. 205–24.

Donnay, J-P., Barnsley, M.J. and Longley, P.A. (eds) (2001) *Remote Sensing and Urban Analysis*, Taylor and Francis, London.

Dorling, D., Mitchell, R., Shaw, M., Orford, S. and Davey Smith, G. (2000) The ghost of Christmas past: health effects of poverty in London in 1896 and 1991, *British Medical Journal*, **321**: 1547–51.

EuroDirect (1999) Promotional material, EuroDirect, Bradford.

Goodchild, M. and Longley, P. (1999) The future of GIS and spatial analysis, in Longley, P.A., Goodchild, M.F., Maguire, D.J. and Rhind, D.W. (eds) *Geographical Information Systems: Principles, Techniques, Applications, and Management*, Second edition, Wiley, New York, pp. 567–80.

Gordon, D. and Forrest, R. (1995) *People and Places 2: Social and Economic Distinctions in England*, University of Bristol.

Gordon, D. and Pantazis, C. (eds) (1995) *Breadline Britain in the 1990s*, Joseph Rowntree Foundation, York.

Gordon, D. and Pantazis, C. (1997a) Appendix I: technical appendix, in Gordon, D. and Pantazis, C. (eds) *Breadline Britain in the 1990s*, Ashgate, Aldershot, pp. 269–72.

Gordon, D. and Pantazis, C. (1997b) Measuring poverty: Breadline Britain in the 1990s, in Gordon, D. and Pantazis, C. (eds) *Breadline Britain in the 1990s*, Ashgate, Aldershot, pp. 5–47.

Gordon, D., Adelman, L., Ashworth, K., Bradshaw, J., Levitas, R., Middleton, S., Pantazis, C., Patsios, D., Payne, S., Townsend, P. and Williams, J. (2000) *Poverty and Social Exclusion in Britain*, Joseph Rowntree Foundation, York.

Green, A. (1997) Income and wealth, in Pacione, M (ed.) *Britain's Cities: Geographies of Division in Urban Britain*, Routledge, London, pp. 179–202.
Hall, P. and Pfeiffer, U. (2000) *Urban Future 21: A Global Agenda for Twenty-First Century Cities*, E & FN Spon, London.
Harris, R. (2001) The diversity of diversity: is there still a place for small area classifications?, *Area*, **33**: 329–35.
Harris, R. (2003) Population mapping by geodemographics and digital imagery, in Mesev, V. (ed.) *Remotely-Sensed Cities*, Taylor and Francis, London, pp. 223–41.
Harris, R. and Frost, M. (2003) Indicators of urban deprivation for policy analysis GIS: going beyond wards, in Kidner, D., Higgs, G. and White, S. (eds) *Socio-economic Applications of Geographic Information Science Innovations in GIS 9*, Taylor & Francis, London (in press).
Harris, R.J. and Longley, P.A. (2000) New data and approaches for urban analysis: modelling residential densities, *Transactions in GIS*, **4**: 217–34.
Harris, R.J. and Longley, P.A. (2001) Data-rich models of the urban environment: RS, GIS and 'lifestyles', in Halls, P. (ed.) *Innovations in GIS 8: Spatial Information and the Environment*, Taylor and Francis, London, pp. 53–76.
Harris, R.J. and Longley, P.A. (2002) Creating small area measures of urban deprivation, *Environment and Planning A*, **34**: 1073–93.
Jarman, B. (1983) Identification of underprivileged areas, *British Medical Journal*, **286**: 1705–9.
Johnston, R.J. (1999) Geography and GIS, in Longley, P.A., Goodchild, M.F., Maguire, D.J. and Rhind, D.W. (eds) *Geographical Information Systems: Principles, Techniques, Applications, and Management*, Second Edition, Wiley, New York, pp. 39–47.
Lee, P. (1999) Where are the deprived? Measuring deprivation in cities and regions, in Dorling, D. and Simpson, S. (eds) *Statistics in Society: the Arithmetic of Politics*, Arnold, London, pp. 172–80.
Lee, P., Murie, A. and Gordon, D. (1995) *Area Measures of Deprivation: A Study of Current and Best Practice in the Identification of Poor Areas in Great Britain*, University of Birmingham.
Longley, P.A. and Clarke, G.P. (eds) (1995) *GIS for Business and Service Planning*, GeoInformation International, Cambridge.
Longley, P.A. and Harris, R.J. (1999) Towards a new digital data infrastructure for urban analysis and modelling, *Environment and Planning B*, **26**: 855–878.
Longley, P.A., Goodchild, M.F., Maguire, D.J. and Rhind, D.W. (2001) *Geographic Information Systems and Science*, Wiley, Chichester.
Longley, P.A. and Mesev, T.V. (2002) Measurement of density gradients and space-filling in urban systems, *Papers in Regional Science*, **81**: 1–28.
Longley, P., Smith, R. and Moore, L. (1989) Cardiff House Condition Survey Phase 1: Inner Areas – Final Report, Keltecs Ltd.
Martin, D. J. (2002) Census population surfaces, in Rees, P., Martin, D.J. and Williamson, P. (eds) *The Census Data System*, Wiley, Chichester, pp. 139–48.
Mitchell, R., Dorling, D,. Martin, D. and Simpson, L. (2002) Bringing the missing million home: correcting the 1991 small area statistics for undercount, *Environment and Planning A*, **34**: 1021–35.
Openshaw, S. (1984) The modifiable areal unit problem, *Concepts and Techniques in Modern Geography*, **38**, Geo Books, Norwich.
Openshaw, S. (1995a) A quick introduction to most of what you need to know about the 1991 Census, in Openshaw, S. (ed.) *Census Users' Handbook*, GeoInformation International, Cambridge, pp. 1–26.
Openshaw, S. (ed.) (1995b) *Census Users' Handbook*, GeoInformation International, Cambridge.
Policy Action Team 18 (PAT 18) (2000) *Report of Policy Action Team 18: Better Information*, The Stationery Office, London.
Raper, J. (2000) *Multidimensional Geographic Information Science*, Taylor and Francis, London.

Shaw, M., Dorling, D., Gordon, D. and Davey Smith, G. (1999) *The Widening Gap: Health Inequalities and Policy in Britain*, The Policy Press, Bristol.

Sleight, P. (1997) *Targeting Customers: How to use Geodemographic and Lifestyle Data in Your Business*, Second edition, NTC Publications, Henley-on-Thames.

Townsend, P., Phillimore, P. and Beattie, A. (1988) *Health and Deprivation: Inequality and the North*, Croom Helm, London.

Voas, D. and Williamson, P. (2001) The diversity of diversity: a critique of geodemographic classification, *Area*, **33**: 63–76.

Yapa, L. (1998) Why GIS needs postmodern social theory, and vice versa, in Fraser Taylor, D.R. (ed.) *Policy Issues in Modern Cartography*, Pergamon, Oxford, pp. 249–69.

7

Assessing Deprivation in English Inner City Areas: Making the Case for EC Funding for Leeds City

Paul Boyle and Seraphim Alvanides

Abstract

One of the key roles of the European Commission (EC) is to encourage a redistribution of wealth throughout the Community to reduce inequalities, both between and within countries. One mechanism for achieving this is the release of Structural Funds to those places that have particularly serious social and economic problems. This chapter reports on the attempt by Leeds City Council to secure Objective 2 status funding from the Commission, as part of a wider bid submitted by the Yorkshire and Humberside region. To be successful, Leeds needed to prove that the socioeconomic situation in its inner city was at least as bad as that in other cities in England. However, to achieve this required some computationally advanced methods beyond the scope of the Council's researchers. In collaboration, a methodology was devised which used some in-house software, embedded within a geographical information system (GIS), to show that Leeds did indeed have a strong case for support from the EC. Here we describe the methods used and the results for Leeds compared with the other cities in England. The chapter ends by commenting briefly on the positionality of researchers who undertake consultancy work of this type and suggests that we rarely spend enough time considering the interesting ethical dilemmas that work of this nature raises.

7.1 Introduction

One of the roles of the European Commission is to stimulate economic growth and encourage a redistribution of wealth that reflects the fact that some parts of the

Applied GIS and Spatial Analysis. Edited by J. Stillwell and G. Clarke
 ISBN: 0-470-84409-4

European Union (EU) are relatively better off than other parts. Rather than simply redistributing funds at the national level, there is increasing recognition of the significant imbalances within countries between cities and towns that need to be addressed. This is particularly so as cities and towns are critical to regional development. In identifying an urban agenda, the Commission states that 'the starting point for future urban development must be to recognise the role of the cities as motors for regional, national and European economic progress' (European Commission, 1998a). In many European cities, however, severe social and economic problems hinder economic development and there is a growing realisation that these need to be addressed before economic growth can be achieved: 'The future of the towns and cities in the EU depends on fighting growing poverty, social exclusion and stemming the loss of certain urban functions' (Committee on Spatial Development, 1999). The EC response has been to distribute central funds to cities across the EU based on uniform criteria designed to identify those most in need of support. The Structural Funds Programmes therefore require that cities and towns with significant social and economic problems can be identified and that 'given the crucial role of towns for regional development and EU regional disparities, it is important for the effectiveness of regional policy that this funding be more explicitly related to urban needs and potential in the regions' (European Commission, 1998b).

The research reported in this chapter was undertaken as part of Leeds City Council's (LCC) attempt to secure Structural Funds from the EU by applying for Objective 2 status funding within a broader bid made by the Yorkshire and Humberside region. For LCC, there was a certain irony in making an Objective 2 application, as in recent times the city has been heralded as one of the fastest growing and most economically successful in the United Kingdom (UK). Thus, on the one hand, the LCC was producing documents aimed at persuading companies to move to what was fast becoming one of the most successful manufacturing, financial and legal services centres in the UK while, on the other, it was attempting to justify the case for significant EU funding, based on small areas within the city that were experiencing severe social hardship. The research described here was undertaken on behalf of LCC with the aim of demonstrating that the problems in Leeds were as significant as those of other competing UK cities that already received EU funding, or were themselves making the case for consideration for funding.

The remainder of the chapter is divided into four sections. In Section 7.2, we describe the nature of the problem and the requirements of LCC. We outline the measures used to identify areas with significant socioeconomic problems and explain that while LCC could speak authoritatively about the most deprived areas in Leeds, it was much more difficult to identify similarly deprived parts of other English cities. In particular, identifying small areas that met certain selection criteria, such as level of deprivation, population size and contiguity, all required a relatively complex methodological approach. Section 7.3 therefore explains the potential measures in more detail and the GIS-based methodology that was developed to solve these problems. Section 7.4 presents the results and compares the area chosen to represent the deprived inner city of Leeds with similarly defined areas for the remaining cities in England. Comparisons are made based on a series of different population cut-offs and using different measures of deprivation. Finally, Section 7.5 concludes by arguing that LCC did indeed have a strong case for Objective 2 status. The Local Authority District of Leeds includes some particularly rich and poor people and, compared with many English cities, ignoring this would be misleading. The geographically refined analysis of socioeconomic conditions in the city that we provided helped to highlight this fact. However, the chapter ends by reminding us that we need to give serious thought to our positionality as researchers when conducting work of this type.

7.2 The Applied Problem and Client Needs

The EC has a series of programmes designed to redistribute resources to those parts of the EU that are most deserving. For example, Objective 1 status areas were defined as those whose per capita gross domestic product (GDP) was less than 75% of the Community average; consequently these were relatively large areas in severe hardship (South Yorkshire was one of the English Objective 1 areas, for example). Objective 3 was available nationally, rather than to geographically targeted areas, and was designed to support education, training and employment policies. Previously, Leeds had been successful in attracting considerable funds under this scheme. Objective 2, which was relevant to this research in Leeds, sought to address the needs of areas experiencing structural difficulties, whether industrial, rural, urban or fisheries-dependent areas, or to promote the restructuring of services (Department of Trade and Industry, 1999). This definition combined Objectives 2 and 5b from previous funding rounds. Leeds did not submit itself for consideration in the previous 1997 round, but was hoping to be considered in the 1999 round.

Before the EC can identify cities, or parts of cities, as deserving of Objective 2 status, a system of measuring the socioeconomic conditions within each is required, such that their relative situations can be compared. Some standard criteria were set out in a lengthy legal document (European Commission, 1999), although these were designed to be indicative rather than a strict list which each application needed to address. Thus, 'urban areas in difficulty' were expected to be densely populated and of a 'reasonable size' (probably over 100 000 population). The application areas were expected to meet at least one of the following criteria:

- a rate of unemployment higher than the Community average;
- a high level of poverty, including precarious housing conditions;
- a particularly degraded environment;
- a high crime rate;
- a low level of education among the population.

Constraints were also imposed by the EC on how the funds would be distributed:

- a ceiling for all types of Objective 2 areas of 18% of the EU population;
- a UK share of that ceiling determined by the population of its industrial and rural areas which qualify as of right and the extent of its structural problems, subject to a maximum of one-third in population coverage at the time;
- a limit of 2% of the EU population to be covered by Urban Areas (the UK hoped to get a larger than average share of this).

It was assumed that applications would be made at NUTS level III (counties or groups of unitary authorities in the UK), but that separate applications should be made by local areas within these broad areas. Thus, Leeds was aiming to be one of a number of areas that was included in a broader submission made by the Government Office for Yorkshire and Humberside.

The criteria identified above that were of particular relevance to this study are those relating to the geographical scale at which areas for support should be identified and the indicators that should be used to identify the level of need. These two criteria are closely related as the choice of geographical units will constrain the data that are available for analysis. Geographically, the two most important criteria were, firstly, that the areas of greatest need should be 'targeted', and secondly, that these areas should be large enough to allow credible and integrated programmes to be mounted. It had been suggested that the most appropriate zones that could be used as building blocks to identify the areas of greatest need across the EU would be NUTS 5 regions or, in UK terms, wards. Whilst the EC did not specify a population size constraint, it was clear that very small areas would not be considered in isolation, and a population of approximately 100 000 was commonly quoted as an appropriate cut-off.

A wide range of indicators, including social, economic, crime and housing measures, were suggested as appropriate for identifying areas of greatest need in the UK, although the manner in which these indicators should be used was not prescribed. Clearly, to be successful, any application for Objective 2 status required that the chosen areas performed badly across a wide range of socioeconomic criteria. Previously within the UK, one of the standard tools that had been used by the Government for identifying need for Objective 2 status was the Index of Local Deprivation (ILD) (Department of Environment, 1995; Department of Environment, Transport and the Regions, 1998). This has been calculated for local authority districts (LADs), of which there were 366 that encompassed the whole of England at the time of the 1991 Census. Through use of the ILD, each district can be ranked according to a standard measure of the material deprivation in that district. The ILD is a useful summary measure that is used in many applied and academic studies to give a feel for the social standing of a particular place. For LCC, the focus of attention was on identifying how Leeds compared with other LADs in England, against whom it was competing for the limited amount of funds.

However, the assumption that LADs are homogeneous areas that can be summarised using a single deprivation measure ignores the diversity of social conditions that exist throughout a city and it is especially biased for cities that are geographically large, encompassing a range of better-off and worse-off areas. The diversity of social deprivation that exists in cities also varies so that an 'average' measure of deprivation will be more biased in some cities than in others. In this round, it was expected that applications would use ward-level deprivation scores as one way of summarising the socioeconomic situation of the local area, partly because this was a useful way of being able to compare submissions from different parts of the country.

Therefore, applicants were encouraged to identify a group of contiguous wards (which aggregate neatly into districts) that was relatively large and that was also an area of great need. LCC believed that the scale of analysis was particularly crucial in Leeds as LAD-level measures masked the reality of inner area deprivation in the city. Two points are relevant here. Firstly, because Leeds LAD is such a large geographical area, it inevitably encompasses a range of places with different levels of deprivation. Other cities like Manchester or Liverpool are defined more tightly with their LAD boundaries encompassing the inner city, but few of the surrounding, often more wealthy, suburban and extra-urban areas. On the other hand, cities like London include a number of LADs within their boundary, each of which is relatively more homogeneous than the LADs that include entire cities. Secondly, it has been argued for some time that the city of Leeds is unusual because it encompasses a heterogeneous mixture of residential environments compared with other UK cities. Put simply, there is considerable variation in the levels of deprivation within the inner city area such that Leeds has been described as 'two cities', one 'working hard to position itself as the capital of the north', and the other 'where unemployment is high, educational attainment is low and where dependency on benefits is almost universal' (Leeds City Council, 1998).

To address this problem, LCC required an analysis of deprivation conducted at a more detailed spatial scale that would reflect these extremes – an analysis of deprivation *within* the city, rather than a summary deprivation score for the city as a whole. The aim, therefore, was to identify a contiguous set of wards within the Leeds LAD which, when aggregated together, represented a relatively large homogeneous population with a high deprivation score. The EC required that areas with relatively large populations be submitted for consideration, as the aim of the Objective 2 status programme is to target relatively large areas of concentrated deprivation that require aid. However, it was not stipulated that this should necessarily include the entire city. By selecting appropriate wards from within the district boundary, it removes the problem that using the entire LAD includes a mixture of richer and poorer areas that, on average, suggest that the city is not particularly deprived.

LCC was therefore keen that an analysis be conducted that compared such a choice of wards in Leeds with similarly defined groups of highly deprived wards in other English cities. As most city councils purchase census and other data that can be used to construct deprivation indices for the wards within their own cities, LCC was able to carry out such an analysis for Leeds. In fact, they also had deprivation scores for the wards across all of England apparently allowing them to compare Leeds with other cities. However, as we demonstrate below, relatively complex methods were required to make sure that the selection of wards in other cities was as equitable as possible.

There were two problems with this approach which were the reasons behind LCC's request for some external consultancy work. First, LCC did not have the digital boundaries required to identify which wards in other 'competing' cities were *neighbouring*. This was important since a requirement of the EC exercise was that the areas submitted for consideration should be compact. Before LCC could argue that their preferred selection of wards was a better reflection of concentrated deprivation within the city, it was necessary to compare the results of their analysis with a similar analysis in other English cities. LCC were convinced that there was a large area of inner city Leeds that was significantly more deprived than most equivalent areas in other English cities, but an analysis was required actually to prove this. The second problem was that, even if contiguous wards could be identified, a strategy for optimising the choice of which wards should be aggregated together was potentially onerous. Given that there are 10 000 wards in England and Wales, deciding how to group wards in every city so that the largest and most deprived combination was achieved was not simple. In Leeds, the planners could do this by hand, knowing well the worst areas of their own city, but they did not have such insights into other English cities. The method we adopted for solving this problem is described below.

7.3 Methods of Identification of Deprived Urban Areas in British cities

The first strategy was to investigate how Leeds LAD compared with other English LADs in terms of the 1998 ILD. This index is based on 12 variables, three of which are derived from the 1991 Census, with the remainder coming from other updateable sources, such as unemployment, mortality and insurance company data. Consequently, this index was relatively up to date and, although we may debate the validity of the index itself, it does at least provide a consistent means of comparing all LADs in England. The ILD can be used as the basis of a number of different comparative measures, including:

- the degree of deprivation, or the overall district level score;
- the intensity of deprivation, or the severity of deprivation in the LAD taken as the average score of the worst three wards in the LAD;
- the ward level extent, or the proportion of the LAD population living in wards that are within the 10% most deprived in England; and
- the ED level extent, or the proportion of the EDs in the LA that fall within the most deprived 7% of EDs in England.

Here we focused mainly on the first of these measures. Of course, as argued above, this method of identifying deprived inner city areas is spatially crude. There is no logical reason why the LAD scale is the most appropriate for analysis and, given the considerable variation in the size and homogeneity of LADs, there is a strong argument that a more considered definition of deprived inner city areas should be sought. In particular, the aim here is to identify *concentrated areas of deprivation*, as these are likely to be especially deprived places.

This objective required analyses that build deprived areas from smaller geographical units. LCC therefore required a series of analyses using aggregations of small areas (wards) for Leeds and other competing LADs in England, which would show that Leeds has a significant area within the city with a higher level of material deprivation compared with other places. These groups of wards would include large populations in most LADs, but in most cases their populations would obviously be smaller than the populations of the entire LADs. Fortunately, a ward-level version of the ILD has been constructed which is based on six variables from the 1991 Census. These six variables were: unemployment, children in low-earning households, households with no car, households lacking basic amenities, overcrowded households, and 17 year olds no longer in full-time education. Although this version of the ILD is not regularly updateable, as it relies on census information which is only collected every 10 years, it does provide an indication of deprivation for small geographical areas.

In consultation with LCC, inner city Leeds was defined based on informed views of the most deprived parts of the city. Of the 12 chosen wards, the least deprived was Armley, with an ILD score of 7.97, which ranked 1062 out of 8620 wards in England. There were 14 wards in Leeds that were at least as deprived as this ward, using the ward-level ILD, although two (Beeston with ILD = 9.60 and Bramley with ILD = 9.50) were excluded from the definition of inner city Leeds used below (Figure 7.1).

Using the deprivation score of 7.97 for Armley as a cut-off, all wards across England with a similarly high level of deprivation could be identified and relevant groups were then aggregated for each city. As can be seen in Figure 7.2, many districts had few wards that reached this level of deprivation. However, it was not a simple matter of aggregating all these wards for each city, as a strategy was required for deciding which to include. Simply using a deprivation score as a cut-off means that isolated wards within LADs would have been included in the list and their

Figure 7.1 *Deprived wards in inner city Leeds © Crown Copyright. Reproduced with permission*

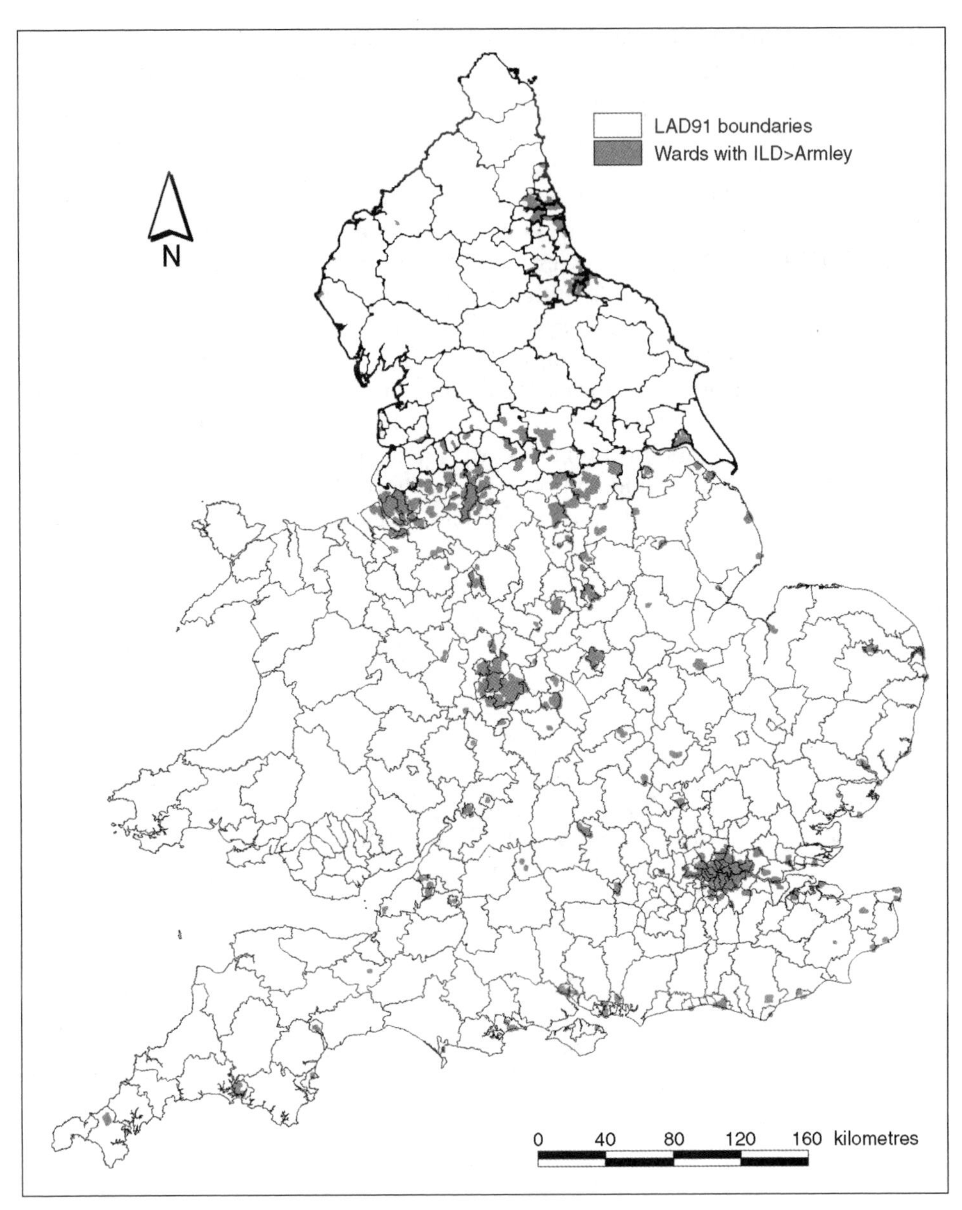

Figure 7.2 *Wards in England that were at least as deprived as Armley ward © Crown Copyright. Reproduced with permission*

inclusion was not acceptable according to the EC guidelines; the group of wards that were chosen needed to be contiguous.

To illustrate this, the districts in West Yorkshire with ILD scores above Armley are mapped in Figure 7.3. Leeds has 14 contiguous wards in total with ILD scores at least as deprived as Armley, including Beeston and Bramley. These 14 wards are clustered to form the contiguous inner area, without any holes or deprived wards outside

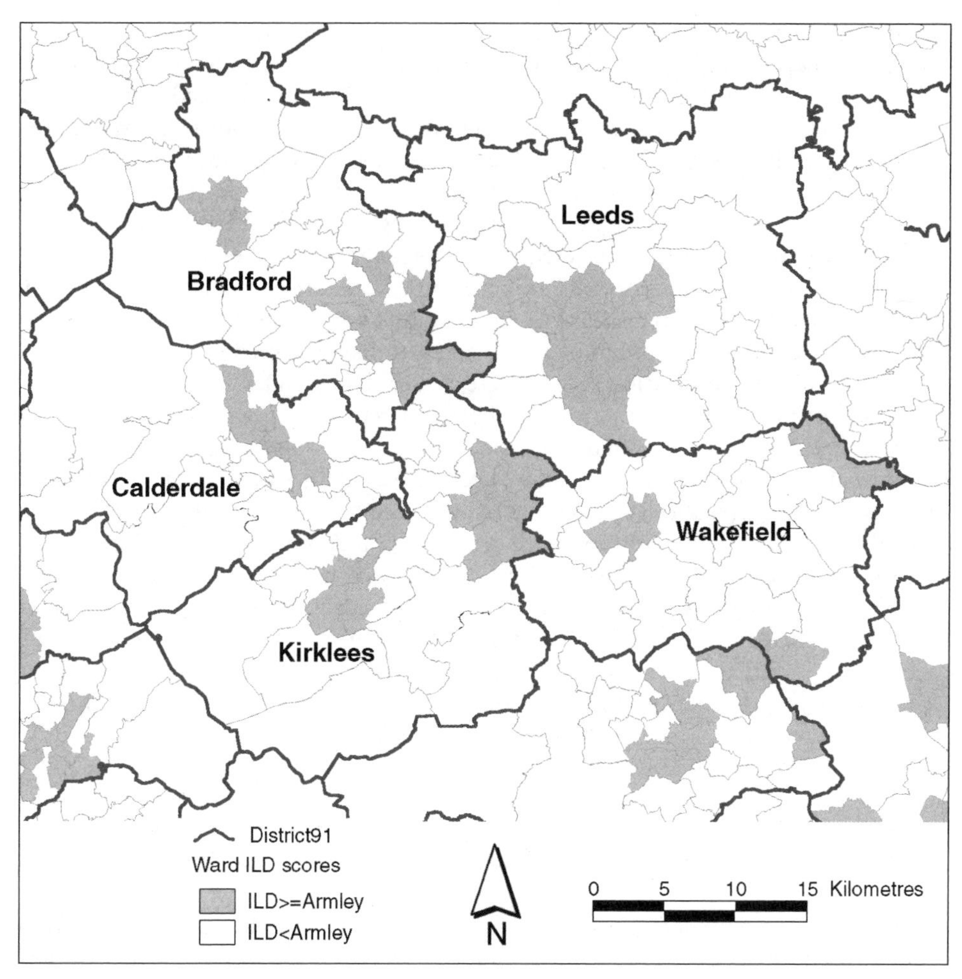

Figure 7.3 *Deprived wards in selected districts in West Yorkshire © Crown Copyright. Reproduced with permission*

this area. This is also the case in Calderdale, where four contiguous wards with ILD scores at least as deprived as Armley cluster to form a deprived area. However, this is not the case in either Bradford or Kirklees, where the deprived wards cluster in two areas, nor in Wakefield where there are three pockets of deprivation. In order to check the contiguity of the selected wards shown in Figure 7.2, our research therefore exploited GIS and some in-house software designed to optimise the selection of contiguous groups of geographical wards based on some pre-determined criteria (maximising deprivation in this case).

The analysis was carried out using semi-automated methods in ArcInfo GIS, utilising the Arc Macro Language (AML) capabilities for topology building, spatial queries and calculations. First, all the wards with ILD scores greater than Armley were selected, as shown in Figure 7.2. Subsequently, the adjacent boundaries of the selected wards were dissolved in order to identify the number of deprived areas within each LAD. In cases like Leeds and Calderdale, the selected wards would only form a single deprived area, with no internal islands and the selection was acceptable. However, in cases like Bradford, Kirklees and Wakefield, the selected wards would form two or more separate areas, within the same district. These districts were automatically identified within ArcInfo GIS and a number of measures were calculated for each deprived area, such as the total population and the degree of deprivation. If the population within one of the sub-district areas was adequate (as described below), then this area was taken as a representative deprived area for the district, otherwise, further GIS analysis was necessary to secure an acceptable population size. Further analysis was undertaken in the form of aggregations, so that wards with relatively low scores were added to the contiguous set if the ward neighboured another ward with a particularly high deprivation score. These 'lower-than-Armley' wards were known as 'bridging wards'. Thus, the addition of both wards to a cluster increased the average deprivation in that cluster, even though the score for the bridging ward may have been lower than the arbitrary cut-off that had been chosen. This procedure therefore identified 187 large aggregations of contiguous wards, in cities across England.

In the analysis, we also introduced a series of population cut-offs enabling us to identify clusters of wards that were above or below certain thresholds. The deprivation cut-off was agreed with LCC as those wards that were at least as deprived as Armley ward (see above). The population cut-offs were also a subjective choice, but one that was guided by the knowledge that the EC was only interested in funding relatively large areas; deprived pockets were not to be targeted. The software allowed us to identify clusters of wards that were above or below certain populations. While there is a strong case that deprived areas with large populations should be targeted for aid, there are also concentrated areas of deprivation that should not be ignored simply because they have a smaller population than other inner cities. The LCC definition of inner city Leeds had a population of approximately 228 000 but, as suggested above, while 100 000 was an oft-quoted figure for an appropriate size of a deprived place, no guidelines specified that smaller areas would not be considered. Four levels of analysis were therefore conducted to reflect the size of the deprivation problem in the various LADs. Initially, we retained only those sets of contiguous wards that had a combined population of at least 200 000. This is useful as it draws attention to the largest areas of concentrated deprivation in England – places that should certainly be considered as worthy of consideration in attempts to redistribute resources. The second and third definitions used clusters of wards with populations of at least 100 000 and 60 000, respectively. Finally, we also considered those clusters with a maximum population of 60 000 as a cut-off, allowing us to identify the most concentrated areas of deprivation in the various inner cities.

The GIS software facilitated the analysis in many ways. Firstly, it was possible to manage the spatial datasets and carry out calculations in a seamless manner. In addition, the AMLs allowed automation of the process, by completing repetitive spatial queries, building the topology, and producing descriptive measures for population and deprivation scores. Finally, the GIS identified the most appropriate bridging wards, restricting the search of appropriate wards considerably. The diagram in Figure 7.4 illustrates the semi-automated process followed in identifying the optimum district ILDs for selected sets of wards.

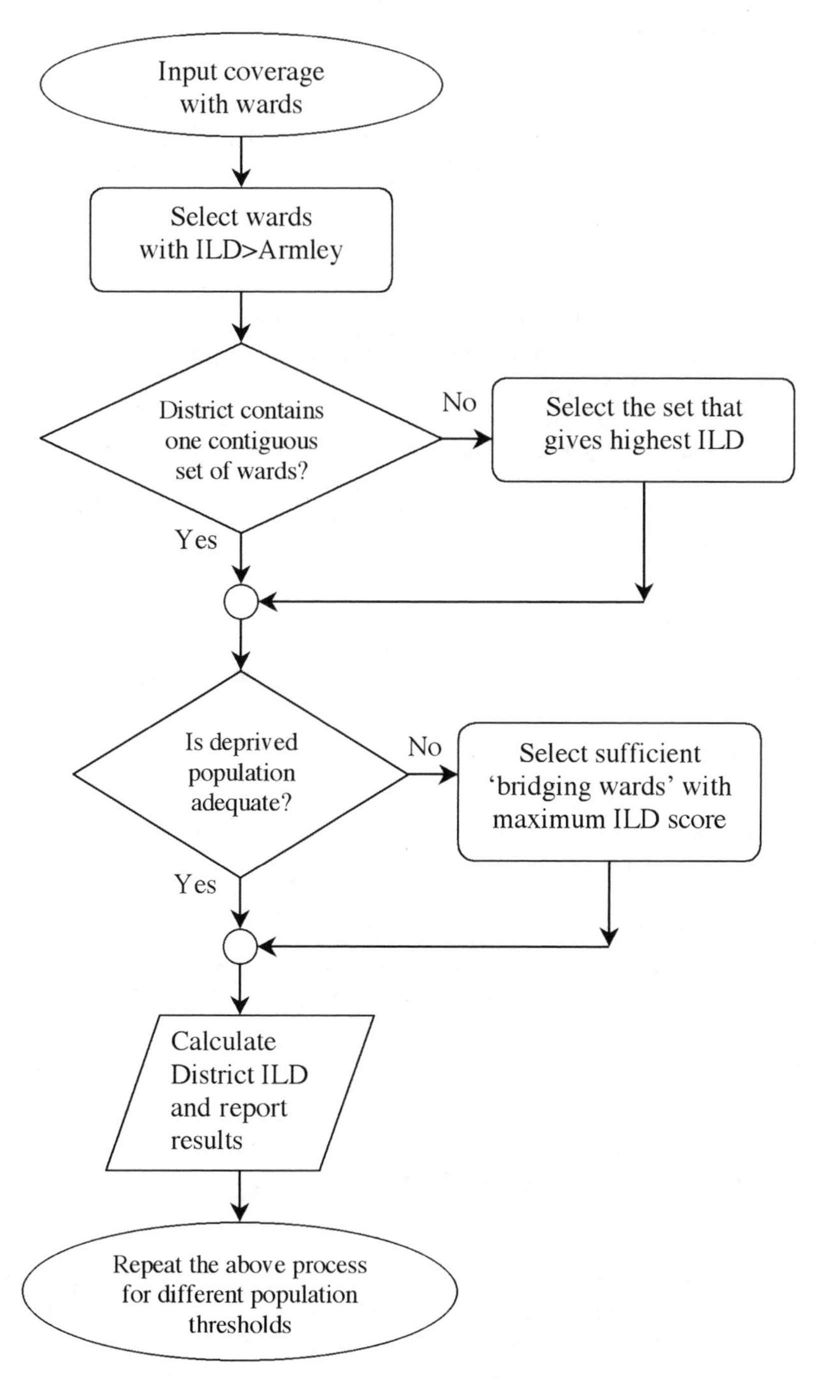

Figure 7.4 *Diagram showing the semi-automated GIS process*

7.4 A Comparison of Concentrations of Deprivation

The results for the *LAD-level* analysis are provided in Table 7.1. The mean deprivation is perhaps the most obvious indicator of overall deprivation. Leeds ranked very low (56th) on this list, demonstrating that it was not among the most deprived LADs in England, using this spatially crude analysis. Note, however, that Leeds ranked seventh in England using the average deprivation in the worst three wards, suggesting that a more sensitive definition of concentrated deprivation may be appropriate for the city.

The next step was to show how Leeds ranked using a simple comparison of *groups of wards* that were included because they had a higher deprivation score than our chosen cut-off value. Table 7.2 provides the cities with the most deprived wards based on this analysis and this is equivalent to the analysis that LCC could have conducted using the information they had to hand. Note, however, that while the wards in Leeds have been chosen because they are contiguous, the wards in the other cities are chosen simply because their deprivation score is higher than a particular cut-off; many of these wards will not have been neighbouring. According to this analysis, Leeds climbed from 56th (Table 7.1) to 11th (Table 7.2). Of particular note is the very large population of the Leeds inner city area compared with other cities. In terms of population size, the group of wards in Leeds was exceeded only by Birmingham, Liverpool, Manchester and Sheffield. Lambeth was the highest-ranking London borough, in sixth position.

The problem with this analysis is that it ignores contiguity for all the cities except Leeds. The aim was to identify a concentrated area of deprivation as the EC had made it known that a smaller number of substantial areas of deprivation were liable to receive funding, rather than a larger number of smaller areas, thus avoiding a 'pepperpotting' of funds. Once a contiguity constraint is introduced it is likely that the deprivation measures will alter for the other cities and, if nothing else, the population of many of the inner city areas will be reduced. For Leeds to make a strong case, it needed to be able to compare its position with that of the other cities in England.

Table 7.3 lists those nine deprived inner city areas with populations of at least 200 000, ranked by the mean deprivation score. In summary, the Leeds inner city area ranked:

- fifth using the ward-level ILD degree score;
- fourth using the ILD ward intensity score;
- third on the maximum index value indicator; and
- fourth on the estimated population size indicator.

Note that the standard deviation of the ILD values is also provided to give an indication of the variation in the deprivation scores for each cluster of wards. Leeds has a fairly high value compared with other clusters (this result is consistent in the following tables).

The second stage of this analysis extended the list of deprived inner city areas to include those that had a population of at least 100 000. These are still large concentrations of deprivation, although many have much smaller populations than the

Table 7.1 *Local authority district rankings on degree, intensity, ward and ED extent measures*

District name	Degree score	Degree rank	Average worst 3 wards	Ward intensity (rank)	% LAD population in worst 10% wards	% LA population in worst 10% wards	% EDs in worst 7%	ED extent (rank)
Liverpool	40.07	1	15.9	2	71.22	11	34.22	6
Newham	38.55	2	14.5	10	100.00	2	55.75	3
Manchester	36.33	3	15.2	4	77.98	8	31.10	8
Hackney	35.21	4	14.7	8	100.01	1	60.79	1
Birmingham	34.67	5	16.6	1	64.61	15	24.88	11
Tower Hamlets	34.30	6	15.1	5	99.99	4	58.26	2
Sandwell	33.78	7	13.3	29	50.02	19	10.95	40
Southwark	33.74	8	14.2	14	88.96	6	39.11	4
Knowsley	33.69	9	14.1	15	72.50	10	38.18	5
Islington	32.21	10	13.8	22	100.00	3	29.93	10
Greenwich	31.58	11	12.2	45	58.27	16	15.24	26
Lambeth	31.57	12	14.7	9	89.42	5	34.09	7
Haringey	31.53	13	14.2	12	79.26	7	30.65	9
Lewisham	29.44	14	13.9	19	53.58	17	18.86	20
Barking and Dagenham	28.69	15	11.0	65	46.40	22	4.51	89
Nottingham	28.44	16	13.8	21	41.30	29	18.94	19
Camden	28.23	17	13.1	32	77.72	9	16.70	23
Hammersmith and Fulham	28.19	18	14.1	17	65.93	14	19.22	16
Newcastle Upon Tyne	27.95	19	12.9	34	39.08	32	19.01	18
Brent	26.95	20	14.1	16	46.86	21	20.76	14
Sunderland	26.90	21	12.2	44	36.65	34	13.06	34
Waltham Forest	26.68	22	14.2	13	66.77	12	16.34	25
Salford	26.64	23	13.3	30	31.43	44	10.72	43
Middlesbrough	26.41	24	12.8	37	53.55	18	22.26	12
Sheffield	26.09	25	14.8	6	36.33	36	13.38	32
Kingston Upon Hull	26.06	26	14.5	11	44.91	24	18.18	22
Wolverhampton	25.94	27	13.5	26	40.87	31	14.66	28
Rochdale	25.13	29	12.3	43	38.63	33	14.29	30
Bradford	25.94	28	15.6	3	42.15	27	21.47	13

Table 7.1 *Continued*

District name	Degree score	Degree rank	Average worst 3 wards	Ward intensity (rank)	% LAD population in worst 10% wards	% LA population in worst 10% wards	% EDs in worst 7%	ED extent (rank)
Wandsworth	25.05	30	11.9	48	36.45	35	8.23	57
Walsall	25.02	31	12.7	39	45.82	23	8.49	55
Leicester	24.95	32	14.0	18	48.46	20	12.95	35
Oldham	24.82	33	13.5	25	34.86	37	14.09	31
Halton	24.69	34	9.7	94	16.52	72	10.26	46
Gateshead	24.58	35	11.7	54	30.30	46	9.47	51
Ealing	24.48	36	11.8	52	29.59	48	7.14	66
Hartlepool	23.72	37	12.0	47	44.85	25	20.57	15
South Tyneside	23.67	38	10.4	76	28.94	50	9.69	50
Doncaster	23.60	39	12.5	40	41.15	30	10.85	41
Coventry	23.48	40	13.4	27	34.77	38	15.00	27
Blackburn with Darwen	23.04	41	12.9	36	32.20	43	19.18	17
Barnsley	22.30	42	10.5	73	23.07	56	6.62	70
Redcar and Cleveland	21.54	43	11.9	49	13.47	81	10.63	44
Wirral	21.25	44	12.8	38	21.40	60	10.81	42
St.Helens	20.98	45	10.9	68	20.91	62	8.70	54
Lincoln	20.70	46	9.3	105	15.57	76	9.88	48
Bolton	20.66	47	13.6	24	29.71	47	11.73	37
Stoke-on-Trent	20.61	48	11.2	60	23.17	55	5.28	81
Stockton-on-Tees	20.41	49	11.9	50	24.90	52	11.51	38
Rotherham	20.23	50	11.3	58	16.71	71	7.05	67
Blackpool	20.14	51	13.7	23	21.99	59	11.99	36
Easington	19.97	52	10.1	82	15.91	75	4.02	99
Tameside	19.78	53	9.8	90	10.11	100	4.08	97
Sefton	19.41	54	11.0	64	32.52	42	9.90	47
Barrow-in-Furness	19.38	55	8.6	117	7.30	122	5.44	80
Leeds	**19.06**	**56**	**14.8**	**7**	**33.30**	**40**	**9.72**	**49**
City of Westminster	19.05	57	13.8	20	66.29	13	18.31	21

Table 7.2 *Inner city area rankings on mean ILD score*

District name	Number of wards	Popn	Popn rank	Index mean	Index mean rank	Average worst 3 wards	Worst 3 wards rank	Index max	Index max rank	Obj 2 status (1 = yes)
Hackney	23	181 248	10	13.33	1	14.73	8	14.91	15	1
Newham	24	212 170	7	13.32	2	14.50	10	14.58	20	1
Tower Hamlets	19	161 064	13	12.91	3	15.06	5	15.66	6	1
Liverpool	26	356 476	2	12.71	4	15.94	2	16.65	3	0
Lambeth	20	222 576	6	12.64	5	14.70	9	15.00	13	0
Haringey	18	158 247	14	12.63	6	14.25	12	14.60	19	1
Birmingham	27	665 336	1	12.54	7	16.60	1	17.12	1	1
Bradford	12	182 938	9	12.25	8	15.64	3	16.79	2	1
Trafford	2	20 260	92	12.23	9	10.60	69	12.83	50	1
Waltham Forest	13	137 821	26	12.21	10	14.22	13	14.47	23	1
Leeds	**12**	**227 747**	**5**	**12.14**	**11**	**14.81**	**7**	**15.77**	**5**	**0**
Oldham	7	75 786	49	12.12	12	13.54	25	14.17	30	1
Sheffield	14	241 960	4	12.09	13	14.81	6	15.55	7	1
Islington	20	164 686	12	12.05	14	13.80	22	14.29	28	0
Southwark	23	201 058	8	11.97	15	14.16	14	14.88	16	0
Hove	2	17 073	96	11.94	16	n/a	n/a	12.08	67	0
Wirral	5	75 181	50	11.90	17	12.76	38	14.00	34	0
Kingston upon Hull	11	139 764	23	11.77	18	14.47	11	15.30	9	1
Blackpool	6	39 837	68	11.63	19	13.69	23	14.33	26	0
Brent	18	141 111	21	11.60	20	14.13	16	14.75	17	0
Manchester	29	355 787	3	11.59	21	15.21	4	15.53	8	1
Salford	8	88 185	39	11.55	22	13.27	30	14.30	27	1
Blackburn	8	52 043	61	11.52	23	12.88	36	13.80	36	1
Bolton	7	90 504	37	11.51	24	13.62	24	16.12	4	1
Knowsley	17	117 525	30	11.46	25	14.13	15	15.13	11	0
Wolverhampton	10	121 095	29	11.32	26	13.47	26	14.23	29	1

Table 7.2 Continued

District name	Number of wards	Popn	Popn rank	Index mean	Index mean rank	Average worst 3 wards	Worst 3 wards rank	Index max	Index max rank	Obj 2 status (1 = yes)
Westminster, City of	15	114 009	31	11.32	27	13.84	20	14.54	22	0
Preston	8	53 087	59	11.31	28	12.99	33	13.22	44	0
Pendle	2	8 959	121	11.26	29	9.71	90	11.51	82	1
Derby	7	76 581	47	11.25	30	12.90	35	13.78	37	0
Kensington and Chelsea	10	65 902	54	11.24	31	13.41	28	14.04	33	0
Coventry	9	147 194	18	11.22	32	13.42	27	15.12	12	1
Sandwell	14	169 220	11	11.13	33	13.30	29	14.12	32	1
Tameside	2	22 782	89	11.11	34	9.77	85	12.00	69	1
Solihull	4	47 026	63	11.10	35	11.45	55	11.99	70	1
Calderdale	4	42 574	66	11.09	36	11.56	53	12.63	54	0
Hammersmith and Fulham	16	103 306	33	11.09	37	14.12	17	14.39	25	0
Hartlepool	7	37 227	72	11.04	38	11.96	46	12.33	61	1
Canterbury	1	4 958	150	11.02	39	8.51	112	11.02	90	0
Camden	20	131 111	27	11.00	40	13.05	32	13.78	38	0
Rochdale	8	80 866	43	10.99	41	12.26	42	13.38	42	1
Portsmouth	5	67 191	53	10.97	42	11.80	49	12.56	57	0
Plymouth	7	85 181	40	10.97	43	13.19	31	15.29	10	1
Nottingham	16	156 161	15	10.97	44	13.81	21	14.69	18	1
Rotherham	4	45 752	65	10.96	45	11.32	56	12.96	47	1
Lewisham	17	151 027	17	10.92	46	13.85	19	14.57	21	0
Leicester	15	144 907	19	10.89	47	13.99	18	14.94	14	0
Milton Keynes	2	16 030	100	10.84	48	8.13	127	11.41	83	0
Shepway	2	7 319	134	10.83	49	9.78	84	12.33	60	0
Newcastle upon Tyne	14	139 753	24	10.75	50	12.92	34	13.20	45	1
Sunderland	12	138 739	25	10.74	51	12.20	43	12.27	63	1

Table 7.3 *Inner city areas with at least 200 000 people: rankings on mean ILD score*

District name	Number of wards	Popn	Popn rank	Area	Index mean	Index mean rank	Standard deviation of index	Max index score	Average worst 3 wards
Newham	24	212 170	7	36 360 712	13.32	1	1.13	14.58	14.50
Liverpool	22	296 891	3	62 447 304	12.81	2	1.81	16.65	15.94
Birmingham	24	602 240	1	137 786 944	12.74	3	2.94	17.12	16.45
Lambeth	20	226 824	5	24 457 256	12.64	4	1.55	15.00	14.70
Leeds	**12**	**227 747**	**4**	**75 948 004**	**12.14**	**5**	**2.25**	**15.78**	**14.81**
Sheffield	14	216 687	6	68 008 216	12.09	6	2.18	15.55	14.81
Southwark	23	201 191	8	23 311 460	11.97	7	1.39	14.88	14.17
Manchester	29	353 203	2	102 427 784	11.59	8	2.07	15.53	15.21

Leeds inner city area. Table 7.4 provides details of all 27 of these deprived inner city areas. In summary, the Leeds inner city area ranked:

- tenth using the ward-level ILD degree score;
- sixth using an ILD ward intensity score;
- fourth on the maximum index value indicator; and
- fourth on the estimated population size indicator.

The list was then extended to include contiguous sets of wards with a population of at least 60 000 (Table 7.5). Even on this much longer list, Leeds ranked high up the list:

- eleventh using the ward-level ILD degree score;
- sixth using an ILD ward intensity score;
- fifth on the maximum index value indicator; and
- fourth on the estimated population size indicator.

These analyses provide a clear indication that the case for Leeds is strengthened by using definitions of deprived inner city areas based on contiguous sets of wards, rather than LADs. We argue that this geographically sensitive analysis is an improvement on simple studies based on the relatively arbitrary definitions of LADs. However, there is a strong argument that the most concentrated areas of deprivation are much smaller than the areas identified in these analyses. While it is important to identify places with relatively large populations, the decision about where such a population cut-off should be is crucial.

The aim in the final analysis was therefore to identify deprived areas with a maximum population of 60 000 – clearly a considerable number of people, but far short of the populations in many of the 'inner city areas' identified above. This strategy identified the most *concentrated areas of deprivation* in the inner cities as the mean deprivation values in Table 7.6 demonstrate. These values are inevitably much higher than those provided in the previous table. This list indicates that the Leeds inner city area ranked:

- third using the ward-level ILD degree score;
- fourth using an ILD ward intensity score; and
- fifth on the maximum index value indicator.

Indeed, both Birmingham and Bradford, which ranked higher than Leeds, already had Objective 2 status at the time of the study.

The populations in this list of contiguous inner city areas were large and, we suggested, should be given serious consideration as zones that may have been deserving of Objective 2 status. While they were much smaller than the populations of entire LADs, these areas were considerably more deprived. The results for this concentrated area of deprivation certainly supported the wider Leeds case for eligibility and it is noteworthy that Leeds also had a high ranking based on the average of the worst three wards indicator.

Table 7.4 *Inner city areas with at least 100 000 people: rankings on mean ILD score*

District name	Number of wards	Popn	Popn rank	Area	Index mean	Index rank	Standard deviation of index	Max index score	Average worst 3 wards
Hackney	23	181 248	9	19 501 144	13.33	1	1.04	14.91	14.73
Newham	24	212 170	7	36 360 712	13.32	2	1.13	14.58	14.50
Tower Hamlets	19	161 064	13	19 733 140	12.91	3	1.46	15.66	15.06
Liverpool	22	296 891	3	62 447 304	12.81	4	1.81	16.65	15.94
Birmingham	24	602 240	1	137 786 944	12.74	5	2.94	17.12	16.45
Bradford	10	163 125	12	50 480 368	12.71	6	2.68	16.79	15.65
Lambeth	20	226 824	5	24 457 256	12.64	7	1.55	15	14.70
Haringey	18	160 269	14	21 242 866	12.63	8	1.30	14.6	14.25
Waltham Forest	13	141 537	15	21 951 868	12.21	9	1.41	14.47	14.22
Leeds	**12**	**227 747**	**4**	**75 948 004**	**12.14**	**10**	**2.25**	**15.78**	**14.81**
Sheffield	14	216 687	6	68 008 216	12.09	11	2.18	15.55	14.81
Islington	20	164 686	11	14 881 020	12.05	12	1.38	14.29	13.81
Southwark	23	201 191	8	23 311 460	11.97	13	1.39	14.88	14.17
Westminster	12	102 430	25	9 238 828	11.83	14	1.94	14.54	13.84
Brent	17	122 595	21	18 418 158	11.70	15	1.87	14.75	14.13
Knowsley	15	104 047	24	63 369 060	11.60	16	1.85	15.13	14.13
Manchester	29	353 203	2	102 427 784	11.59	17	2.07	15.53	15.21
Coventry	8	135 572	17	30 821 952	11.50	18	2.10	15.12	13.43
Wolverhampton	10	122 078	22	32 541 922	11.33	19	2.05	14.23	13.47
Sandwell	14	169 112	10	48 745 228	11.13	20	1.55	14.12	13.30
Hammersmith and Fulham	16	109 860	23	12 270 624	11.09	21	2.07	14.39	14.12
Camden	20	136 438	16	16 145 836	11.00	22	1.27	13.78	13.05
Leicester	13	128 510	20	34 664 192	10.76	23	2.00	14.94	13.69
Newcastle upon Tyne	14	130 489	19	47 679 616	10.75	24	1.89	13.2	12.92
Walsall	10	130 839	18	38 011 604	10.69	25	1.61	13.57	12.73
Doncaster	8	100 040	26	66 884 280	10.60	26	1.90	14.14	12.46

Table 7.5 *Inner city areas with at least 60 000 people: rankings on mean ILD score*

District name	Number of wards	Popn	Popn rank	Area	Index mean	Index rank	Standard deviation of index	Max index score	Average worst 3 wards
Hackney	23	181 248	9	19 501 144	13.33	1	1.04	14.91	14.73
Newham	24	212 170	7	36 360 712	13.32	2	1.13	14.58	14.50
Tower Hamlets	19	161 064	13	19 733 140	12.91	3	1.46	15.66	15.06
Liverpool	22	296 891	3	62 447 304	12.81	4	1.81	16.65	15.94
Birmingham (a)	24	602 240	1	137 786 944	12.74	5	2.94	17.12	16.45
Bradford	10	163 125	12	50 480 368	12.71	6	2.68	16.79	15.65
Lambeth	20	226 824	5	24 457 256	12.64	7	1.55	15	14.70
Haringey	18	160 269	14	21 242 866	12.63	8	1.30	14.6	14.25
Waltham Forest	13	141 537	15	21 951 868	12.21	9	1.41	14.47	14.22
Nottingham	8	72 874	38	17 205 614	12.19	10	1.78	14.69	13.81
Leeds	**12**	**227 747**	**4**	**75 948 004**	**12.14**	**11**	**2.25**	**15.78**	**14.81**
Lewisham	9	85 788	32	11 980 866	12.13	12	1.82	14.57	13.85
Oldham	7	75 493	36	19 797 938	12.12	13	1.61	14.17	13.54
Sheffield	14	216 687	6	68 008 216	12.09	14	2.18	15.55	14.81
Islington	20	164 686	11	14 881 020	12.05	15	1.38	14.29	13.81
Southwark	23	201 191	8	23 311 460	11.97	16	1.39	14.88	14.17
Wirral	5	70 786	41	20 803 394	11.90	17	1.41	14	12.76
Westminster	12	102 430	25	9 238 828	11.83	18	1.94	14.54	13.84
Brent	17	122 595	21	18 418 158	11.70	19	1.87	14.75	14.13
Knowsley	15	104 047	24	63 369 060	11.60	20	1.85	15.13	14.13
Manchester	29	353 203	2	102 427 784	11.59	21	2.07	15.53	15.21

Bolton	7	87730	31	21653956	11.51	22	2.59	16.12	13.63
Coventry	8	135572	17	30821952	11.50	23	2.10	15.12	13.43
Wolverhampton	10	122078	22	32541922	11.33	24	2.05	14.23	13.47
Plymouth	6	68382	43	14981245	11.31	25	2.53	15.29	13.19
Sunderland	8	84662	33	30128508	11.26	26	1.08	12.27	12.16
Derby	7	72873	39	23252992	11.25	27	1.85	13.78	12.90
Kingston Upon Hull	8	94467	28	29991772	11.24	28	2.63	15.3	14.00
Sandwell	14	169112	10	48745228	11.13	29	1.55	14.12	13.30
Hammersmith and Fulham	16	109860	23	12270624	11.09	30	2.07	14.39	14.12
Camden	20	136438	16	16145836	11.00	31	1.27	13.78	13.05
Birmingham (b)	3	67216	44	18791090	10.99	32	0.34	11.39	10.99
Middlesbrough	14	75419	37	21296718	10.97	33	1.29	13.64	12.76
Southampton	5	61648	47	17079916	10.84	34	2.35	14.43	12.06
Leicester	13	128510	20	34664192	10.76	35	2.00	14.94	13.69
Newcastle upon Tyne	14	130489	19	47679616	10.75	36	1.89	13.2	12.92
Walsall	10	130839	18	38011604	10.69	37	1.61	13.57	12.73
Doncaster	8	100040	26	66884280	10.60	38	1.90	14.14	12.46
Brighton	8	71072	40	15052678	10.43	39	1.51	12.82	11.97
Gateshead	8	68843	42	43460340	10.38	40	1.32	11.95	11.68
Sefton	8	94154	29	22979070	10.36	41	0.72	11.93	10.82
Greenwich	16	99902	27	21164610	10.32	42	1.05	12.68	12.18
Wandsworth	6	64785	46	7615645	10.31	43	1.44	12.15	11.48
Kirklees	5	83973	34	40540520	9.67	44	0.84	10.63	10.26
Lewisham	8	67170	45	11015418	9.56	45	0.99	11.04	10.58
South Tyneside	10	75784	35	20240788	9.56	46	0.90	10.57	10.45
Barking and Dagenham	11	88395	30	20086828	9.52	47	1.15	12.05	10.95

Table 7.6 *Inner city areas with a maximum of 60 000 people: rankings on mean ILD score*

District name	Number of wards	Popn	Area	Index mean	Index rank	Standard deviation of index score	Max index score	Average worst 3 wards
Birmingham	2	51 057	14 300 438	16.55	1	0.81	17.12	–
Bradford	3	53 460	13 670 915	15.56	2	1.78	16.79	15.56
Leeds	**3**	**59 399**	**17 355 309**	**14.81**	**3**	**1.00**	**15.78**	**14.81**
Lambeth	4	50 468	4 133 218	14.62	3	0.27	15.00	14.70
Hackney	7	57 588	5 272 163	14.46	4	0.35	14.91	14.73
Tower Hamlets	6	58 134	5 479 824	14.38	5	0.83	15.66	15.03
Newham	7	59 900	8 348 577	14.35	6	0.15	14.58	14.50
Manchester	5	58 517	13 172 278	14.22	7	0.89	15.53	14.72
Sheffield	3	48 659	11 951 534	14.08	8	1.49	15.55	14.08
Liverpool	4	58 774	11 846 305	13.94	9	2.13	16.04	14.84
Haringey	6	57 575	6 879 391	13.71	10	0.62	14.60	14.25
Southwark	6	55 463	4 655 179	13.54	11	0.82	14.88	14.17
Islington	6	58 207	5 044 179	13.49	12	0.44	14.29	13.81
Waltham Forest	5	57 683	8 383 511	13.46	13	1.25	14.47	14.22
Lewisham	5	50 569	6 529 237	13.45	14	0.73	14.57	13.85
Coventry	3	48 796	11 657 353	13.43	15	1.61	15.12	13.43
Westminster	6	59 857	3 930 741	13.34	16	0.85	14.54	13.84
Bolton	4	49 364	11 665 953	13.15	17	2.01	16.12	13.63
Wolverhampton	4	49 358	14 362 921	13.11	18	0.94	14.23	13.47
Knowsley	8	55 099	45 252 499	13.10	19	1.00	15.13	14.13

Hammersmith and Fulham	7	54818	6661414	12.95	20	1.49	14.39	14.12
Brent	8	58936	8991411	12.93	21	1.14	14.75	14.13
Nottingham	6	53166	11644620	12.88	22	1.16	14.69	13.81
Oldham	5	55414	12821654	12.80	23	1.26	14.17	13.54
Sandwell	5	57821	13426594	12.45	24	1.35	14.12	13.30
Salford	5	45142	12107229	12.41	25	1.39	14.30	13.15
Wirral	4	54909	17531268	12.24	26	1.37	14.00	12.76
Derby	5	51926	16319104	12.12	27	1.29	13.78	12.90
Walsall	4	49586	14565825	12.08	28	1.46	13.57	12.73
Sunderland	5	51216	14480702	12.01	29	0.29	12.27	12.21
Blackpool	5	32315	4822156	12.00	30	2.44	14.33	13.69
Blackburn	7	43996	14785446	11.94	31	1.09	13.80	12.88
Plymouth	5	58340	13461113	11.83	32	2.44	15.29	13.19
Kensington and Chelsea	6	42163	3230210	11.79	33	1.69	14.04	12.93
Doncaster	4	49482	38684399	11.73	34	1.93	14.14	12.46
Newcastle upon Tyne	6	52710	14727652	11.57	35	1.74	13.20	12.78
Camden	9	59032	7156480	11.53	36	0.84	13.08	12.53
Kingston Upon Hull	4	48313	13078984	11.52	37	2.51	13.76	12.67
Leicester	6	59304	11352824	11.51	38	1.56	14.07	12.47
Rochdale	5	52913	16486902	11.37	39	1.41	13.38	12.23
Preston	8	53421	19178555	11.32	40	1.48	13.22	12.99
Bristol	5	58863	12828471	11.30	41	1.37	12.60	12.27
Middlesbrough	11	58932	15249053	11.27	42	1.28	13.64	12.76
Southampton	4	47751	12570230	11.14	43	2.60	14.43	12.06

Finally, we should not ignore the fact that the geographical size of deprived areas will also influence the relative inequalities experienced by people resident there. Living in a small area of deprivation, where the distance to various resources such as cheap shops, good quality food, parks or medical facilities, may not be as detrimental to well-being as living within a large area of deprivation where the distance to such facilities is greater. Leeds ranked third in geographical area among those clusters of wards with a population larger than 200 000 (Table 7.3) and fifth among those clusters with a maximum population of 60 000 (Table 7.6). Even based on this criterion, Leeds ranked high compared with most other LADs.

7.5 Conclusion: A Tale of Two Cities

The analysis that we were asked to conduct formed part of the LCC application for Objective 2 status (Leeds City Council, 2000). The technique showed that Leeds had a strong case and that a substantial area of the inner city could be described as suffering from severe material deprivation. We believed that the LCC was correct in its assertion that the inner city area was relatively more deprived than would be indicated by a crude analysis based on data collected for LADs. Redefining Leeds based on clusters of wards in the inner city area better reflected the relative needs of Leeds compared with other English cities. Indeed, the final results in Table 7.6 demonstrate that Leeds ranks third among the most deprived inner city areas with populations less than 60 000, while Table 7.3 showed that it ranked fifth among the cities with more than 200 000 people. We concluded that Leeds was indeed a 'two-speed economy' and that despite the strong record of economic growth, the benefits were failing to 'trickle down' to the inner city residents.

The analysis was crude in the sense that it relied on a single measure of deprivation to assess the position of Leeds relative to other English cities. Little attention was given to the extent of relative inequalities that may be relevant to the Leeds inner city case. Leeds can be described as having 'two cities' where the relative inequalities between two broad groups – the rich and poor – are quite extreme. There are good reasons to suppose that people take the standards of others around them as the basis for self-appraisal and evaluation (see, for example, Boyle *et al.*, 1999, 2001; Elstad, 1998). Those living in deprived conditions may find it harder to cope if they are surrounded by people that are better off than themselves, rather than being surrounded by people in similar circumstances. This appears to be particularly pertinent to the Leeds case; the analyses presented above demonstrate that there is considerable variation among those living in the Leeds LAD – more so than in most other LADs. There is good reason to suppose, therefore, that this may cause the deprivation in this area to have even worse effects than might otherwise be the case.

Finally, as contracted researchers, we should comment on our own positionality in the context of this research. Obviously, we were employed as neutral researchers who could provide an objective insight into the relative position of the Leeds inner city. It was important for LCC to be able to argue that the research was independent and this was the case. We introduced a methodological approach that was implemented as fairly as possible in every English city, and the results did indeed demonstrate that Leeds' position was worse than a simple district-level analysis would indicate. However, while our choice of population cut-offs was not influenced by LCC, it is unlikely that the results from each would have been reported by them in their application for funding had the results not supported their general argument. This raises an interesting ethical dilemma. Do we as researchers demand that all of our results are used in any publications associated with the work, or is this an unreasonable expectation given that our services have been paid for? In this case, we felt that the use of our analysis was reasonable and that our conclusions were robust enough for LCC to use them in arguments concerning their claim for Objective 2 status. However, consultancy work of this type does raise questions that are perhaps given little attention in many studies.

Epilogue

Leeds City Council were part of the successful Yorkshire and Humberside application for Objective 2 status following the submission of its report which demonstrated, among other things, that Leeds had a significant cluster of deprived people living in its inner city, compared with other cities and towns in England. As a result, the region has been awarded a total of approximately £320 million from the EC, which when matched with UK and private funds that were also secured as part of the bid, brought the total to £870 million. Leeds, and other areas within Yorkshire and Humberside that were part of the successful bid, are now and have been, competing for these funds between 2000 and 2006 under a series of programmes designed to support business, community regeneration and physical developments.

References

Boyle, P.J., Gatrell, A.C. and Duke-Williams, O. (1999) The effect on morbidity of variability in deprivation and population stability in England and Wales: an investigation at small-area level, *Social Science and Medicine,* **49**: 791–9.

Boyle, P.J., Gatrell, A.C. and Duke-Williams, O. (2001) Do area-level population change, deprivation and variations in deprivation affect self-reported limiting long-term illness? An individual analysis, *Social Science and Medicine,* **53**: 795–9.

Committee on Spatial Development (1999) *European Spatial Development Perspective: Towards Balanced and Sustainable Development of the Territory of the European Union,* Office for Official Publications of the European Communities, Luxembourg.

Department of Environment (1995) *1991 Deprivation Index: A Review of Approaches and a Matrix of Results,* HMSO, London.

Department of Environment, Transport and the Regions (1998) *1998 Index of Local Deprivation: A Summary of Results*, DETR, London.

Department of Trade and Industry (1999) *EU Structural Funds: Determining Areas for Objective 2 Eligibility*, DTI, London.

Elstad, J.I. (1998) The psycho-social perspective on social inequalities in health, in Bartley, M., Blane, D. and Davey-Smith, G. (eds) *The Sociology of Health Inequalities,* Blackwell, London.

European Commission (1998a) *Towards an Urban Agenda in the European Union*, Communication from the European Commission.

European Commission (1998b) *Sustainable Development in the EU: A Framework for Action*, Communication from the European Commission.

European Commission (1999) *Council Regulation No 1260*, Communication from the European Commission.

Leeds City Council (1998) A tale of two cities, *New Statesman*, 22–3.

Leeds City Council (2000) *The City of Leeds – A Two Speed Economy*, Leeds City Council, Leeds.

8

GIS for Joined-up Government: the Case Study of the Sheffield Children Service Plan

Massimo Craglia and Paola Signoretta

Abstract

This chapter gives a practical example of the use of GIS in supporting public policy at the local level. The issues it raises in respect to the sharing of data among multiple agencies, and the methodology it puts forward to develop a composite indicator able to identify priority areas for intervention will be of interest to many practitioners, particularly because the project it discusses is set in the context of the preparation of the Children and Young People Service Plan, which is a statutory requirement on all local authorities under Section 17(4) of the Children Act 1989. Moreover, the range of agencies involved in this project reflects many of the partnerships being established at the local level to deliver public policy in a range of fields, from urban regeneration to crime reduction. Therefore, much of the discussion will be applicable beyond the specific application discussed here. The chapter is divided into five sections. The first introduces the project and the local setting, the second discusses the use of GIS to identify children's needs using data from different local sources; the third reports on the case study and the methodology used; the fourth introduces the results obtained, and finally the fifth draws some conclusions.

8.1 Introduction

This chapter examines the result of the *Needs Analysis and Mapping Project* supporting the preparation of the Children and Young People Service Plan (CSP)

Applied GIS and Spatial Analysis. Edited by J. Stillwell and G. Clarke
 ISBN: 0-470-84409-4

2000–03 in Sheffield. The formulation of CSPs, and their yearly updates, is a statutory requirement on all local authorities (LAs) established under Section 17(4) of the Children Act 1989. The project was requested in 1998–9 by the Joint Commissioning Group (JCG), a local partnership in Sheffield, to the Sheffield Centre for Geographic Information and Spatial Analysis to support data collection and analysis as input to the preparation of the plan. The Partnership led by the City Council included representatives of South Yorkshire Police, the Department of Social Services (DSS), the Department of Education, the Young Children's Service (YCS), Community Health Sheffield, Sheffield Health Authority, and the City Council's Central Policy Unit and Department of Housing.

The chapter demonstrates the value of GIS in bringing together data coming from different agencies for policy purposes, and puts forward a simple methodology for constructing composite indices that capture different dimensions of need, and enable the identification of key priority areas for intervention. The use of composite indices is now well established for identifying need and targeting resources (see Chapters 6 and 7). These include, for example, the Townsend Index in the health field (Townsend *et al.*, 1988), and the Index of Multiple Deprivation regularly updated by the Department of Transport, Local Government and the Regions (DTLR). However, the methodological approach put forward in this chapter has three advantages. Firstly, that a simple methodology can be more easily understood and adopted by the government agencies themselves. This is important to facilitate the wider use of GIS in practice, and enable regular updates to monitor policy and assess outcomes. Secondly, by using enumeration districts as spatial units instead of the wards, which are used both by the DTLR Index and other recently proposed relevant indices (Prince's Trust, 2000), it is possible to identify some of the pockets of needs that are otherwise masked by the ward frame. Finally, for all its simplicity, we did explore the extent to which the results also stood against other methodologies, which whilst more complex, had the advantage of being grounded on firmer statistical bases. Having carried out the necessary tests, we are satisfied that the proposed methodology yields comparable results, with the bonus of simplicity and adaptability.

This project was carried out in the context of children's welfare but has a number of aspects that make it of wider interest for policy analysis: in the first place, there has been an increasing emphasis in Britain, but also in other countries, on a more integrated approach to policy making, which recognises the increasing complexity of many social issues, and therefore the need for a collaborative multi-agency approach. This often requires the integration of data coming from different agencies on a spatial basis, and the methodology to construct a composite index discussed here has wider applicability. In the second place, whilst inter-agency collaboration is fine in principle, it does not always come easily or naturally to many of those involved. Moreover, there may be just as many difficulties of communication within each agency as there are between agencies. This was well illustrated in the course of this project in relation to the challenges of collecting the data necessary from each agency. These challenges are discussed briefly in Section 8.2 and more fully in Signoretta and Craglia (2002).

The project is set in Sheffield, a city of approximately half a million people located in the north of England. Its economy has been traditionally based on steel production and cutlery, while its hinterland was also strongly based on coal mining. The restructuring of these industries in the 1980s has had a major impact in the city. Employment in the steel industry alone fell by approximately 90%, from 45 100 in 1971 to 4700 in 1993, while another 10 000 jobs were lost in the mining industry in the surrounding region between 1985 and 1994 (Taylor *et al.*, 1996). These job losses have particularly affected the inner city wards where much of the traditional manual labour was concentrated. According to a study based on the use of a deprivation index similar to Townsend's (Woodhouse, 1999, p. 38), 'the most deprived wards, Manor, Park, Castle and Burngreave, are clearly concentrated in the inner city. They consist of both nineteenth-century industrial housing and post-war local authority estates. The next most deprived category consists of areas that have similar housing stock to the most deprived wards, such as Netherthorpe and Firth Park' (Figure 8.1). Regardless of the period of construction, many of these areas share the common characteristics of having been built when the local authority was under enormous pressure to build fast and re-house population after slum clearance or in the aftermath of World War II. Hence the quality of the construction was often low, and physical decay set in relatively quickly (Crook, 1993). Moreover, parts of these estates have been plagued for more than 30 years by high levels of crime, relative to the otherwise generally lower rates for Sheffield as a whole compared with other conurbations

Figure 8.1 *Wards of Sheffield*

in England (Baldwin and Bottoms, 1976). The cumulative impacts of economic restructuring, physical decay and social malaise, including crime, make it all the more important to identify areas where children are most vulnerable, target intervention across many agencies in a coordinated fashion, and monitor outcomes.

With these considerations in mind, the chapter is divided into four further sections. The first sets the project in context and discusses the use of GIS in performing small areas analysis to identify children's needs using data from different local sources to support the Joint Commissioning Group's requirements; the second section reports on the case study and the methodology used; the third introduces the results obtained, finally the fourth draws some conclusions.

8.2 Small Area Analysis for the Identification of Children's Needs

Since New Labour came into power in 1997, there have been numerous policy initiatives aimed at improving children and young people's lives. These policy actions are very important as recent research work points out that experiencing financial disadvantage in childhood has severe implications in adult life (Prince's Trust, 2000). The Children's Services Plans (CSP) which are the focus of this chapter provide the general framework for the strategic planning of children and young people's services. Their importance has therefore increased as a way to help coordinate the many initiatives that are now underway in this field. With this in mind, the Joint Commissioning Group in Sheffield had two aims in setting up the Needs Analysis and Mapping Project with the University. The first was to pool together data about children from the different participating agencies, and set the framework for the yearly updating process; the second was to use the expertise at the University to have a series of analyses made that could inform the preparation of the CSP and in particular help identify areas of priority for intervention. The first aspect of the project highlighted the difficulties that those involved in joined-up initiatives face in trying to build a common base of data to support policy making. These difficulties included organisational issues (such as communication difficulties within agencies, varying organisational priorities among the agencies, and resource constraints), uncertainty on what could be shared in the light of the Data Protection Act, technical difficulties relating to the characteristics of the computer systems (both hardware and software) existing in each agency, and the degree to which they could respond to demands unforeseen at the time of their installation. These problems are discussed more fully in Signoretta and Craglia (2002), whilst the remainder of this chapter focuses on the methodological aspects related to the analysis of the data, and in particular to the development of composite spatial indices that bring together different dimensions of 'need', and help to prioritise areas for investment or additional analysis.

The 1996 Guidance (DfEE, 1996) indicates the content of the CSP for children in need[1] while the latest non-statutory guidance aimed at vulnerable children[2]

[1] Defined as children who will not achieve a reasonable standard of health and development unless they receive services under part III of the Children Act 1989.

[2] Defined as children whose life chances will be jeopardised unless action is taken to meet their needs and reduce the risk of social exclusion.

(http://www.doh.gov.uk/scg/childplan.htm) gives a more strategic role to the plan (defined as the Children and Young People Strategic Plan). The Plan should link vertically to more general plans that set out strategies concerning for instance environment and transport issues, and to more focused centrally prescribed plans such as *Sure Start* aimed at improving the health and well-being of children and their families and area-based programmes such as *Education Action Zones* aimed at raising the achievement of pupils tackling the underlying causes of poor performance and behaviour. Relevant for this work are the latter area-based programmes. These stem from the recognition that areas of deprivation and social exclusion need a multi-agency and cross-sectoral approach for their regeneration and also that the identification of these areas is vital to pursue a programme of local regeneration as proposed in *A New Commitment to Neighbourhood: National Strategy Action Plan*, the Government's strategy for the socioeconomic renewal of the most deprived neighbourhoods in England.

The Children and Young People Strategic Plan contains elements such as information about levels and types of needs and gaps between needs and service provision, which allows the preparation and coordination of focused joined-up intervention among the local agencies. A recent report on disadvantaged young people from the Prince's Trust (2000) highlights that mapping is an essential tool for an approach that is based on finding local solutions to local problems. Small area analysis is an important tool to help identify pockets of deprivation. The report looks at a variety of factors to define multiple deprivation for young people and describes several indices to measure disadvantage using broad measures of poverty together with measures more specific to young people. However, the analysis is based on very broad geographical areas (regions, counties, districts and wards) within which severe pockets of deprivations may be overlooked. This aspect severely limits its use for policy intervention at the local level.

The need for small area analysis has also been recognised by the Social Exclusion Unit of the Cabinet Office, and in particular by the Policy Action Team 18 Report on Better Information (PAT 18, 2000). The report recommended that a set of neighbourhood statistics at ward level should be established nationally covering access to services, community well-being, crime, economic and work deprivation, education, health, housing, and the physical environment. Following this recommendation, a new service *Neighbourhood Statistics* was established by the Office for National Statistics in collaboration with central, and local government and other agencies. These small area indicators are used at national level to compare areas targeted by a specific initiative and at local level to evaluate local conditions, to target resources, and to monitor and assess outcomes (http://www.statistics.gov.uk/neighbourhood/patinitiatives.asp). However, the choice of the ward as the basic geographic unit for the dissemination of data about neighbourhoods is problematic. As recognised by a sub-group of PAT18, which focused on Geographic Referencing, wards vary greatly in size and population and their boundaries are continuously modified.

In the UK, the use of deprivation indices is well established and a detailed review was produced by PAT 18 (for more details on indices and associated methodologies, see http://www.cabinet-office.gov.uk/seu/2000/pat18/depindices.htm). For instance, the Townsend Index of Material Deprivation used for health analysis is the sum total of the standardised Z scores of four census variables. Other indices such as those developed by the former Department of the Environment, now ODPM, for 1991, 1998 and 2000 are based on various approaches. As an example, the latest Index of Multiple Deprivation 2000 (IMD 2000) (DETR, 2000) comprises six ward-level domain indices which were first standardised to indicate the departure of each value from its mean, then transformed to a common distribution and finally weighted. Another example especially relevant for the present work is the already mentioned study by the Prince's Trust (2000) which measures disadvantage for young people using five indices. Each index relates to a different theme (namely population, income and work, care, education and crime) with sub-components each of which is ranked according to a defined classification (for more details, see Prince's Trust, 2000, p. 33). The total score is obtained by adding up the single rankings. The advantage of this methodology is in its simplicity that makes it easier to be employed by government agencies; the downside being that the indices are developed for large units (districts, wards) which may not be very useful for policy purposes at the local level.

With these considerations in mind, this project was set up to use fine-grained socio-economic data available in the city, and use GIS as a way of integrating these data sets, and analyse them over small geographical units to support local area planning. The methodology developed and the data utilised are described in the following section.

8.3 Methodology

8.3.1 The Needs Analysis and Mapping Project

According to the project design, three main questions about children and young people needed exploring; firstly, the significant needs of under 18s in the city which affect their educational achievement, health and welfare; secondly, the distribution of these needs according to age, gender, ethnicity and geography; thirdly, the geographical distribution of children in areas affected by multiple deprivation. The project was organised in two phases: the first for data collection and input in GIS, the second for data analysis and discussion with the partnership.

The project brief contained a list of 33 data sets to collect from the participating agencies reflecting several dimensions of children and young people's lives. The list was based on the *Children's Services Planning: Guidance* (DfEE, 1996) and included information on children and young people relating to the following issues: disabilities, mental health, distribution of under 8 years olds, looked-after children, runaways, adopted children, children leaving care, protected children, young carers and children in conflict with the law. The project steering group also decided that the information was to be collected at the most disaggregated level, focusing on the post-code unit as the most appropriate scale. The time period to which the data referred was limited to one particular period (1997–8) to facilitate the collection due to the restricted time scale of the project. In some instances adjustments had to be made in relation to specific data sets. For instance, the time period chosen for education data was the academic year, rather than the calendar year. In other cases, as detailed below, it was not possible to have a whole year's worth of data but only a snapshot at a particular time. However, in all cases, questions about particular trends affecting the data were asked of the data owners to ensure that the data sets were not affected by any particular bias.

In spite of the support verbally committed by the partners in the Joint Commissioning Group, who also 'owned' the data, it proved impossible to collect all 33 data sets. After three months of efforts, only 18 data sets were made available, even though all 33 existed, mostly already in digital form. Four data sets were dropped because they were only available in paper format or they were too time consuming to collect; for two variables only sample data were available, while nine data sets were not received. As outlined before, the problems associated with data sharing were largely due to organisational issues and the different level of priority accorded to the project in the participating agencies. Following an assessment of the overall quality and robustness of the 18 data sets received, and discussions with the steering group representing the agencies involved, six data sets were selected for analysis to detect areas where children are affected by multiple needs, and thus identify areas for priority action and funding.

For the present work it was decided that a simple but effective methodology should be employed. Thus the use of statistical tests usually included in specialised software, which may not be available to local government data analysis and research units, was avoided. The six variables utilised represent different dimensions of children's life in terms of health, education, crime and social background, these are:

- children with disabilities and special needs;
- children on the Child Protection Register;
- children in families on income support;
- children with low weight at birth;
- children excluded from school; and
- children and young people in conflict with the law.

Each variable represents a group of young people with specific needs, and their combination points out areas where children's needs are more acute.

Initially each data set was checked for invalid postcodes, and then the remaining postcodes were georeferenced using the ED/Postcode directory provided by Manchester Information & Associated Services (MIMAS, www.mimas.ac.uk) at the University of Manchester. The postcoded data were aggregated at the enumeration district level (using the 1991 boundary provided by EDINA Edinburgh University Data Library, http://edina.ac.uk), since this level would preserve the anonymity of single individuals but at the same time was fine enough to pick up pockets of deprivation within Sheffield. The six variables were combined to produce two indices: one was derived using count data while the other rates, thus reflecting different policy needs, both operational (where are the largest number of children in need?) and strategic (where do we have more or less children in need than we would expect other things being equal?). The indices measure needs of children and young people in Sheffield, not material deprivation as for instance in the Townsend Index, and are *indirect* indicators of deprivation as they represent the victims of particular disadvantaged circumstances (for more details on this issue, see http://www.cabinet-office.gov.uk/seu/2000/pat18/depindices.htm). A description of the six variables and the denominators used is given below.

8.3.2 Description of the Data Sets Utilised

The data set on *Children with Disabilities and Special Needs* was delivered by Sheffield Social Services and includes four data sets (Children registered disabled with Sheffield Social Services, Children with a user group for disabilities, Children referred during the required period to the Social Work Team for Children with Disabilities, Children and Families User Group with a referral reason of mobility problems, disability problems or mental health). The steering group decided to include all four data sets, as this would give a reliable picture of the problem in Sheffield. Ninety-two per cent of the records could be mapped while the remaining 8% had an invalid postcode. *Children on the Child Protection Register* was also delivered by Sheffield Social Services. Ninety per cent of the records were mapped while the remaining 10% had an invalid postcode. Data on *Children aged 0–17* in families on income support was a snapshot at December 1998. This was because the data set was derived from a system designed for administrative purposes which could not easily respond to the requirements of this project. The data was supplied by the Central Policy Unit at Sheffield City Council. *Children born less than 2.5 kg* was derived from a file supplied by Sheffield Health Authority on the children born in Sheffield in 1998 with details of their weight. The steering group selected the threshold of 2.5 kg because this is the value at which health practitioners define 'under weight'. Ninety-nine per cent of the records delivered were mapped. The data set *Children excluded from school* was based on data delivered by the Department of Education and includes children aged 3–16 permanently excluded from school during the academic year 1997–8. A permanent exclusion applies when pupils are excluded from the school on a permanent basis because the school does not wish them to return. Ninety-nine per cent of the total number of records were mapped. Finally, the Youth Offending Team supplied the data set on *Children aged 10–17 in conflict with the law*. The data set contained the cases brought to court during April 1997 to March 1998. From this initial data set information on young offenders was extracted and 90% of the records were geocoded.

As far as the denominators are concerned four data sets were used: total number of children aged 0–17 (disabled children, protected children, children in families on income support), total number of children aged 3–16 (children excluded from school), total number of children aged 10–17 (children in conflict with the law) and total number of children born in Sheffield during 1998 (babies born less than 2.5 kg). The data were delivered by the Central Policy Unit; 98% of the records were geocoded. The method to combine the variables to obtain the indices is explained in the next section.

8.3.3 Construction of the Two Indices

The methodology deployed included the following steps. Firstly, for each variable we computed rates and counts at the ED level. Rates were calculated as the proportion of observed cases in each ED over the total population at risk in that ED (e.g., number

Table 8.1 Score attributed to each class

Class	Score
Worst 20%	5
Second worst	4
Mid-range	3
Second best	2
Best 20%	1
No cases	0

of disabled children aged 0–17 in ED/Total number of children aged 0–17 in ED). Secondly, we classified each variable in quintiles to identify the worst 20% of EDs. This was done for both sets of counts and rates. For all variables, only the EDs with cases were included in the classification and a score from one to five was assigned to the EDs falling in each group (Table 8.1), with the worst 20% having the highest score.

A score of zero was given to the EDs with no cases. In this way, each ED ended up having a score assigned for each variable for both sets of counts and rates. Thirdly, the scores given to each ED were summed and two total scores were derived, one for counts and one for rates. Finally, the two total scores were grouped in quintiles and then mapped. This methodology makes it easy to assign different weights to each variable prior to the construction of the composite index, and thus could be applied in the context of multi-criteria evaluation (see for example Henig, 1996; Hill, 1972). In the case of this project, however, the stakeholders represented in the steering group decided that differential weighting was not warranted, and therefore none was applied.

8.4 The Spatial Distribution of Child Deprivation

As far as the initial three questions are concerned, the project investigated the overall multiple needs of children and young people and their geographical distribution in Sheffield. Gender and ethnicity issues were not incorporated in the analysis due to the incompleteness of the data sets collected but will be part of the Plan update. Out of a theoretical maximum score of 30, which would indicate that an ED was consistently among the worst 20% on all six variables, the maximum score obtained in practice was 27 on counts, and 26 on rates (Figures 8.2 and 8.3).

As far as the total score on rates is concerned the worst EDs (first two categories), appear to be concentrated to the east of the city centre, with a similar pattern when looking at the counts. The figures also highlight the outliers. The outliers common in both indices are located in Southey Green, Firth Park, Manor, Norfolk Park and Park Hill, i.e. many of the areas that have been significantly impacted by economic restructuring as discussed in Section 8.1. It is important though to acknowledge that even within an estate or neighbourhoods there can be significant variations, with problematic areas sitting right next to others that are not. Here lies the value of working with fine-scale data. This issue is all the more apparent if one used wards, as suggested by PAT 18, as the basic unit for neighbourhood data collection and

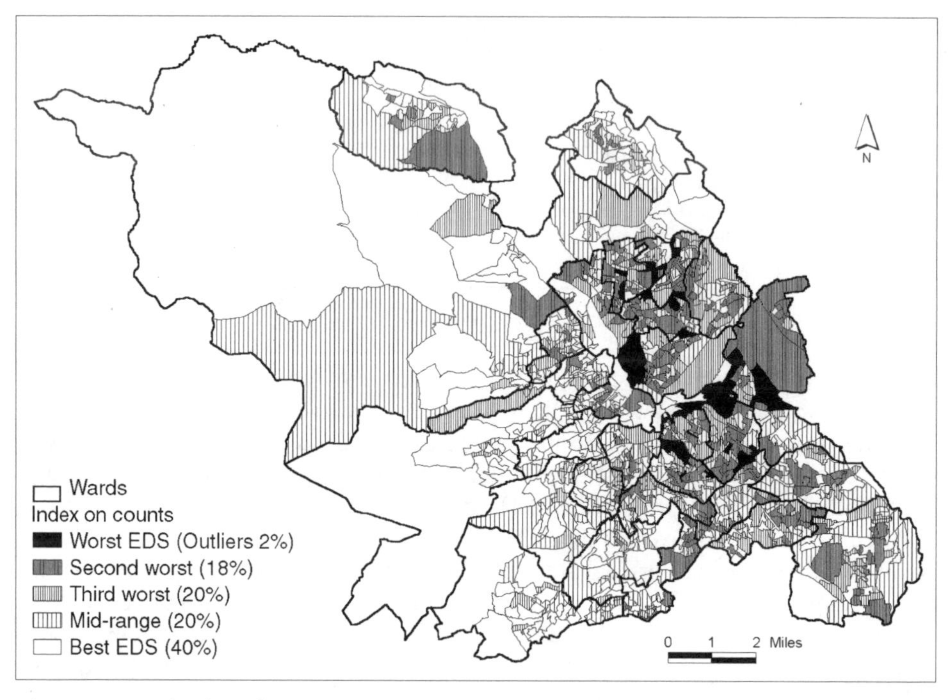

Figure 8.2 *Index based on counts*

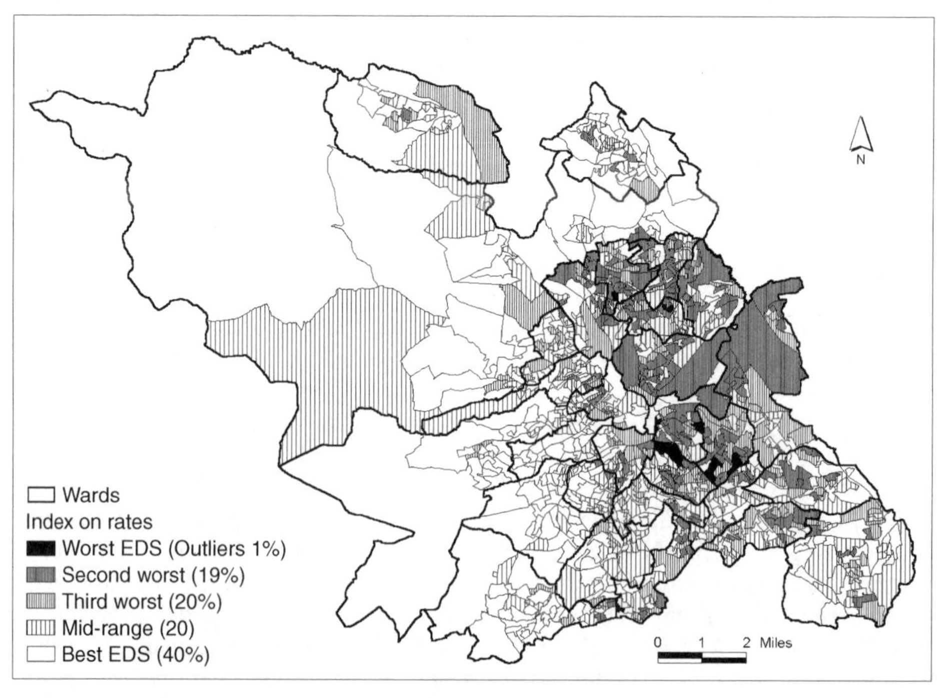

Figure 8.3 *Index based on rates*

dissemination. The figures show this by superimposing the ward boundaries onto the ED framework thus confirming the unsuitability of aggregating data at the ward level for local planning and service delivery purposes. This is yet another example of the Modifiable Areal Unit Problem (MAUP), i.e., of the extent to which the outcome of spatial analysis depends on the nature of the chosen spatial framework. Whilst well known to academics, highlighting this type of problems to the partnership funding the project was particularly useful. Partly as a result of that, the City Council is currently in the process of revising both ward and corporate area boundaries to maximise the homogeneity of key social indicators.

The advantage of the method described above to combine different variables, and identify areas affected by multiple needs lies in its arithmetic transparency, which makes it ideal for use by local authorities and agencies engaged in needs assessment and planning. Moreover, the simplicity of the method makes its outcomes more easily understood by local stakeholders and community groups. The potential drawback is that this methodology disregards any spatial structure in the data and gives no measure of statistical significance. Cognisant of these potential limitations, we undertook further analysis using the Besag–Newell test (Besag and Newell, 1991). This is a test for detecting clusters in rare events over a geographical area that has been divided into small zones (in the present study enumeration districts). For each zone, data on the population at risk must be available and a coordinate assigned to represent the location of the zone – such as the population-weighted centroid. Each event occurring in the ED is then allocated to the centroid. For each ED a population-centroid is available and for each variable and each ED a population at risk (which as in the first method will depend on the particular variable) is available and a count of the number of cases. In this application all EDs that were included in a cluster test attained the 5% significance level (for more details on the procedures followed see Craglia *et al.*, 2003). The results confirmed that the simpler method described in this chapter to create composite indices was more than adequate in identifying the key problem areas in the city. This finding is important as the methodology adopted can be easily employed for similar analysis especially in the context of local government, with some confidence that the results are similar to those obtained with more sophisticated, and complex, statistically based methods.

The results of this project proved very helpful for the planning and programming of priority actions for all the agencies involved, and gives a benchmark to evaluate the effectiveness of strategies over time. The results formed the basis for consultation by the area groups of the Young Children Service with the local communities, and were also utilised for subsequent projects. For example, the data on disabled children supported the Residential Respite Service project and the provision of additional services for disabled children in the northern part of the city. A final result was the setting up of 14 multi-agency groups each with the task to investigate issues relating to children and young people's welfare.

8.5 Conclusions

This chapter has presented a case study based on the *Needs Analysis and Mapping Project* within the context of the preparation of the Children Service Plan 2000–03 in Sheffield. In order to identify priority areas for intervention, the methodology developed has produced two indices based on a total score for counts and rates using six variables. The analysis carried out identified areas where children and young people are most affected by multiple needs. One advantage in the context of local government is that this methodology overcomes the difficulties associated with sophisticated statistical tests that are usually included in specialised software developed and mainly utilised in academic environments. One additional advantage of the present work was the small level (ED) of analysis adopted that allowed the identification of pockets of deprivation within wards which would not have emerged as affected by multiple needs.

This project demonstrated the high value that data collected from several local sources and integrated through GIS can offer to policy making in tackling local problems. The difficulties encountered in sharing the data among agencies have raised awareness among all the participant organisations of the practical difficulties of 'joined-up' working. As a result, more formalised arrangements have been put in place among all the stakeholders with written protocols committing each partner to provide key data sets to the University at regular intervals. In this context, the role of the University has been crucial as a neutral broker, able to collect highly disaggregated data, and develop small area analyses that add value to the data for policy making without infringing the requirements of the Data Protection Act. From a policy perspective, the analysis pointed out that the most deprived areas in terms of the six variables selected are still, not surprisingly, the same areas that started suffering from economic decline over 30 years ago. As this analysis concerned young people, the future for these areas and consequently for the people that live in them looks unpromising unless sustained action is taken. There are now new opportunities to help the local economy through the structural funds awarded to South Yorkshire and Sheffield from the European Union for the period 2001–06. Hence, the added importance of having in place an analytical framework to target intervention and monitor outcomes. GIS has a key role to play in this framework, as shown in this chapter. More practical applications like this will do much to raise further the awareness among policy makers of the opportunities of this technology and method of analysing spatial data.

Acknowledgements

The authors would like to thank the Young Children's Service and the Joint Commissioning Group in Sheffield for providing the data used in this analysis. However, the views expressed are those of the authors and not necessarily of the agencies involved. This work is based on data provided with the support of the ESRC and JISC and uses boundary material which is copyright of the Crown and the ED-LINE Consortium.

References

Baldwin, J. and Bottoms, A. (1976) *The Urban Criminal,* Tavistock, London.

Besag, J. and Newell, J. (1991) The detection of clusters in rare diseases, *Journal of the Royal Statistical Society*, **154**(1): 143–56.

Craglia, M., Haining, R. and Signoretta, P. (2003) Identifying areas of multiple social need: a case-study in the preparation of Children Service Plans, *Environment and Planning C*, **21**(2) April: 259–76.

Crook, A.D.H. (1993) Needs, standards and affordability: housing policy after 1914, in Martin, D. and Binfield, J.C.G. (eds) *The History of the City of Sheffield 1843–1993*, Society Vol. 2, Sheffield Academic Press.

Department for Education and Employment (1996) *Children's Service Planning: Guidance*, G15/004 3756 1P 10k March 1996 (03).

Department for Education and Employment (2000) *Connexions, The Best Start in Life for Every Young Person*, DfEE Publications, London.

Department for Transport, Local Government, and the Regions (DTLR) (formely DETR, currentlt ODPM) (2000) Index of Multiple Deprivation. Office for National Statistics, http://www.statistics.gov.uk/.

Department of Environment, Transport and the Regions (2000) Indices of Deprivation 2000, *Regeneration Research Summary No 31*, DETR, London.

Henig, M. (1996) Solving MCDM problems: process concepts, *Journal of Multicriteria Decision Analysis*, **5**: 3–21.

Hill, M. (1972) A goals–achievement matrix for evaluating alternative plans, in Robinson I. (ed.) *Decision-Making in Urban Planning*, Sage, Beverly-Hills, CA.

Policy Action Team 18 (2000) *Better Information*, http://www.cabinet-office.gov.uk/seu/2000/pat18/.

Prince's Trust (2000) *Mapping Disadvantage*, Greenhouse, London.

Signoretta, P. and Craglia, M. (2002) Joined-up government in practice: a case study of the children needs in Sheffield, *Local Government Studies*, **28**(1): 59–76.

Taylor, I., Evans, K. and Fraser, P. (1996) *A Tale of Two Cities*, Routledge, London.

Townsend, P., Phillimore, P. and Beattie, A. (1988) *Health and Deprivation: Inequality and the North*, Croom Helm, London.

Woodhouse, P. (1999) Deprivation and health in Sheffield, *Geography Review*, November: 36–41.

9

The Application of New Spatial Statistical Methods to the Detection of Geographical Patterns of Crime

Peter Rogerson

Abstract

The evolution of geographic information systems has spawned a resurgence in the development of spatial statistical methods. In this chapter, several recent developments in the field of spatial statistical analysis are summarised and applied to burglary data for census tracts in Buffalo, New York. The methods employed here are concerned with the detection of spatial patterns in geographic data. Some of the methods are concerned with finding regions with rates that are significantly higher than those that could be expected under the null hypothesis of no geographic pattern, relative to prior expectations. Others have as their goal the detection of temporal change in spatial data.

9.1 Introduction

As the value of geographic information systems (GIS) has become more evident to researchers and practitioners in an expanding array of disciplines, the demand for spatial analysis, spatial modelling and spatial statistics has witnessed substantial growth. Although the early benefits of GIS derived primarily from their ability to store, retrieve and display spatial information, GIS is now increasingly used in the analysis of spatial data. The addition of modules such as Spatial Analyst and

Applied GIS and Spatial Analysis. Edited by J. Stillwell and G. Clarke
 ISBN: 0-470-84409-4

Geostatistical Analyst to ArcView, and Spatial Statistics to S-PLUS represent examples of how vendors are beginning to meet the needs of the new marketplace for geographical analysis. The effects of these changes vary from one subfield to another. In epidemiology, the use of spatial modelling and spatial statistics has soared in recent years. Many new methods for the statistical assessment and modelling of spatial patterns have been developed recently (e.g., Elliot *et al.*, 2000; Lawson *et al.*, 1999; Lawson, 2001). In crime analysis, the development of new statistical methods has perhaps been less apparent, but the use of GIS by police agencies as well as by other local government agencies has been much more widespread than in public health and epidemiology.

This chapter demonstrates how recent advances in spatial statistical analysis may be used in the context of crime analysis. Section 9.2 provides a brief description of the burglary data (from Buffalo, New York) that will be used to illustrate the various methods. Section 9.3 describes several alternative approaches to detecting geographic patterns in the data, and in each case the approach is illustrated by using the burglary data.

9.2 Burglary Data

Data on burglary incidents in the city of Buffalo, New York, for 1998 is used to illustrate the methods. Buffalo has a population of approximately 300 000 residents. Its population has declined steadily over the last several decades as it has coped with the difficulties associated with transforming its economy and becoming less reliant on what was a once-strong manufacturing base. The maintenance of infrastructure and the provision of services are, of course, difficult for declining manufacturing cities. In part reflecting these difficulties, the Buffalo Police Department (BPD) has had to operate with a crime analysis unit that is understaffed. Consequently, the crime analysis unit has a greater focus on database development and reporting, and less of a focus on statistical analysis.

The BPD has simultaneously maintained strong links with the University at Buffalo, and the two have been involved with several joint projects. Within this context, and with the funding of a grant from the National Institute of Justice to (a) develop new methods, and (b) work with local agencies, an effort was made to examine geographic aspects of crime. In particular, we worked with the Crime Analysis Unit to develop new methods for monitoring temporal changes in the geographic patterns of crime.

Information is available on the number of burglaries by census tract, as is information on the coordinates of tract centroids. In addition, the data for 1998 is disaggregated by quarter; a summary of the number of incidents in each quarter is given in Table 9.1. Since there are 90 census tracts in the city, there is an average of approximately 15 burglaries per tract, per quarter.

The data are spatial in the sense that regional frequencies are available for the 90 census tracts that comprise the city of Buffalo. Consequently, the methods discussed in this chapter focus on those that are useful for spatial data. Other methods are relevant for situations where coordinates are available for the point locations of events; these latter methods are not covered in this chapter.

Table 9.1 Burglary incidents by quarter in 1998, Buffalo

Period	Number
January–March	1364
April–June	1540
July–September	1645
October–December	1564
Total	6113

Table 9.2 Chi-square test comparing observed frequency of burglaries with that expected under the null hypothesis of no spatial pattern

Period	χ^2
January–March	810.3
April–June	925.6
July–September	1027.8
October–December	875.5

Note: The critical value of chi-square, when $df = 89$ and $\alpha = 0.05$, is approximately 112.

9.3 Analysis of Burglary Data Using Alternative Spatial Statistical Methods

9.3.1 Expected Versus Observed Frequencies

The expected numbers of burglaries in each tract were calculated by assuming that there was no spatial pattern in the burglary data, and by conditioning on the total number of burglaries in each quarter. Specifically, expected values were determined by multiplying the total number of burglaries in a quarter by each tract's share of city-wide population. One straightforward way to compare observed and expected frequencies is via the common, aspatial, chi-square test:

$$\chi^2 = \sum_i \frac{(f_{obs,i} - f_{exp,i})^2}{f_{exp,i}} \tag{9.1}$$

where the fs refer to tract i-specific observed and expected frequencies, and the sum is over all tracts. This statistic has a chi-square distribution with degrees of freedom equal to the number of tracts minus one, under the null hypothesis that the number of burglaries observed in tracts is generated by distributing the total number of burglaries randomly across tracts, in accordance with tract population shares.

The results, shown in Table 9.2, indicate a very high degree of statistical significance. In all cases the observed value of the chi-square statistic exceeds greatly the critical value of 112, and the pattern of crime is thus substantially different from the underlying pattern of population. This is not particularly surprising; it is well known that crime tends to cluster geographically, and the results are consistent with the visual impression one obtains from Figure 9.1. Of interest is the fact that the fit is different in each quarter; the burglary pattern most closely resembles population distribution in the first quarter (even though it is very different from it), and the

Figure 9.1 *Burglary incidences in Buffalo by census tract, 1996*

pattern of burglaries is most different from the spatial distribution of population in the third quarter. We will investigate these differences in more depth in the following sections.

9.3.2 Changes in Spatial Pattern: Redefining Expectations

After finding that the spatial pattern of burglaries differs significantly from that which would be expected if the true probability of a burglary were proportional to population, we now turn our attention to the changes in spatial patterns that occur from one quarter to the next. An alternative way to derive expected frequencies is to use the spatial pattern of crime itself, rather than the spatial pattern of population. One might, for example, choose to use the spatial pattern of burglaries that prevailed during the previous year as a basis for expectations. Then, any significant difference between what is observed in the current year and what is expected (on the basis of the past year) would pertain to *changes* that had occurred in the spatial pattern. Since we do not have access to 1997 burglary data for Buffalo, we will simply use the complete set of data for 1998. Thus the expected frequency of burglary in a census tract, in a given quarter, is now derived as the total number of burglaries in the quarter, multiplied by the fraction of all 1998 burglaries that occurred in that particular census tract. Note that this particular calculation of expectations uses some of the actual data, and hence judgements of statistical significance based on these expectations might be viewed as conservative. An alternative might be to calculate expectations for each quarter based upon the spatial pattern that prevailed during the other three quarters, but that seems to be even more *ad hoc*. Clearly the question of what data to use in the definition of expectations is an important one. Most often one will want to use previous, historical data. Even then, the question of how much historical data (e.g., last quarter, last year or last three years) is a difficult one.

Table 9.3 reveals that only the first quarter, January–March, has a spatial pattern that differs significantly from the pattern that prevailed for the entire year. Of course it is of interest to know *how* the January–March pattern differs from the remainder of the year (e.g., where are burglaries higher or lower than expected?). This question will be addressed in section 9.3.4.

Table 9.3 *Chi-square test comparing observed frequency of burglaries with that expected under the null hypothesis of no spatial deviation from the annual pattern*

	χ^2	Significance (p)
January–March	129.9	0.003
April–June	63.9	0.979
July–September	106.4	0.101
October–December	77.6	0.800

9.3.3 A Spatial Chi-Square Test

When the categories in a chi-square test represent a set of geographic regions, the observed frequencies in each category/region may be spatially dependent, and one approach to account for this dependence is to deflate the usual chi-square statistic in an appropriate manner (Fingleton, 1983, 1986). Cerioli (1997) has indicated how chi-square tests should be modified in the presence of spatially autocorrelated data for 2 × 2 tables. A specific weakness of the chi-square goodness-of-fit statistic, when applied to spatial data, is that it does not account for *where* the greatest deviations between observed and expected values are, relative to one another. Figure 9.2, taken from Rogerson (1999), shows that both (a) and (b) have the same chi-square statistic, and yet the spatial pattern of deviations is quite different. In (a), there is a distinct cluster of regions where the fit is poor, while in (b), the spatial distribution of deviations is more random. Suppose that the usual, aspatial chi-square statistic is computed for (a) and (b), and it is found that it is almost, but not quite, significant. It would be desirable to have a statistic with which we would conclude that in (a), the combination of the aspatial deviations *and* the spatial pattern of deviations could not have occurred by chance (that is, where the observed *spatial* pattern of frequencies is not likely under the null hypothesis of independent, expected regional frequencies).

Likewise, Moran's I statistic might be used to examine the spatial pattern of deviations but, by convention, the diagonal elements of the spatial weight matrix are zero, implying that deviations between the observed value and the expected value within

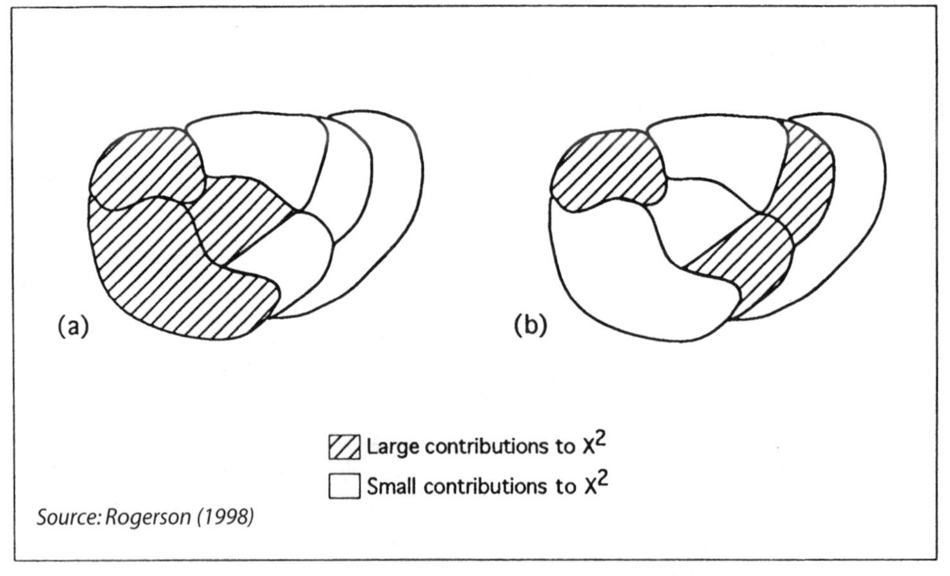

Figure 9.2 *Two patterns of deviations (observed–expected) yielding identical χ^2 values. From Rogerson (1998)*

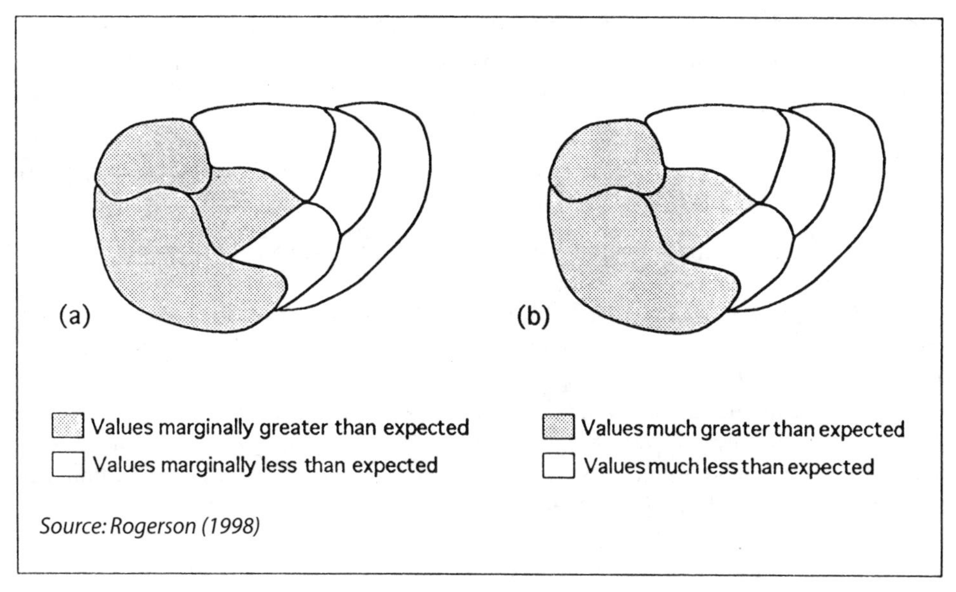

Figure 9.3 *Two patterns of deviations (observed–expected) yielding identical I values. From Rogerson (1998)*

each of the regions are not accounted for. Figure 9.3, also from Rogerson (1999), shows two situations that could have identical I values, and yet in (a) the pattern is consistent with the null hypothesis, while in (b) the deviations are unlikely to have arisen under the null hypothesis. Whereas the chi-square goodness-of-fit test focuses upon individual regions and ignores the geographical pattern of deviations, Moran's I focuses upon the pattern of deviations, but ignores the fit *within* regions.

Rogerson (1998, 1999) introduced the spatial chi-square statistic, R, to construct a test of the null hypothesis that a set of m observed regional frequencies, N_i, $i = 1, 2, \ldots, m$, could have occurred under the following null hypothesis:

$$H_0 : E[N_i] = \lambda \xi_i, \quad i = 1, 2, \ldots, m. \tag{9.2}$$

For example, in the case of the spatial pattern of crime, the population of each region is given by ξ_i, the overall crime rate in the population is λ, and the expected number of cases in each region is $E[N_i]$.

Rogerson defines the spatial version of the chi-square goodness-of-fit statistic as

$$R = (\mathbf{r} - \mathbf{p})' \mathbf{W} (\mathbf{r} - \mathbf{p}) \tag{9.3}$$

where there are m regions, and the observed frequencies are used to construct the $m \times 1$ vector of proportions $\mathbf{r}$, where $r_i = N_i / N$ and N is the total number of observed cases. The values $\xi_1, \xi_2, \ldots, \xi_m$ are known quantities, which, for example, might be taken as the populations of the regions. The nonrandom $m \times 1$ vector of expected proportions is **p**. A natural definition is to have $p_i = \xi_i/\xi$, where $\xi = \xi_1 + \xi_2 + \ldots + \xi_m$ and $\lambda = N/\xi$. Also, **W** is an $m \times m$ matrix containing elements w_{ij} (including elements along the diagonal, where $i = j$):

$$w_{ij} = \frac{a_{ij}}{\sqrt{p_i p_j}} \tag{9.4}$$

The elements of a_{ij} measure the 'connectedness' of region i to region j, and are discussed further below. The spatial chi-square statistic tests the null hypothesis by combining terms that focus on within-region deviations (as does the chi-square goodness-of-fit statistic) and the spatial pattern of deviations found in nearby regions (as does Moran's I).

High values of the statistic indicate that there may be single regions where observed and expected proportions differ significantly, and/or there may be areas where positive (or negative) deviations between observed and expected proportions exhibit significant spatial autocorrelation. Since R may also be written as

$$R = \sum_i \sum_j w_{ij}(r_i - p_i)(r_j - p_j) = \sum_i \frac{(r_i - p_i)^2}{p_i} + \sum_{i \neq j}\sum a_{ij} \frac{(r_i - p_i)(r_j - p_j)}{\sqrt{p_i p_j}} \tag{9.5}$$

it may be conceptualised as the sum of two terms – the aspatial chi-square statistic, and a term that accounts for the spatial pattern of deviations between observed and expected values. High values of R will lead to the rejection of the null hypothesis for at least one of two reasons: either there are large deviations between observed and expected regional values (in keeping with the usual interpretation of the aspatial chi-square test), or the deviations exhibit a significant spatial pattern, or both.

The spatial chi-square statistic is closely related to Tango's (1995) statistic:

$$C_G = (\mathbf{r} - \mathbf{p})' \mathbf{A} (\mathbf{r} - \mathbf{p}) \tag{9.6}$$

where **A** is a $m \times m$ matrix containing the elements a_{ij} that measure the degree to which regions are connected. Tango uses

$$a_{ij} = e^{-d_{ij}/\tau} \tag{9.7}$$

$$a_{ij} = 1 \tag{9.8}$$

where d_{ij} is the distance between regions i and j, and τ is a measure of the importance of distance (where larger values of τ correspond to larger 'spheres of influence' around each region). Tango gives the expected value and variance of C_G as

$$E[C_G] = \frac{1}{N}\mathrm{Tr}(\mathbf{AV}_p) \tag{9.9}$$

$$V[C_G] = \frac{2}{N}\mathrm{Tr}(\mathbf{AV}_p)^2, \tag{9.10}$$

where

$$\mathbf{V}_p = \Delta(\mathbf{p}) - \mathbf{pp}' \tag{9.11}$$

with $\Delta(\mathbf{p})$ defined as an $m \times m$ diagonal matrix containing the elements of $\mathbf{p}$ on the diagonal. Furthermore, the null hypothesis may be tested using the chi-square distribution. Specifically, the quantity

$$\nu + \frac{C_G - E[C_G]}{\sqrt{V[C_G]}} \sqrt{2\nu} \tag{9.12}$$

has an approximate chi-square distribution, with ν degrees of freedom, where

$$\nu = \left[\frac{(\mathrm{Tr}(\mathbf{AV}_p))^{1.5}}{\mathrm{Tr}(\mathbf{AV}_p)^3} \right]^2 \tag{9.13}$$

The significance of the spatial chi-square statistic, R, may be assessed by replacing $\mathbf{A}$ with $\mathbf{W}$ in equations (9.9)–(9.13).

Results for the burglary data are shown in Table 9.4 for various choices of τ (which can be roughly interpreted as proportional to cluster size). For a Type I error of $\alpha = 0.05$, the statistic is significant only for the January–March data (though note that the statistic would be barely significant at $\alpha = 0.10$ for the July–September data). For January–March, the significance is highest ($p = 0.002$) when $\tau = 0.9$. As τ increases, the results approach those of the aspatial chi-square test. The results here are similar to those for the aspatial chi-square test (Section 9.3.1); when used in an exploratory fashion, by examining different values of τ, the spatial chi-square test has the additional capability of better capturing the scale on which clustering is occurring.

9.3.4 Outliers and Local Tests

When the global chi-square test (spatial or aspatial) is significant, it is of interest to know which categories (in this case, census tracts) contributed to the significant value. Perhaps it was simply one or two tracts that might be then considered as potential

Table 9.4 *Spatial chi-square tests*

	Jan–Mar		Apr–Jun		Jul–Sept		Oct–Dec	
τ	*R*	Sig.	*R*	Sig.	*R*	Sig.	*R*	Sig.
0.5	.108	.006	.0508	.496	.0590	.195	.0522	.429
1	.097	.002	.0429	.962	.0644	.107	.0535	.623
1.5	.0953	.003	.0416	.978	.0648	.099	.0501	.781
2.	.0952	.003	.0416	.979	.0647	.101	.0497	.799
Optimal	.098	.002	.0408	.147	.0648	.099	.0286	.356
τ	(0.9)		(0.2)		(1.3)		(0.2)	

outliers. Also, when the global test is insignificant, it is still possible that an examination of individual categories will uncover instances where the observed value differs significantly from the expected value (and these might then be considered local hotspots). Fuchs and Kennett (1980) propose the M test to find outlying cells in the multinomial distribution. The test may be used in conjunction with the global, aspatial chi-square test (and the related quadrat test). Their statistic is the maximum adjusted residual. For cells $i = 1 \ldots, m$ define

$$z_i = \frac{N_i - Np_i}{\sqrt{Np_i(1 - p_i)}} \tag{9.14}$$

A one-sided test is based on max z_i; $i = 1 \ldots, m$, and bounds for the critical value are

$$\Phi^{-1}\left\{1 - \frac{m - [m^2 - 2\alpha m(m-1)]^{1/2}}{m(m-1)}\right\}, \Phi^{-1}\left\{1 - \frac{\alpha}{m}\right\} \tag{9.15}$$

where Φ is the cumulative distribution function for the standard normal distribution. The upper bound can be recognised as the common Bonferroni adjustment. Fuchs and Kennett (1980) show that the test is more powerful than the global chi-square test in the presence of a single outlier. Thus the M test can be used in conjunction with the global test – when the global test is significant, the M test may be useful in determining whether the global significance is achieved via contributions from many regions, or via contributions from only a small number of outliers. Similarly, when the global test is not significant, the M test may still uncover anomalous values that remain hidden in the global statistic, since it is more powerful at detecting outliers.

To determine those census tracts that contributed significantly to the overall chi-square value in the burglary data, the tract-specific local z-statistics were calculated using equation (9.13). Using (9.12), with $k = 90$, $\alpha = 0.05$, bounds for the critical values of $M = \max z$ are $\Phi^{-1}(.99943, .999444) = \{3.253, 3.261\}$. The observed maximal z-values for each quarter are shown in Table 9.5. In each of the first three quarters, the observed value of $M = \max z$ is significant (though only marginally so in quarters two and three). The most significant value is found in the first quarter – a result that is consistent with the global analysis summarised in Table 9.3. Table 9.5 also depicts the tract number associated with the maximal z value; in all four quarters, only one of the 90 tracts had a z-value exceeding the critical value, and thus there were no other tracts that contributed significantly to the overall chi-square.

Table 9.5 *Results of M test*

Period	Maximum z	Tract
January–March	3.68	69
April–June	3.29	87
July–September	3.30	26
October–December	2.67	75

9.3.5 A Local Version of *R*

A recent development in spatial statistics has been the advent of 'local' spatial statistics that focus more specifically upon the possible existence of spatial association as viewed from a particular local region (e.g., Getis and Ord, 1992; Ord and Getis, 1995; Anselin, 1995). Thus 'hotspots' of significant local positive spatial autocorrelation may be detected, despite the absence of pattern on a global scale. Also, as Anselin has indicated, local statistics are useful in detecting outliers when the global indicator is significant. For example, in their application to international trade, O'Loughlin and Anselin (1996) found that a significant global measure of autocorrelation was due to a small number of outliers, and when these were removed, the spatial pattern of the remaining values was essentially random. The local version of the spatial chi-square statistic, *R*, for a given location *i*, is:

$$R_i = \frac{(r_i - p_i)}{\sqrt{p_i}} \sum_j \frac{a_{ij}(r_j - p_j)}{\sqrt{p_j}} \tag{9.16}$$

Under the null hypothesis, the quantity

$$1 + \frac{R_i - \mathrm{E}[R_i]}{\sigma[R_i]}\sqrt{2} = \frac{R_i}{\mathrm{E}[R_i]} \tag{9.17}$$

can be approximated by a chi-square distribution, with one degree of freedom. The expectation is defined as

$$\mathrm{E}[R_i] = \frac{a_{ii}(1 - P_i)}{N}$$

This local test can be thought of as a spatial generalisation of the aspatial *M* test, so that the assessment of fit in individual sub-regions also includes an assessment of how similar the fit is in nearby sub-regions as well.

Table 9.6 displays the maximum value of *R* in each of the four time periods. This is done for that value of τ which yields the highest local value of *R*. The results are similar to those shown in Table 9.5 – the first time period has the most significant local spatial association, around census tract 69. The next two quarters are

Table 9.6 *Local R tests*

Period	Local statistic	Max τ	Observation number
January–March	13.53	τ = high and flat	69
April–June	10.85	τ = 1.7 (flat)	87
July–September	12.358	τ = 0.46	26
October–December	10.175	τ = 0.52	75

Note: ($R/\mathrm{E}[R]$) is approximately chi-square with 1 degree of freedom; under the null hypothesis we use $3.26^2 = 10.63$ as the critical value (where 3.26 is the critical value of a standard normal random variable with $n = 90$ tests).

marginally significant, and the last quarter fails to reject the null hypothesis that any single tract significantly differs from the expected number of burglaries (where expectations are based here upon the annual pattern). For the first two quarters, the value of τ leading to the maximum value of R is quite high; the results of the maximum local R test approach those of the M test as τ gets large.

9.3.6 The Q^2 Statistic for Detecting Change in Multinomial Distributions

When spatial data are available for a number of time periods, it is of interest to test the null hypothesis of no change in spatial pattern, versus the alternative that a shift in pattern occurred at a particular point in time. Srivastava and Worsley (1986) discuss how changes may be detected retrospectively in data that come from a multinomial distribution. Let $\mathbf{x}_t$ be a $1 \times (m-1)$ row vector with elements indicating the number of crimes in each of $m-1$ tracts during period t. There are m tracts in the study region, and so one region is arbitrarily omitted from the vector. For n time periods, the data may be given as an $n \times (m-1)$ matrix, $\mathbf{X}$, consisting of n of these row vectors. Srivastava and Worsley (1986) derive a likelihood ratio statistic that is asymptotically equivalent to the maximum of χ_r^2 over r. Here χ_r^2 is the usual chi-square statistic used for testing for association in the $2 \times (m-1)$ table formed by first segmenting the observations into the two sets $t = 1, 2, \ldots, r$ and $t = r+1, \ldots, n$ and then forming the two aggregate row vectors corresponding to the two sets of observations. The distribution of the test statistic, $Q^2 = \max\chi_r^2$, under the null hypothesis of no change, is given by

$$p(Q^2 > c) = 1 - F_m(c) + q_1 \sum_{r=1}^{n-2} t_r - q_2 \sum_{r=1}^{n-2} t_r^3 \tag{9.18}$$

where $f_m(c)$ and $F_m(c)$ are, respectively, the probability density and cumulative density function of chi-square random variables with m degrees of freedom. Furthermore,

$$q_1 = 2(c/\pi)^{1/2} f_m(c) \tag{9.19}$$

$$q_2 = \frac{q_1}{12}\left[\frac{(m^2-1)}{c} + c - 2m + 1\right] \tag{9.20}$$

and

$$t_r = \sqrt{1-\rho_r} \tag{9.21}$$

$$\rho_r = \left(N_r / N_r^*\right)^{1/2} / \left(N_{r+1} / N_{r+1}^*\right)^{1/2} \tag{9.22}$$

where N_r is the total number of observations in the first r rows of the table, and $N_r^* = N - N_r$, with the total number of observations equal to N.

Extramultinomial variation may be due either to situations where (a) the probabilities of falling into particular categories are themselves subject to uncertainty, or

(b) where there are correlations between successive rows of the contingency table. To model this, one first calculates

$$\hat{\sigma}_r^2 = \frac{Q - Q_r^2}{(n-2)m}; \quad n \geq 3 \tag{9.23}$$

where Q is the value of χ^2 measure of association between rows and columns for the entire $n \times (m + 1)$ table. Then define $K_r^2 = Q_r^2/\hat{\sigma}_r^2$ as the test statistic. Under the null hypothesis, K_r^2/m has an F distribution with m and η degrees of freedom, where $\eta = (n - 2)m$.

Results are shown in Table 9.7. These results were derived by omitting the last census tract from the analysis; again the results are consistent with those found previously. The most likely 'change point' is between the first quarter and the remainder of the year (p-value = 0.004); no other break-point yields a greater distinction between 'before' and 'after' patterns. The geographic pattern of burglaries during the fourth quarter is very similar to the pattern during the first three quarters ($p = 0.94$).

Once a change point has been found, it is then of interest to see whether other change points exist within either the 'before' or 'after' segments of the data. The bottom part of Table 9.7 reveals that further disaggregation of quarters 2, 3 and 4 does not uncover additional, significant change in the spatial pattern of burglaries.

9.3.7 Spatial Outliers

As we have seen, local spatial statistics may be used to pick out local regions with values that are significantly higher or lower than expected. They are defined for individual regions as a function of the value of the variable in that region, and values of the variable in nearby regions. Local statistics are designed primarily for testing hypotheses of spatial association for particular localities; the issue of multiple testing arises when one wishes to test more than one local statistic for significance. The M test applied to spatial data and the local R test were implemented with a Bonferroni adjustment. However, since there is correlation among tests of local statistics that are near to one another in space, the Bonferroni adjustment is conservative, and a more refined adjustment is desirable. The method summarised in this section addresses this issue.

Bailey and Gatrell (1995) and others have described the use of kernel-smoothing as one way to represent the spatial variability in the mean of a variable of interest. The value at any particular location is taken to be a weighted function of the values

Table 9.7 *Srivastava and Worsley's maximal chi-square*

	χ_r^2	σ_r^2	$K = Q_r^2/\sigma_r^2$	K/m	p-value
Quarter 1 vs Quarters 2, 3, 4	167.2	1.16	143.9	1.62	0.004
Quarters 1, 2 vs Quarters 3, 4	126.6	1.39	91.1	1.02	0.44
Quarters 1, 2, 3 vs Quarter 4	101.3	1.53	66.1	0.74	0.94
Quarters 2, 3 vs Quarter 4	107.3	1.13	94.8	1.07	0.38
Quarter 2 vs Quarters 3, 4	85.8	1.37	78.1	0.70	0.95

in the neighbourhood of the location, with closer locations receiving higher weights. The result is a surface portraying the regional variation in the underlying value, smoothed enough to eliminate the roughness of the image that would result if the original data were used, but not so much that underlying geographic variability is eliminated. Although these images represent a useful visual way to explore data, one often wishes to assess the significance of peaks in the surface. Attempts at such hypothesis testing have been limited to Monte Carlo simulation (e.g., Kelsall and Diggle, 1995, for an application to point data) or to more formal statistical methods that do not control properly for the likelihood of a Type I error (e.g., Bowman and Azzalini, 1997).

In this section, there is a summary of a statistical method for the detection of geographic clustering that is based upon Worsley's (1996) work on the maxima of Gaussian random fields and developed further in Rogerson (2001). The method provides a way to assess the significance of the maximum of a set of local statistics. It also may be viewed as a method that allows for the assessment of the statistical significance of a kernel-based, smoothed surface. Finally, the method is similar in concept to spatial scan statistics (e.g., Kulldorff and Nagarwalla, 1995) since, like the scan statistic, it considers many possible sub-regions and evaluates the statistical significance of the most extreme value. In addition, the method yields a calculable critical value that may be derived without resorting to Monte Carlo simulation methods.

The method assumes that the underlying data are normally distributed, and a Gaussian kernel is then used to smooth individual values and to construct local statistics. The steps involved may be described as follows:

1. Standardise the original observations so that the regional values may be considered as observations from a normal distribution with mean zero and variance one.
2. Construct local statistics z_i for each region using the standardised original observations (denoted y), and weights defined as

$$w_{ij} = (\sqrt{\pi}\sigma)^{-1} \exp\left(-d_{ij}^{2}/2\sigma^{2}\right) \tag{9.24}$$

   where d_{ij} is the distance from cell i to cell j and σ is chosen as the standard deviation of a normal distribution that matches the size of the hypothesised cluster. Then define the local statistic as

$$z_i = \sum_j w_{ij} y_j \Big/ \sqrt{\sum_j w_{ij}^2} \tag{9.25}$$

   To account for edge effects, it can be useful to add a guard area around the study region, with values for sub-regions in the guard area assigned from a standard normal distribution.
3. Find the critical value M^* such that $p(\max z_i > M^*) = \alpha$ by using that value of M^* that leaves probability $(1 + .81\sigma^2)\alpha/A$ in the tail of the standard normal distribution, where the study region has been subdivided into a grid of A square cells, each having side of unit length. Alternatively, M^* may be approximated by

$$M^* = \sqrt{-\sqrt{\pi}\ln\left(\frac{4\alpha(1+.81\sigma^2)}{A}\right)} \tag{9.26}$$

Alternative improvements to the Bonferroni adjustment are addressed in Rogerson (2002).

Figure 9.4 displays the areas where the observed number of burglaries is significantly greater than expected (where expected numbers are based upon the background population). Dots on the maps represent census tract centroids. Of interest is the degree to which the shaded regions are changing over time. Although there are no strong visual changes from one figure to the next, the results are consistent with the tests described above, since the clustering shown in Figure 9.4a (January–March) is, visually, somewhat different from Figures 9.4b–d. The clustering shown in Figures 9.4b–d occupies a greater sub-area, and the intensity of clustering (which can be determined from the scale) is stronger in these figures than it is in Figure 9.4a. This is consistent with the findings of the global, aspatial chi-square test, which found that the pattern of crime was most similar to the underlying population in the first quarter.

9.4 Summary

In this chapter, we have reviewed some recent developments in spatial statistical analysis, and their applications to crime analysis. The methods discussed here are for spatial data, and they have as their objective the detection of any deviations from expected geographic patterns. A key element of their use is to pay important attention to the definition of expectations. It is well known that patterns of crime exist, over and above their relation to background population, or to, say, the size of the male population in the young adult age groups. One approach would be to introduce other, known covariates (such as measures of socioeconomic status) in the definition of expectations. If the interest is in changes in patterns over time, the geographic patterns that held during some previous period may be used for the expectation.

The focus in this chapter on the temporal differences in geographic patterns has been one that assumes such detection is retrospective; i.e., a given set of data covering multiple time periods is searched for potential change points. When interest is in the *prospective* detection of change, the goal is to uncover any changes in spatial patterns as quickly as possible. If the time series of data describing the geographic pattern is of sufficient length, and if data on the areal location of successive observations is available, Rogerson (1997) suggests how quick detection may be carried out by combining Tango's statistic with the cumulative sum methods used in industrial process control (see also Montgomery, 1996). When the objective is similar and the data are for point locations, Rogerson and Sun (2001) demonstrate how quick, prospective detection of changes in spatial patterns may be achieved by combining the nearest neighbour statistic with cumulative sum methods.

As the development of spatial statistical analysis and spatial modelling continues to evolve, there will be a continuing need to embed the tools of spatial analysis within GIS. The emergence of packages such as *CrimeStat* (Levine and Canter, 1998) and

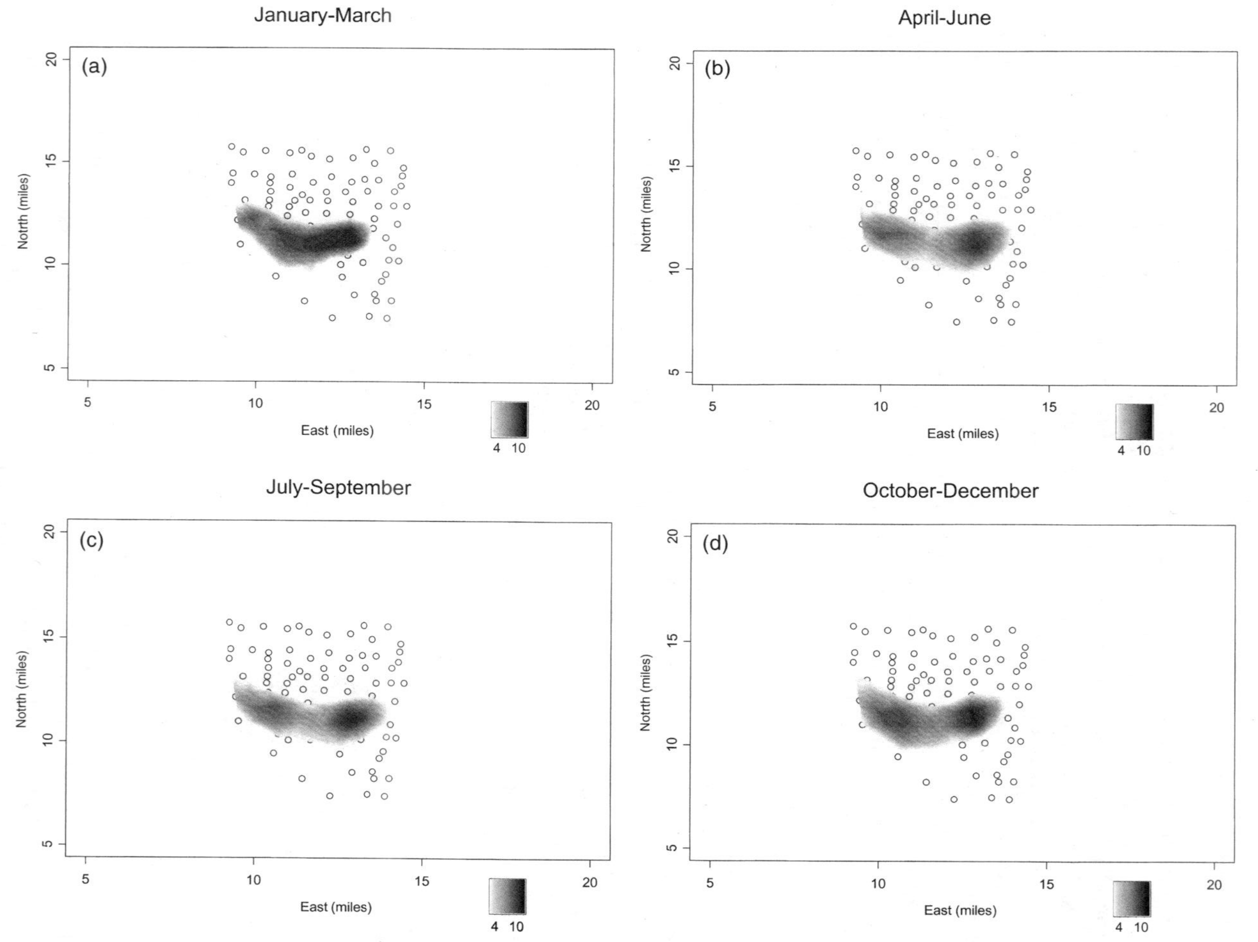

Figure 9.4 *Burglary clusters in Buffalo in (a) January–March, (b) April–June, (c) July–September and (d) October–December*

the Spatial Statistics package associated with SPLUS is an encouraging sign that these needs of spatial analysts will be met.

Acknowledgements

This chapter is based upon work supported by Grant No.98-IJ-CX-K008 awarded by the National Institute of Justice, Office of Justice Programs, U.S. Department of Justice. It was also supported by the National Science Foundation under Award No. BCS-9905900. The points of view in this document are those of the author and do not necessarily represent the official position or policies of the U.S. Department of Justice or the National Science Foundation. The support of a Guggenheim Fellowship is also gratefully acknowledged. I would also like to thank Daikwon Han for his assistance in producing several of the figures.

References

Anselin, L. (1995) Local indicators of spatial association-LISA, *Geographical Analysis*, **27**: 93–115.

Bailey, T. and Gatrell, A. (1995) *Interactive Spatial Data Analysis,* Longman, Essex (published in the US by Wiley).

Bowman, A. and Azzalini, A. (1997) *Applied Smoothing Techniques for Data Analysis,* Oxford University Press, New York.

Cerioli, A. (1997) Modified tests of independence in 2 × 2 tables with spatial data, *Biometrics*, **53**: 619–28.

Elliot, P., Wakefield, J.C., Best, N.G. and Briggs, D.J. (eds) (2000) *Spatial Epidemiology: Methods and Applications*, Oxford University Press, Oxford.

Fingleton, B. (1983) Independence, stationarity, categorical spatial data and the chi-squared test, *Environment and Planning A*, **15**: 483–500.

Fingleton, B. (1986) Analyzing cross-classified data with inherent spatial dependence, *Environment and Planning A*, **18**: 48–61.

Fuchs, C. and Kennett (1980) A test for detecting outlying cells in the multinomial distribution and two-way contingency tables, *Journal of the American Statistical Association*, **75**: 395–98.

Getis, A. and Ord, J.K. (1992) The analysis of spatial association by use of distance statistics, *Geographical Analysis*, **24**: 189–206.

Lawson, A. (2001) *Statistical Methods in Spatial Epidemiology*, Wiley, New York.

Lawson, A., Biggeri, A., Bohning, D., Lesaffre, E., Viel, J.-F. and Bertollini, R. (eds) (1999) *Disease Mapping and Risk Assessment for Public Health*, Wiley, New York.

Levine, N. and Canter, P. (1998) *CrimeStat*: a spatial statistical program for crime analysis: a status report, paper presented at the National Institute of Justice Cluster Conference of the Development of Spatial Analysis Tools, Washington, DC (http://www.icpsr.umich.edu/NACJD/gis_data.html).

Montgomery, D. (1996) *Introduction to Statistical Quality Control*, Wiley, New York.

Kelsall, J.E. and Diggle, P.J. (1995) Non-parametric estimation of spatial variation variation in relative risk, *Statistics in Medicine*, **14**: 2335–42.

Kulldorf, M. and Nagarwalla, N. (1995) Spatial disease clusters: detection and inference, *Statistics in Medicine*, **14**: 799–810.

O'Loughlin, J. and Anselin, L. (1996) Geo-economic competition and trade bloc formation: United States, German, and Japanese exports, 1968–1992, *Economic Geography*, **72**, 131–60.

Ord, J.K. and Getis, A. (1995) Local spatial autocorrelation statistics: distributional issues and an application, *Geographical Analysis*, **27**: 286–306.

Rogerson, P. (1997) Surveillance systems for monitoring the development of spatial patterns, *Statistics in Medicine*, **16**: 2081–93.

Rogerson, P. (1998) A spatial version of the chi-square goodness-of-fit test and its application to tests for spatial clustering, in Amrhein, C., Griffith, D. and Huriot, J.M. (eds) *Econometric Advances in Spatial Modelling and Methodology: Essays in Honor of Jean Paelinck*, Kluwer, Dordrecht, pp. 71–84.

Rogerson, P. (1999) The detection of clusters using a spatial version of the chi-square goodness-of-fit test, *Geographical Analysis*, **31**: 130–47.

Rogerson, P. (2001) A statistical method for the detection of geographic clustering, *Geographical Analysis*, **33**: 215–27.

Rogerson, P. (2002) The statistical significance of the maximum local statistic, in Getis, A. (ed.) *Advances in Spatial Econometrics*, Palgrave, London.

Rogerson, P. and Sun, Y. (2001) Spatial monitoring of geographic patterns: an application to crime analysis, *Computers, Environment, and Urban Systems*, **6**: 539–56.

Srivastava, M.S. and Worsley, K.J. (1986) Likelihood ratio tests for a change in the multivariate normal mean, *Journal of the American Statistical Association*, **81**: 199–204.

Tango, T. (1995) A class of tests for detecting 'general' and 'focused' clustering of rare diseases, *Statistics in Medicine*, **14**: 2323–34.

Worsley, K. (1996) The geometry of random images, *Chance*, **9**(1): 27–40.

Part Three
TRANSPORT AND LOCATION

10

Modelling and Assessment of Demand-Responsive Passenger Transport Services

Mark E.T. Horn

Abstract

An investigation of public transport options was carried out for the central Gold Coast region of Queensland, Australia, using the LITRES-2 public transport modelling system. Several road-based demand-responsive passenger transport services were modelled in combination with conventional buses, trains and taxis, under a range of assumptions about future demand. The demand-responsive options included multiple-hire taxis, 'roving buses' and a 'smart-shuttle' local service between railway stations and their immediate suburban localities. These were shown to be valuable and operationally credible as complements to the conventional timetabled modes. Besides substantive conclusions of this kind, the project raised some methodological issues concerned with project planning and execution, and the capabilities and usage of a modelling system such as LITRES-2.

10.1 Introduction

The responsibility for public transport in Australia lies with state rather than with local governments; nevertheless, many local councils take an active interest in the provision of such services, working in consultation with state transport departments. Such was the case in the project described in this chapter. An Integrated Regional Transport Plan was prepared by the Queensland Transport Department for the south-eastern section of Queensland during the mid-1990s. The Integrated Plan

Applied GIS and Spatial Analysis. Edited by J. Stillwell and G. Clarke
 ISBN: 0-470-84409-4

included ambitious targets for increased usage of public transport services, and the achievement of those targets was one of the considerations leading to the initiation of a Public Transport Passenger Study by the Gold Coast City Council (GCCC) in 1997. The Passenger Transport Study included the modelling project described in this chapter, and was in turn one of the main inputs to the Gold Coast City Transport Plan, a joint undertaking of the GCCC and the Queensland Departments of Transport and Main Roads (GCCC, 1998).

The main task of the Gold Coast modelling project was to assess the viability of several proposed road-based demand-responsive transport modes, which were envisaged as supplements to anticipated bus, rail and taxi services in the near and medium-term future: that is (approximately) the period 2000–2010. The work was carried out by the Commonwealth Scientific and Industrial Research Organisation (CSIRO), in collaboration with Booz Allen and Hamilton (BAH). BAH's main role was to supply estimates of key parameters concerned with current and forecast demand, service levels and fares. BAH subsequently also used the results of the project in formulating a wider range of recommendations for the enhancement of passenger services and infrastructure (BAH, 1998).

The main analytical tool used in the project was a specialised transport modelling system called LITRES-2, which embodies research carried out by the author and his CSIRO colleagues over several years (Horn, 2002; see also Smith, 1993; Rawling *et al.*, 1995). No proprietary GIS software was used: all data were stored in ordinary Unix files, and all data preparation was carried out using standard Unix tools, or using editing and display programs developed as adjuncts to LITRES-2. The supplementary software includes a graphical program for browsing spatial inputs (see Figures 10.5 and 10.6); there is also a graphical version of the modelling system itself, which has proved useful in demonstrating and validating the system's operation (see Figures 10.3 and 10.4). Besides the substantive issues under investigation, the project involved extension and refinement of the software, and an effort to assess the software's capabilities in practice.

Section 10.2 of this chapter surveys urban patterns in the Gold Coast, states the objectives of the project, and introduces the transport modes under investigation. Section 10.3 describes the LITRES-2 modelling system and the main categories of data involved. Section 10.4 outlines the scenarios tested in the project and the results obtained in each case. The chapter concludes in Section 10.5 with a summary of substantive conclusions from the project, a discussion of methodological issues, and an assessment of LITRES-2 in comparison with alternative modelling packages.

10.2 Project Objectives

The City of the Gold Coast comprises a stretch of Pacific seashore and estuarine hinterland commencing some 32 km south of the Brisbane central business district, and extending another 60 km south to Coolangatta on the New South Wales border. Figure 10.1 covers the more densely settled parts of the region, running 40 km from north to south between Oxenford and Coolangatta. During the past 50 years the region has evolved from a collection of small farming, fishing and holiday towns,

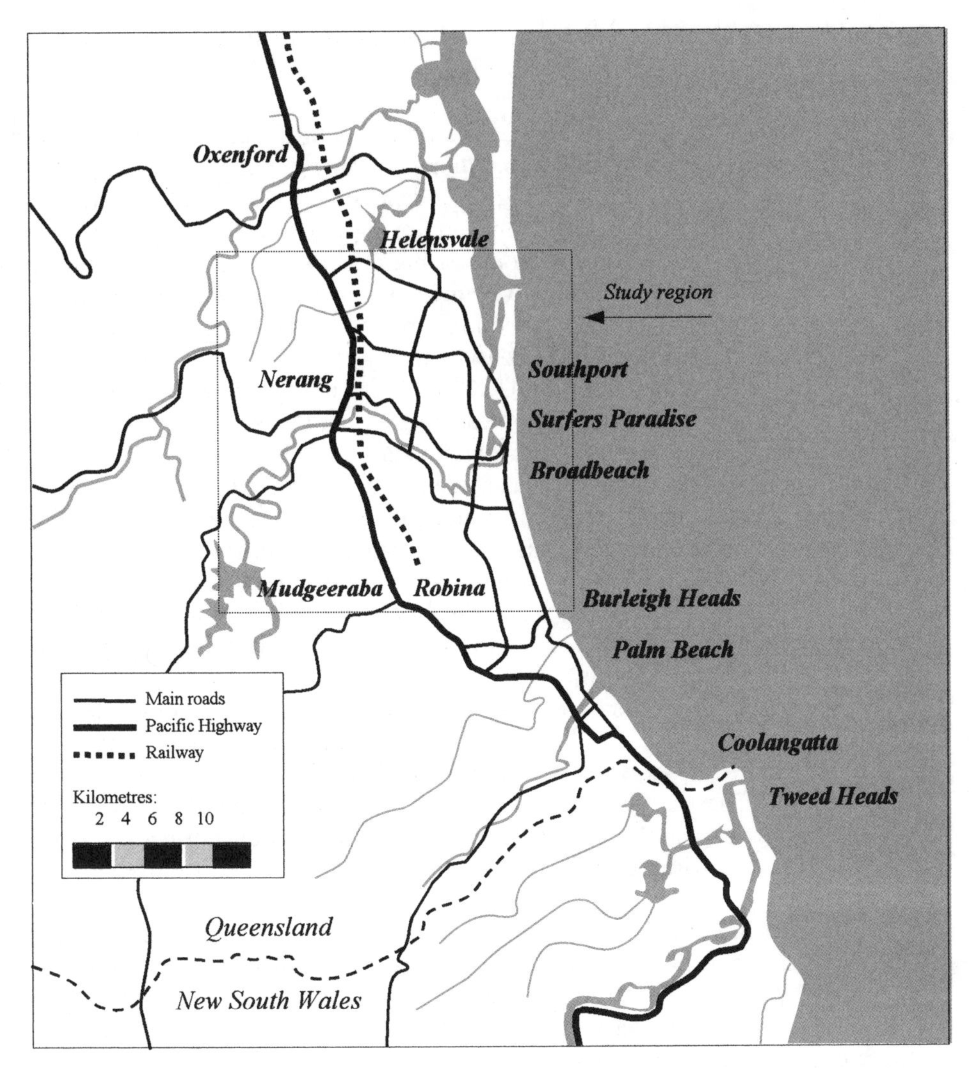

Figure 10.1 Sketch map of the Gold Coast region

into a leading tourist destination, favoured also as a residential location by many elderly people on account of its mild climate. Substantial numbers of residents commute to work in Brisbane, and the region now houses a diversified range of industries in its own right. It is one of the fastest-growing urban areas in Australia, with a permanent resident population already exceeding 425 000.

A chain of high-density urban development (e.g., apartment buildings, hotels and other tourism facilities) follows the main north–south coastal road, concentrating especially in the north around the regional centres of Southport and Surfers Paradise. West of the coast there are extensive low-density residential areas built on the banks of estuaries and canals. This waterfront zone is bounded on its western edge by the Pacific Highway, which traverses regional centres at Helensvale, Nerang and Mudgeeraba, from which suburban developments extend farther westward into the foothills of the coastal ranges.

The Gold Coast City Council initiated the project with the general objectives of reducing growth in car traffic and congestion, improving accessibility within the region for residents and visitors, reducing dependence on the private car and the need for new roads and parking facilities, and improving environmental sustainability. These objectives were to be pursued by increasing the range of choice, convenience and integration of the public transport system, improving the accessibility and affordability of public transport and, in general, improving the balance between private and public modes of transport (BAH, 1998).

In preliminary discussions with the GCCC and BAH, it was decided that the analysis should concentrate on a study area measuring approximately 21 by 15 km. This covers the central part of the Gold Coast region, and includes all the main regional centres mentioned earlier. In some respects the region is well served by public transport: there are high-frequency bus services along the coastal strip, and a railway line has been built from Brisbane to Helensvale, with extensions to the south under way at the time of the project. Even so, current levels of passenger transport usage are very low (GCCC, 1998, Section 4.3), and the Gold Coast's generally dispersed settlement pattern constitutes a difficult environment in which to sustain geographically comprehensive bus or rail services.

The public transport modes covered by the project (see Table 10.1) included the existing bus network, the railway line with its projected extensions, and a variety of demand-responsive services. The attraction of the latter in the context of a region such as the Gold Coast is that they can be operated more flexibly than fixed-route services, and thus should be more effective in serving spatially dispersed urban areas. They may be envisaged as variants of conventional taxi services, designed to achieve economies of scale through the use of larger, shared vehicles. The demand-responsive modes are characterised by the use of recently developed technologies to handle communications between vehicles and a central dispatching facility, and to allow prospective travellers to make enquiries and bookings. Beyond that, their viability will often depend on the use of automated, centrally located scheduling software to obtain an effective and efficient deployment of vehicles.

Table 10.1 *Characteristics of transport modes*

Transport mode	Pickup and setdown	Advertised service	Routing
Walking	Stop to stop	–	Euclidean or Manhattan path
Bus, Rail	Stop to stop	Timetabled	Fixed
Taxi	Door to door	Free-range	Demand-responsive
TaxiMulti	Door to door	Free-range	Demand-responsive
RovingBus	Stop to stop	Free-range	Demand-responsive
SmartShuttle	Stop to stop	Timetabled, stops or zones	Demand-responsive

The main demand-responsive modes modelled in the Gold Coast project are outlined below.

- *Taxis*: single-hire, door-to-door taxi service. As with the other demand-responsive service modes, provision is made for advance bookings. The traditional 'step-in' style of taxi service was not modelled in the Gold Coast project.
- *TaxiMulti*: multiple-hire taxis guaranteeing service within prescribed travel-time standards.
- *Roving Bus*: multiple-hire services like TaxiMulti, but picking up and setting down passengers at designated stops.
- *SmartShuttle*: a bus service based on an advertised timetable but run on demand. That is, the actual route taken by a SmartShuttle bus needs to traverse only the locations for which demand has been registered, and the service can be omitted altogether on occasions when no demand is registered. The timetabled locations may be points (e.g., bus stops), or sets of points (in spatially defined zones), or a combination of the two.

10.3 Methodology

The LITRES-2 modelling system provides a fine-grained representation of demand and transport services, with particular focus on the operational characteristics of demand-responsive modes (Horn, 2002). The system's modelling architecture is outlined below, before proceeding to the data framework and modelling strategy used in the Gold Coast project.

10.3.1 Modelling Framework

LITRES-2 applies a microsimulation strategy to the representation of events and processes evolving over a given period of time (some alternative approaches are discussed at the end of this chapter). By contrast with more statistically based approaches, the parameters are all empirically based, and a formal calibration of the system is thus not required. A primary distinction between simulation and control modules (Figure 10.2) provides a clear conceptual basis for the system, and is consistent with the intended future usage of the control modules in a real-time context. In particular, the request-brokering and journey-planning modules embody algorithms that are applicable in a public transport information system, while the fleet-scheduling module comprises procedures designed to manage bookings and deployment for a fleet of demand-responsive vehicles in real time. The simulation modules embody representations of real-world processes, and are the source of the control systems' knowledge of critical real-world phenomena such as passenger requirements and vehicle movements. By contrast, the role of the control modules is to influence the simulated processes, this influence being exerted by means of messages (e.g., dispatch commands) sent to the simulation modules.

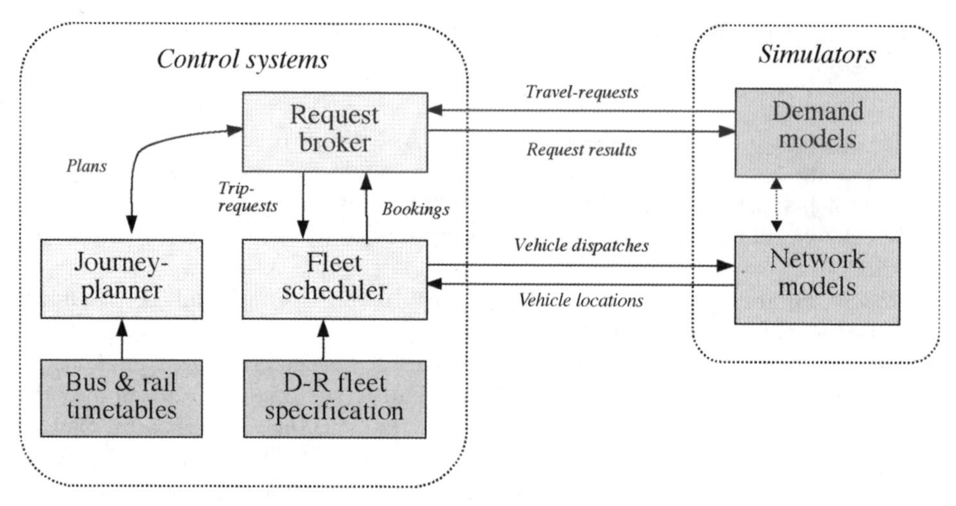

Figure 10.2 LITRES-2 modelling framework. Reprinted from Horn, M., Multi-modal and demand-responsive passenger transport systems, Transportation Research A, ***36****(2): 167–88. Copyright 2002, with permission from Elsevier Science*

The main simulation modules are demand simulators and a network simulator. The demand simulators convert predefined models of aggregate demand into a stream of travel-requests, which is sent to the control modules. The network simulator tracks the movements of demand-responsive vehicles through the road network, and interacts mainly with the fleet-scheduling module. Several additional simulation models (not shown in Figure 10.2) provide for the modelling of randomly occurring contingencies, such as vehicle breakdowns and delays, and passenger bookings and no-shows. The simulation models are executed in a temporal context provided by the C++Sim toolkit (Little and McCue, 1993), which ensures that timed processes evolve in a chronologically consistent time-sequence. Outside the simulation modules, the time-sequence is manifest as time-stamps on messages sent to the control modules.

Aggregate demand for passenger transport is represented as a collection of predefined demand models, each associated with a particular market segment, parameters for party size (i.e., the number of people travelling together in a given journey), and a given time period (e.g., 8:00 to 9:00 a.m.). Two types of demand model were used in the Gold Coast project. An origin–destination (OD) model is a matrix of which a given element specifies the total number of journeys from one zone to another during the period covered by the model. A source–sink (SS) model consists of an array of which a given element specifies the aggregate demand for travel from a given point to a given zone, or vice versa: such a model provides a straightforward way to represent demand associated with a major regional facility, such as a shopping complex or tourist venue, as an overlay on the more uniformly distributed patterns associated with an OD model.

The task of a demand simulator is to disaggregate a demand model in a randomised fashion, so as to generate a stream of travel-requests: each such request refers to a person (or party of people) belonging to a particular market segment, requesting travel between a given origin and a given destination point, within a given journey-envelope specifying an earliest departure and a latest arrival time. As a typical example of the simulation techniques employed in LITRES-2, we shall consider the demand-simulation process for an element (i,j) of a given OD model. This element specifies the total number of journeys k_{ij} to be made from zone i to zone j during the period (tlo,thi) covered by the model.

In summary, there are k_{ij} travel-requests to be generated for element (i,j), consistent with the aggregate parameters mentioned above but differing from each other in a random fashion. To generate the starting-point for one such request, we refer to the bounding rectangle of zone i (i.e. the smallest rectangle enclosing all parts of the zone), with width and height w_i and h_i respectively. We use a random-number generator to obtain a pair of numbers (p,q), both in the range (0,1), and hence generate a point (x,y), with x located at $p \cdot w_i$ units from the left-hand edge and y at $q \cdot h_i$ units from the bottom of the bounding rectangle. We check that (x,y) lies within the boundaries of the zone and, if not, we obtain a new pair of random numbers and repeat the process. The same technique is used to obtain a destination point in zone j for the travel-request. The commencement-time *tstart* for the request is interpolated at random within the model period (tlo,thi); that is, given a random number r $(0 \leq r < 1)$, $tstart = tlo + r\ (thi - tlo)$. The end-time *tterm* is found by reference to walking-time and travel-delay parameters of the market segment s associated with the OD model: $tterm = tlo + wmax_s + ttmin_{ij} \cdot kmax_s$, where $wmax_s$ and $kmax_s$ are obtained by uniform random interpolation in ranges specified for market segment s, and $ttmin_{ij}$ is the shortest possible travel time by road from i to j.

10.3.2 Journey Planning

The travel-requests generated by the demand modules are passed, one by one, to a request-broker module, whose task is to plan an optimal journey that satisfies the requirements of each incoming request. A journey consists of a sequence of one or more legs; a journey-leg carried by a vehicular mode is called a trip. A single-leg journey may be walked all the way or carried by a door-to-door vehicular service (i.e., taxi or TaxiMulti). A multi-leg journey involves a sequence such as 'Walk → Bus → Train → SmartShuttle → Walk'; or more precisely, 'Leave origin-point at 10:10 a.m., walk to bus stop, catch 290A bus at 10:15, alight at Nerang at 10:25, take the train from Nerang to Robina at 10:30, alight at Robina at 10:40, take SmartShuttle at 10:45, alight at local stop at 10:50, walk to destination, arriving at 10:56 a.m.'. An example is shown in Figure 10.3.

The planning of multiple-leg journeys is carried out by a journey-planning module, an optimisation procedure embodying a branch and bound algorithm. The journey-planning process is subject to behavioural constraints specified for each market segment, such as the maximum distance that the traveller is prepared to walk at a single stretch. The availability of service is determined by timetable data, or in the case of demand-responsive modes, by reference to the fleet-scheduling module. The objective in planning a journey is to minimise the generalised cost of travel, defined as the sum of any fares, together with the imputed cost of time for the person or persons making the journey. This imputed cost is estimated using valuation factors for the various time components of the journey, namely waiting at the origin, waiting

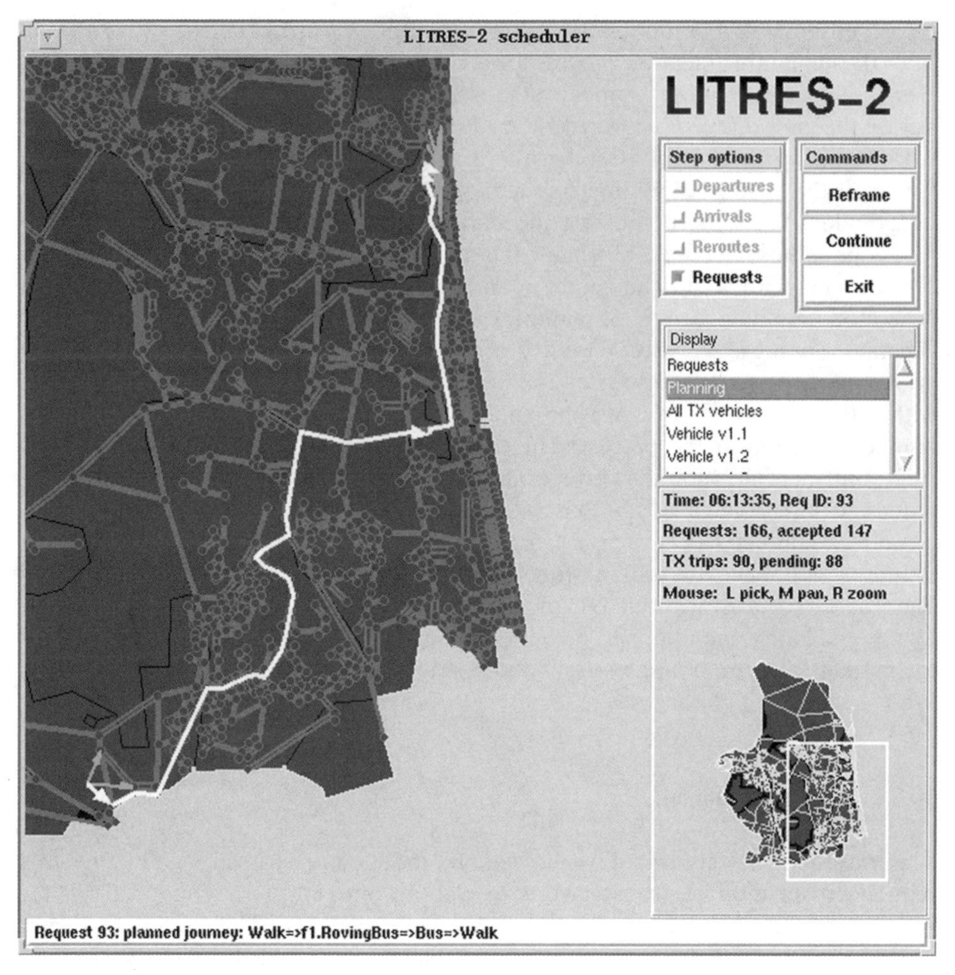

Figure 10.3 *Planning a journey*

at interchanges, walking and travelling. Travel times are estimated as follows. For a walked leg, the time is simply the length of the leg divided by the walking speed associated with the market segment associated with the traveller; while for a leg carried by a timetabled mode, the travel time is obtained from the timetable. For a leg carried by a non-timetabled mode (i.e., RovingBus, TaxiMulti or taxi), travel time is calculated by reference to the mode's service-quality standards (see Section 10.3.4 below).

10.3.3 Fleet Scheduling

The LITRES-2 fleet scheduler manages the deployment of a fleet of demand-responsive vehicles, by planning each vehicle's itinerary in response to trip-requests received over time from the request-broker. Figure 10.4 shows one such itinerary,

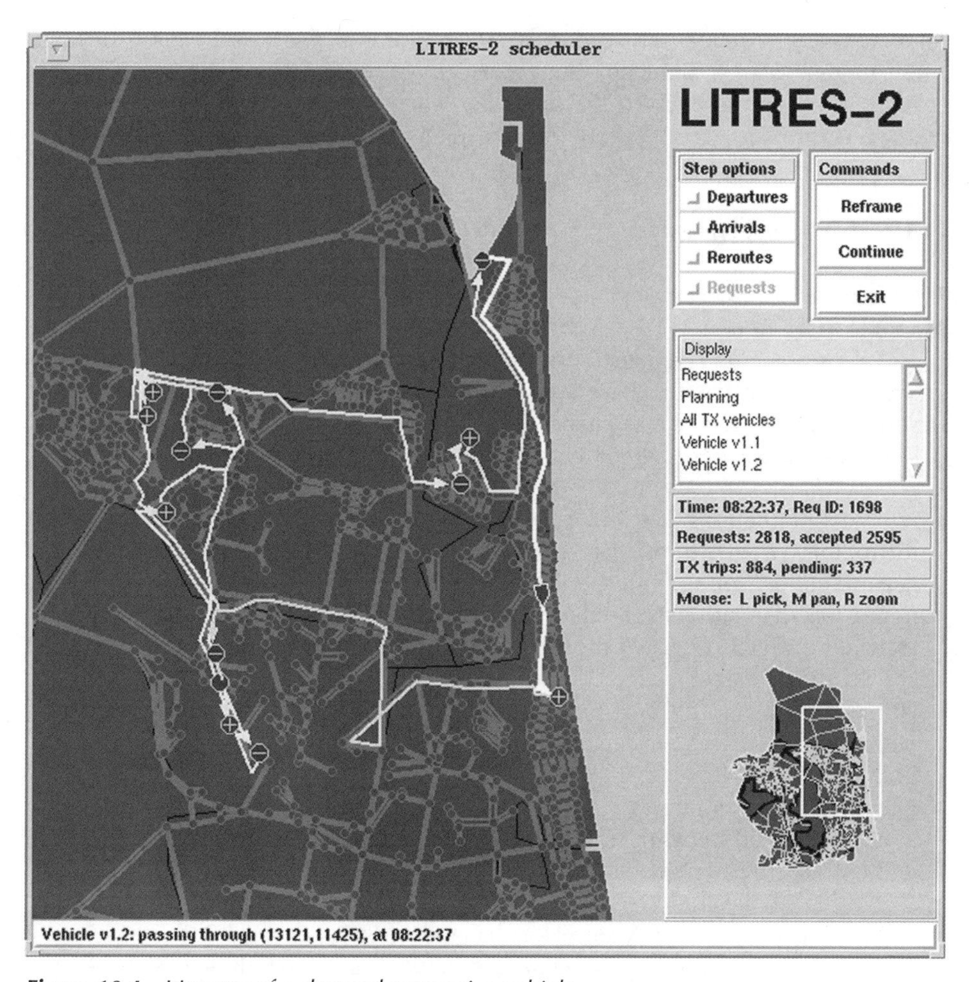

Figure 10.4 *Itinerary of a demand-responsive vehicle*

including the vehicle's current location and its planned pickup and setdown points ('+' and '–' respectively). A trip-request refers to a particular transport mode, and has explicit time-windows specifying early and late limits for departure from the start and arrival at the end of the trip: these limits are defined by the journey-planning module by reference to the service-quality standards pertaining to the mode. The overall objective when choosing an implementation for a trip is to produce favourable operational outcomes for the fleet operator (e.g. by minimising fuel costs). The finding of shortest-time paths between each successive pair of stops in a vehicle's itinerary is a scheduling sub-task, and is discussed below in connection with the road network model. A distinction is made between the routes and timings planned by the scheduling module, and on the other hand the actual trajectory of a vehicle as it moves through the road network. In particular, the fleet scheduler sends dispatch commands (e.g., 'Cab 99, proceed at 10:10 to location (*x*,*y*)') to the network simulation module, which simulates the movement of each vehicle through the road network, and reports back to the scheduler when a vehicle arrives at a pickup or setdown point or is affected by an incident such as a breakdown.

10.3.4 Inputs and Outputs

The Gold Coast project was planned as a series of scenarios, each specified as a set of inputs to a run of LITRES-2. In particular, a Base scenario provided a check of LITRES-2's predictions against current conditions, while other scenarios were constructed as variants of the Base scenario, so as to investigate the performance of various postulated sets of passenger service options over different levels of demand. A LITRES-2 scenario is defined by input data of the following kinds.

- A model of the road network is used as a basis for estimating travel times, and for routing vehicles in the fleet-scheduling and network simulation modules. The network model comprises a set of nodes and a set of directed links connecting the nodes, together with link speeds and distances. This is a format used in many commercially available GIS, but the intensive use of network calculations has required special attention to the algorithms and storage strategies used for network calculations. For the Gold Coast project, a single speed was defined for each link and path durations were precalculated: subsequent extensions allow for fluctuation in link speeds over the course of a day. The techniques used to find shortest-time paths and their timings are variants of Dijkstra's label-setting algorithm, with allowance for off-network travel between the (Euclidean) starting point of a path and a nearby network node, and vice versa at the end of the path.
- A division of the study region into polygonally bounded small-area units called zones provides spatial reference for demand models. Zonal decompositions are used also as bases for defining fare structures and zone-based SmartShuttle services.
- Behavioural characteristics of travellers are defined in market segments. A market segment defines travel parameters for a particular set of travellers, such as elderly people. These parameters include a set of acceptable modes, valuations of time pertaining to generalised costs, a walking speed and a maximum walking distance. In addition, there are parameters for acceptable travel duration, and for the 'notice period' between the time when travel is requested and when it is to take place.
- As indicated earlier, aggregate demand is represented in the form of origin–destination (OD) and source–sink (SS) models, each of which is associated with a market segment, party-size parameters, and a time-range.
- Specifications of timetabled services include locations of the points served (e.g., bus stops and railways stations), and service routes (sequences of points). The specifications for each route include a transport mode (e.g., bus), a fare structure, and a timetable.
- Demand-responsive services are defined by specifying the vehicles in a demand-responsive fleet or fleets. Each vehicle has a passenger-carrying capacity; a list of

the modes offered; and an operating shift, specified as the period when the vehicle is in service, with a corresponding pair of terminal points. Associated with each demand-responsive mode are parameters specifying a fare structure, pickup and setdown times, and service quality. The service-quality standards comprise a waiting-time limit and a travel factor, the latter specifying an upper limit on the ratio of actual to minimum possible travel time: this ratio is one for single-hire taxi, and more than one in the case of the multiple-hire modes.

The main outputs from LITRES-2 are outlined below. The volume of these outputs makes them somewhat unwieldy in their raw form, and their interpretation in practice requires considerable care in the selection and examination of a smaller number of key indicators (for this purpose the outputs are now provided in a coded format, to facilitate transfer to spreadsheets).

- *Travel statistics* include the total numbers of passengers carried and journey-legs completed, with average travel times, distances and costs, for the whole simulation period and for nominated time-intervals. These statistics are broken down by leg-sequence, by market segment, and by transport mode.
- *Scheduling statistics* include the total numbers of passengers carried, with mean trip time and distance per passenger, for the demand responsive fleet(s), for each mode, and for each vehicle. As with the travel statistics, statistics are aggregated for the whole simulation period and also over nominated time-intervals. They include measures of efficiency (e.g., the ratio of travel time to the least possible travel time, relevant in the case of multiple hire modes) and of vehicle occupancy (i.e., the average number of passengers on board at any given time).
- *Route-usage statistics* include the total numbers of passengers carried on each timetabled mode, and the totals for each route and each service run along each route, with average trip times and occupancy ratios. Again, statistics are aggregated for the whole simulation period and for nominated time-intervals.

10.4 The Case Study

The Base and other scenarios were planned in advance, and then adjusted and extended in the course of the project. The most demanding and time-consuming task was to fine-tune the considerable range of data inputs in the Base scenario, so as to obtain a credible match between model outputs and known conditions (Horn *et al.*, 1998). The simulations themselves incurred little real delay: the run time for each scenario was generally between 20 to 40 minutes, on a Sun UltraSparc workstation.

10.4.1 The Base Scenario

An eclectic approach was required in the preparation of the 'Base scenario', involving the use of sources whose applicability to the project was often somewhat indirect. Outlined below are the main sources for the data categories described earlier, and the adjustments applied to them in preparing the Base scenario.

The Road Network

The GCCC provided a road network data set covering nearly all roads in the region, and comprising 4166 nodes and 9472 links. This was somewhat excessive in terms of the computational resources available; furthermore, it included a great deal of geometric detail which would be irrelevant to the analysis carried out by LITRES-2. The network was therefore reduced in size through the application of a filtering program, followed by manual correction of a few remaining anomalies. The filtering program was developed especially for this purpose. It removed 'superfluous' dead-end links (i.e., each link to a node with no other links inward or outward), and 'superfluous' intermediate nodes (i.e., each node with exactly one inward and one outward link). Aggregate distances were preserved in this process (e.g., where a chain of links representing a curved section of road was reduced to a single link). The reduced network comprises 1686 nodes and 4515 links, and is illustrated in Figure 10.5.

Link speeds were defined initially by applying speed limits corresponding with the road classifications provided with the network. These speeds were then systematically scaled down in the light of a preliminary simulation of taxi system performance, and ground speed checks carried out by BAH. For off-network vehicular travel, a uniform speed of 25 km per hour was assumed.

Zones

A zonal division of the study region was needed to provide spatial reference for bus fares and demand models. The zones were defined as those used by the Surfside Bus Company in determining fares, with unpopulated areas removed. These excisions – mainly in the estuarine parts of the region – were made in order to minimise distortions in the representation of demand, which is assumed within LITRES-2 to be distributed uniformly within each zone.

Market Segments

BAH identified six market segments as a basis for modelling demand. The market segments may be characterised informally as follows:

- *Mainstream*: regular users of public transport when service is readily available.
- *Captive old and captive young*: people who have no real alternative to public transport.
- *Tourists*: tourists who choose to use public transport.
- *Car-oriented*: people who currently travel by car but may decide to switch to public transport if service of sufficiently high quality is available.
- *Rail commuters*: people who live within the corridor served by the railway and its planned extensions, and are likely to commute by train to workplaces in Brisbane.

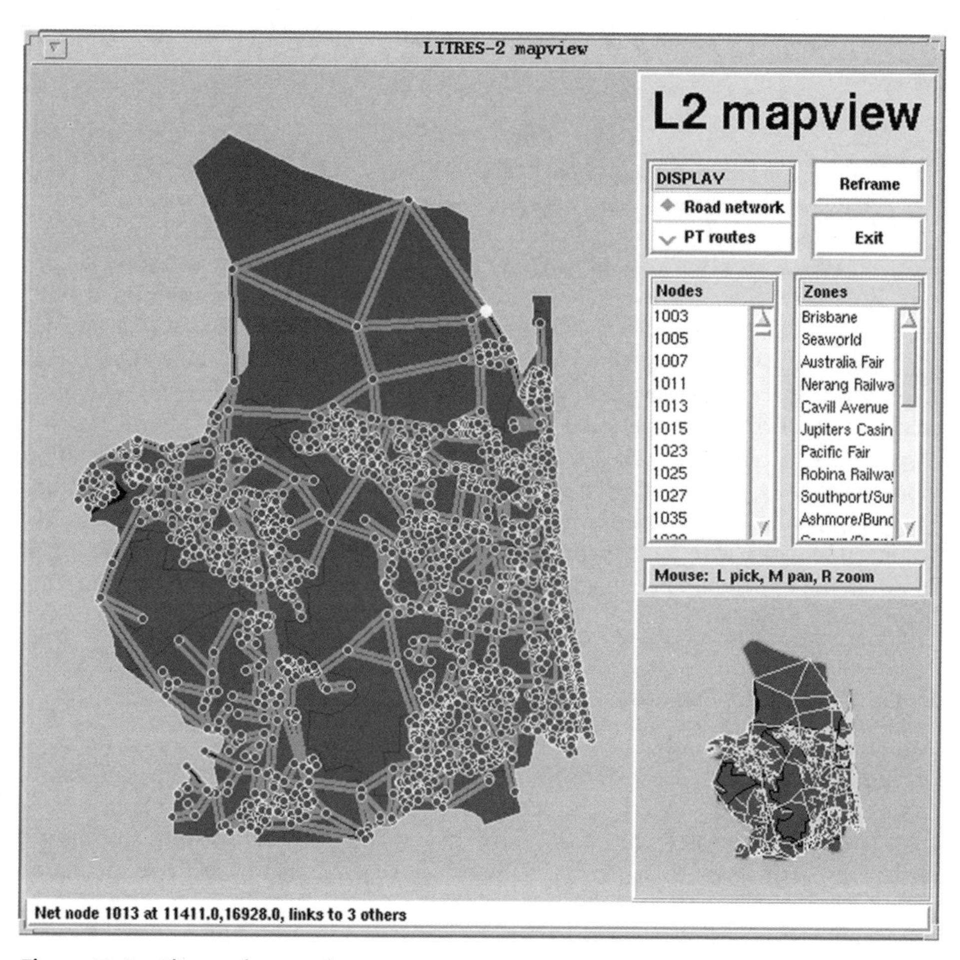

Figure 10.5 *The road network*

Demand Models

Estimates of current demand were obtained from a variety of sources (BAH, 1998). Demand specifically associated with regional attractors (e.g., shopping malls, casinos and funfairs) was modelled in the form of source–sink models, using estimates from economic studies of these facilities. Other demand was modelled in several series of origin–destination matrices, specified (as for the source–sink models) at hourly intervals over a 24-hour period (notionally a weekday). These OD series were compiled by GCCC staff from the Surfside Bus Company's fare-collection data, Queensland Rail passenger data, and estimates of private car journeys from an earlier study in that region. The Base demand amounted to 21 148 travel-requests (31 722 travellers) during a 24-hour period. In temporal terms the patterns of demand were generally similar to those seen in other metropolitan areas, but with smaller peaks around the start and end of the conventional working day, and with a continuation of activity into the evening (or the early morning, in some of the entertainment districts).

Timetabled Transport Services

The geographic locations of bus stops and railway stations were digitised, and a small proportion of them were designated as interchange points for transfer on multi-leg journeys (this restriction proved to be unnecessary, since subsequent testing yielded very similar results with transfer allowed at every bus stop). The published timetables were transcribed and then elaborated, as follows. First, each route was defined as a bi-directional pair (e.g. an advertised north–south route 290 yielded two routes, 290N and 290S); second, variant routes implied by the timetables (e.g., with skip-stop service during peak periods) were generated explicitly; third, arrival times at stops without explicitly advertised times were interpolated in proportion to distances along the routes; and fourth, services on the projected extension of the railway line were extrapolated from existing timetables. Figure 10.6 shows a snapshot of the graphical browser, with a list of the routes connecting a designated pair of bus stops shown in the right-hand panel, and with one of those routes highlighted on the main map display window.

Demand-Responsive Services

About 220 taxis currently operate in the wider Gold Coast region, under the aegis of Regent Taxis Limited. Some of the drivers specialise in traditional 'step-in' service from cab-ranks, and some – with larger vehicles – specialise in multiple-hire work from specially designated cab-ranks; most drivers, however, rely mainly on Regent's central dispatching centre for most of their work (Horn *et al.*, 1999). The simulation involved a taxi fleet about half the size of the actual Regent fleet, reflecting the approximate proportion of taxi activity encompassed by the study region. As with the timetabled modes, service details and fares for taxis were matched to existing conditions as closely as possible. The new demand-responsive services (TaxiMulti, RovingBus and SmartShuttle) were conceived as intermediate in terms of service quality and fares between taxis and buses. The fare structure in each case was defined as a base fare and a distance-based component, with group discounts for parties of two or more people.

10.4.2 Future Transport Scenarios

Low, medium and high demand scenarios were constructed by scaling up demand from the Base scenario to represent (notionally) the range of conditions possible in the study region within the next decade. Different scaling factors were applied to the various demand components, yielding estimates of 42 291, 53 997 and 65 799 passengers respectively for the three scenarios, compared with 31 722 in the Base scenario. To avoid excessive optimism regarding the prospects for new transport services, the

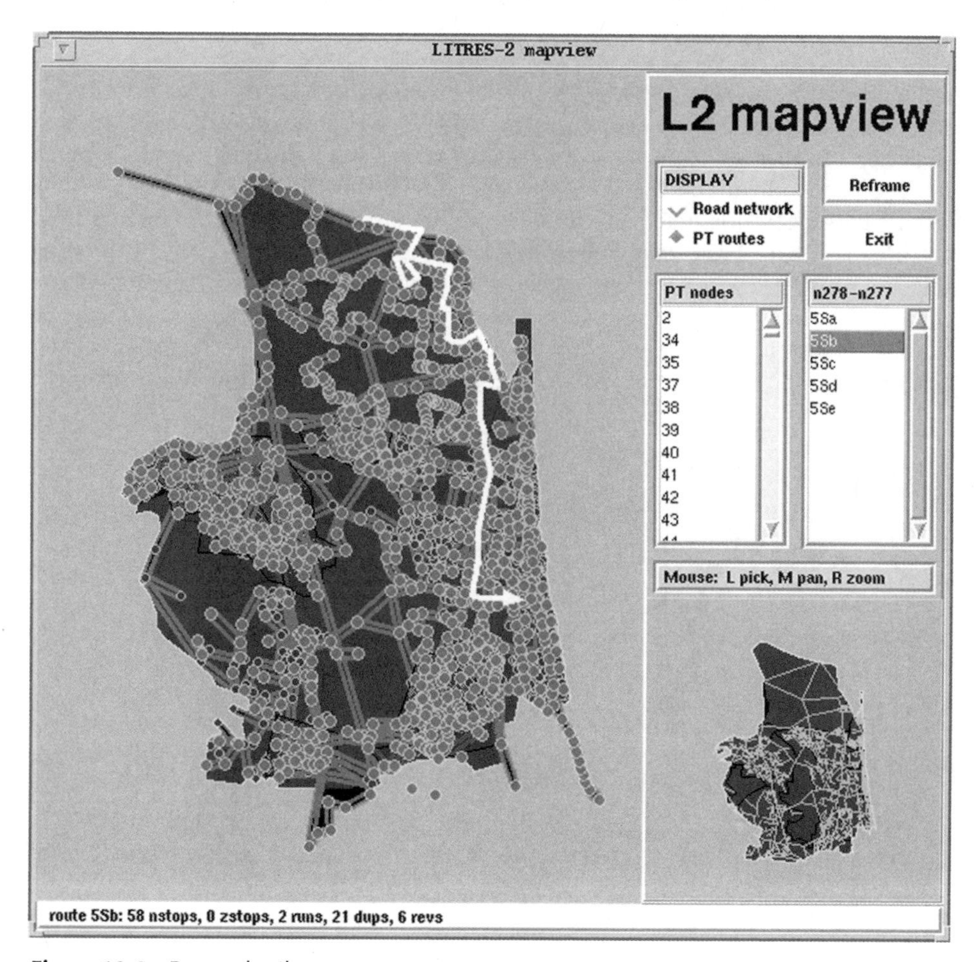

Figure 10.6 *Bus and rail routes*

analysis focused mainly on the *low* demand scenario. A more extensive study might well encompass a wider range of possibilities in this respect, for example with geographical differentiation in the rates of growth in demand.

Fixed-Route Services

The assumptions in the future scenarios regarding rail, bus and taxi services were the same as those in the Base scenario. As might be expected, the simulations indicated that bus usage would increase substantially under increasing demand. In fact, there is probably scope for augmentation of bus services and routes in some areas: an investigation of such possibilities was beyond the scope of the project, but – with the Base scenario already available – would require little additional effort.

Multiple-Hire Services

Demand-responsive services were modelled in each scenario as a single fleet of vehicles, with each vehicle dedicated to a single mode. The low, medium and high demand scenarios included 40, 60 and 80 mini-buses respectively, divided equally between TaxiMulti and RovingBus service. The simulations showed these vehicles carrying several passengers for much of the time, although they were rarely close to their maximum carrying capacity (Figure 10.7). High occupancy ratios are obviously desirable from a fleet operator's point of view, but saturation of fleet capacity prevents additional passengers from using the fleet; thus if the fleet were larger, it might be chosen by passengers who currently use other modes. Consequently, the simulated patronage of multiple-hire services in the main scenarios does not fully reflect the

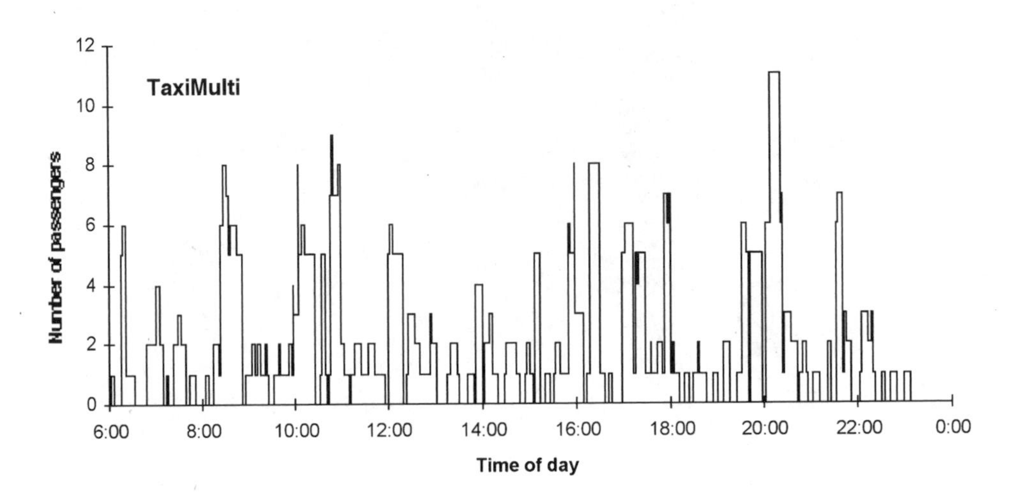

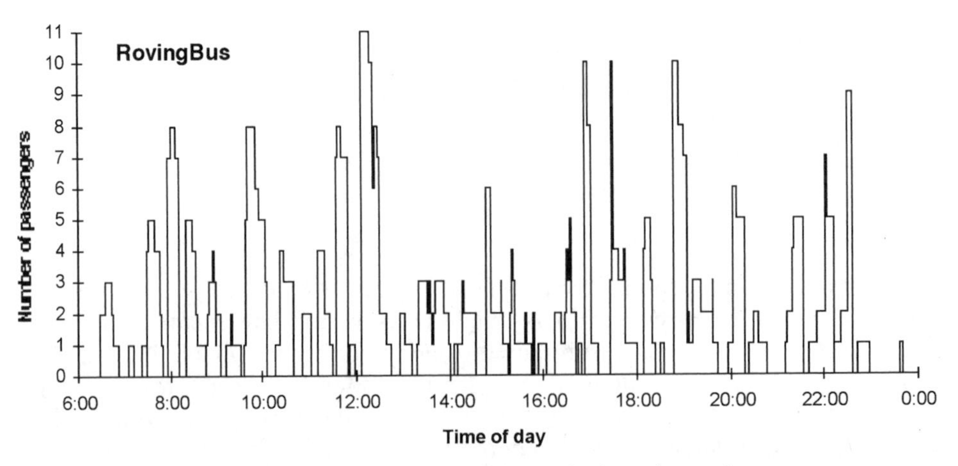

Figure 10.7 *Occupancy patterns for two vehicles under low demand*

attractiveness of these services to the travelling public, a deduction that was borne out by supplementary simulations with larger fleet sizes. More generally, the 'carrying capacity' of a fleet is a somewhat ambiguous concept, since 'heavily loaded' or 'saturated' conditions are characterised by a gradual decline – rather than a complete disappearance – in the additional trips that can be accepted under marginal increases in demand: the limit, where no further passengers can be accepted, is arbitrary, fluctuating over time and sensitive to arbitrary variations in current fleet deployment. A reasonably precise estimate of fleet capacity may be obtained, however, by allowing a progressive stretching of service standards under saturated conditions (Horn *et al.*, 1999; Horn, 2002).

Results from the *low* demand scenario (Figure 10.8) indicate that the TaxiMulti and RovingBus services should generate gross earnings per vehicle greater than those obtainable from single-hire taxis, and carry roughly twice as many passengers, provided that fleet sizes are matched to demand as discussed above. These and other results also suggest that trips on multiple-hire modes are on average comparatively long. The trip-makers highlighted here presumably are those who wish to travel between extremities of the region, and are deterred from using alternative services by their unavailability or excessive slowness; thus the multiple-hire modes may be particularly useful in spatially dispersed conditions.

Analyses of usage for RovingBus and TaxiMulti suggested how these modes might be managed in practice. For example, given the temporal patterns illustrated in Figure

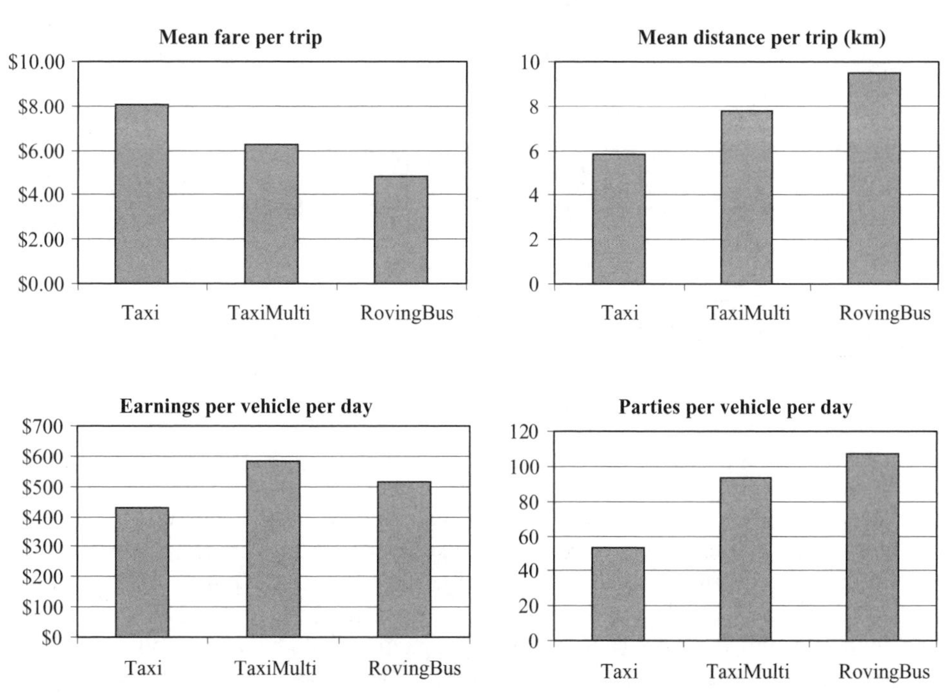

Figure 10.8 *Total trips on demand-responsive modes under* low *demand*

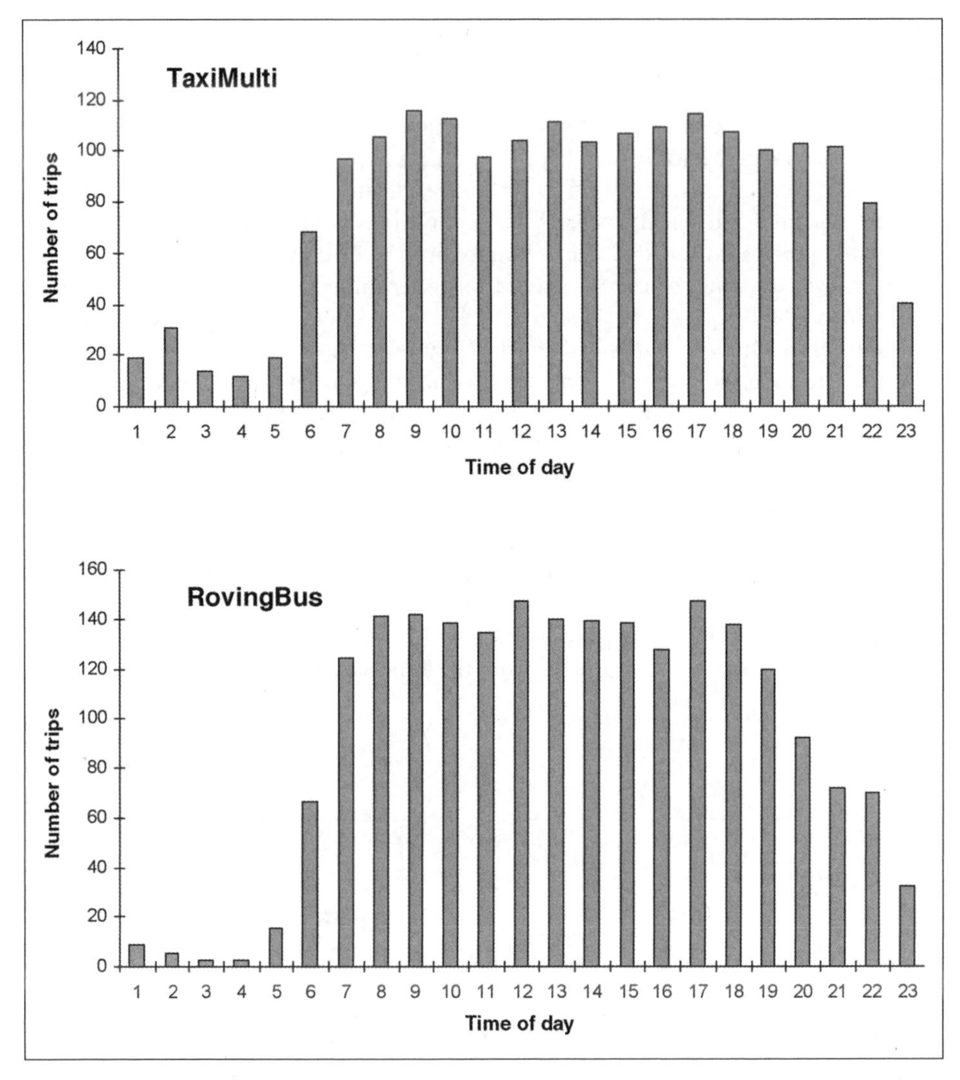

Figure 10.9 *Trips by TaxiMulti and RovingBus services under low demand*

10.9, an adequate service might be maintained by replacing RovingBus with TaxiMulti late at night, the latter perhaps run with a fleet reduced to one-third its daytime size. It may be observed also that although each vehicle was assumed to be able to accommodate 11 passengers in the low, medium and high demand scenarios, the actual occupancies rarely reached this level (Figure 10.7). A separate series of tests indicated that with capacity reduced to seven passengers per vehicle, the reductions in fleet performance and fleet-wide carrying capacity would be fairly small (of course this ignores less tangible considerations such as the sense of crowding that passengers would experience under pervasively nearly full conditions).

SmartShuttle

As indicated earlier, a SmartShuttle service is advertised in the form of a timetable, but the actual routes taken by vehicles are defined on demand. A 'feeder' role was envisaged for SmartShuttle in the Gold Coast, with vehicles plying between local railway stations and their residential catchment areas. The following account concerns a series of tests focused on the new Nerang railway station and its catchment area, which was divided into five SmartShuttle zones. One SmartShuttle vehicle was provided per zone, and timetables were designed in coordination with the railway timetable so that (notionally) each vehicle could tour its zone between visits of the mainline train, generally at 30-minute intervals. For comparison, an approximately equivalent set of fixed-route bus services was designed for the same area.

Two series of simulations were run to test these arrangements, one with SmartShuttle feeders and one with buses, the level of service in each case being held constant while the 'rail commuter' component of demand was scaled up incrementally from the Base scenario. These simulations showed that the SmartShuttle vehicles could serve nearly all prospective passengers under Base scenario assumptions while satisfying the journey-planning conditions discussed in Section 10.3 above; by contrast, fixed-route buses would leave about 8% of travellers beyond walking distance. With multipliers of 100–500% applied to base commuter demand, the proportion of requests that could be feasibly satisfied by SmartShuttle would decline steadily from 100% to 85%; the simulations suggested that the corresponding proportion for buses would be approximately constant across the same range of demand. On closer examination, however, it became apparent that the latter figure significantly overstated the performance of buses, which LITRES-2 currently models in less detail than demand-responsive vehicles; in particular, no allowance is made for passenger-capacity constraints or for delays incurred when picking up and setting down passengers. Thus on local routes and under moderate demand, small buses running in SmartShuttle mode would almost certainly be more convenient for travellers than the same vehicles run in conventional fixed-route mode.

Mode Usage by Market Segment

Figure 10.10 shows aggregate mode choice results for each market segment in the *low* demand scenario. In summary, *car-converts* are relatively heavy taxi users because they place high value on their own time. *Tourists* are heavy bus users, their activities being concentrated in the coastal strip where the bus service is frequent and convenient. The *captive old* are willing to walk only short distances, and so are the heaviest users of TaxiMulti.

10.5 Conclusions

10.5.1 Effectiveness of Transport Services

The effectiveness of the new demand-responsive service modes as demonstrated in the project has led to a recommendation that they be implemented on a pilot basis

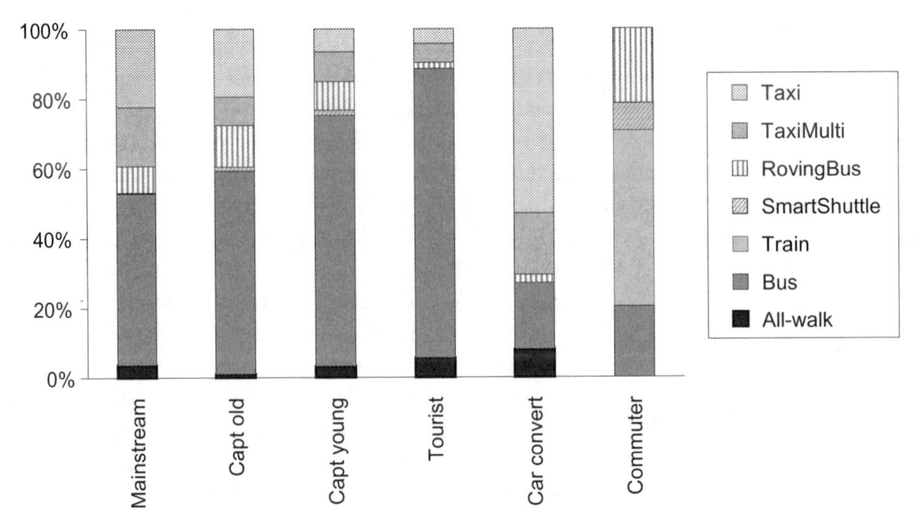

Figure 10.10 *Mode usage by market segment under* low *demand*

(GCCC, 1998, Section 5). The recommendation has not so far been implemented, presumably due to reluctance on the part of the Queensland Government to commit the substantial resources that would be required. In general, however, the project confirmed the benefits of a coordinated, multi-modal array of transport services. Those benefits were apparent in the way that the various modes complemented each other with respect to service quality, geographic coverage and market segments: even RovingBus and taxi, which are closely related in logistical terms, were mutually competitive only to a marginal extent.

Multiple hiring of vehicles in demand-responsive fleets was shown to be effective in terms of service quality, fleet economics and market share. It appears that TaxiMulti and RovingBus may best cater to longer journey-legs for which bus services are either unavailable or too slow to meet travellers' requirements, while taxis are more suited to shorter legs. SmartShuttle was investigated only to a limited extent, but as a feeder to railway stations and other regional interchanges, this mode appears to have advantages over a conventional scheduled bus service.

A critical requirement for TaxiMulti and RovingBus is the achievement of reasonably high occupancies in order to obtain substantial economies of scale over conventional taxi services. Occupancy rates in turn are directly sensitive to the travel factor, which defines the maximum allowable deviation from direct routing (see Section 10.3.4 above, and Rawling *et al.*, 1995). Increasing the travel factor above the value of 130% assumed in the Gold Coast project would provide broader time-windows and hence greater flexibility for scheduling, making it easier to achieve high vehicle occupancies. This would be a desirable outcome, but a reduction in fares might then be needed to compensate travellers for longer travelling times.

Taxi services were not a major concern in the Gold Coast project reported here, but were investigated more thoroughly in a related study that concentrated on the Regent taxi fleet over its full geographical range (Horn *et al.*, 1999). That study was concerned mainly with operational issues such as the balance between phone bookings and traditional 'step-in' services at cab-ranks, the impact of advance notice on carrying capacity and waiting times, and temporal staggering of shift changeovers. The availability of historical booking records made data preparation much easier than in the wider-ranging project reported here; the methodology was also more straightforward, since the focus was on individual initiatives and changes, rather than broader combinations of assumptions and proposals like those discussed elsewhere in this chapter.

10.5.2 Methodological Considerations

Data Preparation and Validation

For CSIRO and its collaborators the most difficult challenges posed by the Gold Coast project were those arising in the preparation of a credible Base scenario, especially with respect to data items – market-segment parameters, demand models, and road speeds – defining behavioural characteristics of transport system users. Here there was no alternative but to make the most of whatever data sources were available, often in an indirect way or by reference to prior experience in other locations. The empirical foundation of the project cannot therefore be regarded as definitive, and the conclusions are best seen as indicative rather than strictly predictive. Even so, detailed examination of Base scenario outcomes, and sensitivity analyses of key variables (Horn *et al.*, 1998), showed a satisfying conformity with known aspects of transport system operation.

Apart from these substantive concerns, the effort involved in data preparation constituted a large proportion of that invested in the project as a whole, and it seems clear that an effort of this kind might be difficult to justify in a once-off consultancy exercise. Conversely, however, the project suggests a framework within which public authorities might accumulate information on transport demand and system usage, the long-term benefits in this respect lying in the ease with which new proposals can be investigated once a suitable database is established.

Investigative Strategy

Some other methodological issues were concerned with the relations between modelling – the central task of the project – and other phases of the planning process. In a naive view, design or planning may be regarded as proceeding in a linear sequence from the enunciation of a problem (e.g., as a set of objectives and conditions), to the formulation of a proposal (e.g., by proposing new transport services), to assessing the implications of that proposal (e.g., by means of a LITRES-2 simulation), and then to evaluating those implications (e.g., by interpreting simulation outcomes). However, in any substantial planning exercise it can be expected that there will be backtrackings or feedback cycles to complicate a simple sequence like this, and it must be admitted that the extent of these complications was underestimated when the Gold Coast project was planned. The lesson here is that modelling should not be treated as an isolated task: although the broad outlines of transport options could be sketched in advance, the development of those options as credible proposals was intertwined with the modelling process and with the interpretation of its outputs.

For example, the main simulations of multiple-hire demand-responsive services involved the testing of a small number of proposals, each comprising specifications of fleet size, the carrying capacities of vehicles, the fares to be charged, service-quality standards, and so on. In fact these proposals themselves emerged from a preliminary series of simulation runs, in which a range of plausible values was explored for each of the main parameters. There was a similarly close relationship between design and simulation in the tests of SmartShuttle feeder services, which involved the additional design task of formulating 'benchmark' bus services for comparative purposes.

A further point concerns the criteria applied in assessing simulation outcomes. Besides convenience, accessibility, and so on, it emerged during the course of the project that the clients believed that new passenger services should be able to operate with little or no public subsidy. The desire for operationally attractive performance entailed considerable effort in tuning the various proposals, and led to the exploration of demand-responsive services on a smaller scale than might have been undertaken otherwise. In retrospect, the rationale for this approach seems questionable, given the emphasis elsewhere on the benefits of consistency and integration in the provision of transport services. Operational viability is certainly a significant planning consideration, but to require that demand-responsive services should be fully self-supporting could prejudice their availability at the lower end of the socioeconomic spectrum. It would be unfortunate indeed if an effort to modernise transport services were to involve the relegation of traditional fixed-route modes to a second-class or 'safety-net' role.

Modelling Tools

Experience in the Gold Coast project shows that effective usage of LITRES-2 requires a considerable level of interpretive intelligence, as indeed with any modelling system of comparable scope. This is illustrated by the counter-intuitive results encountered in the SmartShuttle tests, which highlighted an anomaly in the level of detail applied to fixed-schedule as opposed to demand-responsive vehicles (see Section 10.4 above). The anomaly can be removed by applying the same operational constraints and detailed simulation to buses as to demand-responsive vehicles, and further attention might be given also to the system's representation of passenger behaviour (Horn, 2000). In addition, an interface between LITRES-2 and a GIS might facilitate the preparation of input data and the analysis of modelling outputs.

There appears to be no immediate prospect of overcoming LITRES-2's reliance on external estimates of passenger transport demand (Horn, 2002). In this respect it is interesting to contrast the microsimulation approach adopted in LITRES-2 (see also Tong and Wong, 1999), with network-equilibrium models such as EMME/2, in which probabilistic methods are used to estimate the distribution of *all* journeys, including those made by private car (Spiess and Florian, 1989). From a planning perspective, the main differences between the two approaches have to do with granularity and scope: where the equilibrium models are coarse-grained and comprehensive, microsimulation is typically fine-grained and specialised (consider for example the practical difficulties attending a microsimulation of all road traffic).

It is interesting also to consider the use of GIS buffering capabilities in estimating 'catchment areas' and latent demand in the vicinity of transport routes, bus stops, railways stations and so on. This is clearly pertinent to analyses concerned with accessibility, especially where bus or rail routes are to be planned in conformity with official standards (e.g. maximum walking distances to bus stops). But accessibility by itself must be reckoned a weak indicator of transport system quality, since it is concerned merely with identifying the transport services that are physically available to people in a given area, rather than the services that those people are likely to use in practice. This criticism applies even to some interesting recent research involving the generalisation of temporal and operational aspects of accessibility (O'Sullivan *et al.*, 2000).

The foregoing discussion has highlighted some strengths and weaknesses of LITRES-2 in the context of projects such as the one described in this chapter. In summary, LITRES-2 emerges as a very suitable means for assessing passenger transport operations and performance, with unique capabilities with respect to demand-responsive services. Considerable effort must be devoted to data collection and validation; nevertheless, once that investment is made, it is possible to explore a wide range of passenger transport planning options, to a level of detail that is not available elsewhere.

Acknowledgements

Acknowledgement is due to the author's collaborators in the work described in this chapter: Ken Deutscher, Melissa Doueihi, Garry Glazebrook, Mac Hulbert, Bella Robinson and John Smith.

This chapter includes some material adapted from the report cited as Horn, Robinson and Smith (1998), and included here by permission of the Gold Coast City Council. Enquiries with respect to the contents of the report or their dissemination should be directed to the Council's Transportation Planning Branch.

This chapter includes some other material that has been adapted from an earlier article written by the author. The material is reprinted from Horn, M.E.T. (2002) Multi-modal and demand-responsive passenger transport systems: a modelling framework with embedded control systems, *Transportation Research A*, **36**(2): 167–88. Copyright with permission from Elsevier Science.

References

BAH (Booz Allen and Hamilton) (1998) *Gold Coast Public Transport Study: Proposed Strategy and Implementation Plan*, Report submitted to Gold Coast City Council.

GCCC (Gold Coast City Council) (1998) *Gold Coast City Transport Plan.*

Horn, M.E.T. (2000) A framework for modelling passenger movement in urban public transport systems, in *Proceedings, 22nd Conference of Australian Institutes of Transport Research*, Canberra.

Horn, M.E.T. (2002) Multi-modal and demand-responsive passenger transport systems: a modelling framework with embedded control systems, *Transportation Research A*, **36**(2): 167–88.

Horn, M.E.T., Robinson, B. and Smith, J.L. (1998) *LITRES2 – Public Transport Simulation Research Project*, Report submitted to Gold Coast City Council, CSIRO Australia.

Horn, M.E.T., Smith, J.L. and Robinson, B. (1999) Taxi fleet performance under flexible operating conditions and with improved scheduling procedures, in *Proceedings, Fourth International Conference of ITS Australia*, Adelaide.

Little, M.C. and McCue, D. L. (1993) Construction and use of a simulation package in C++, *Computer Science Technical Report #437*, University of Newcastle upon Tyne, July 1993 (also in *C Users' Journal*, **12**(3), March 1994).

O'Sullivan, D., Morrison, A. and Shearer, J. (2000) Using desktop GIS for the investigation of accessibility by public transport: an isochrone approach, *International Journal of Geographical Information Science*, **14**(1): 85–104.

Rawling, M., Smith, J.L. and Davidson, I. (1995) Modelling of personal public transport in a multi-mode service environment, in *Proceedings International Conference on Application of New Technology to Transport Systems*, ITS Australia, Melbourne, v1, pp. 167–78.

Smith, J.L. (1993) An activity oriented urban passenger transport model, Proceedings, *Australian Transport Research Forum*, **18**(2): 993–1006.

Spiess, H. and Florian, M. (1989) Optimal strategies: a new network assignment model for transit networks, *Transportation Research B*, **23**(2): 83–102.

Tong, C.O. and Wong, S.C. (1999) A stochastic transit assignment model using a dynamic schedule-based network, *Transportation Research B*, **33**(2): 107–21.

11

The South and West Yorkshire Strategic Land-use/ Transportation Model

David Simmonds and Andy Skinner

Abstract

A major multi-modal transport study is currently being carried out in South and West Yorkshire. This chapter outlines the land-use/transport/economic modelling system that has been developed for the study. This system is being used at the strategic level to forecast future land uses and travel demands, and to examine the impact of alternative transport and development plans.

11.1 Introduction and Client Needs

The South and West Yorkshire Multi-Modal Study (SWYMMS) is one of a series of studies which were proposed in *A New Deal for Trunk Roads in England*, published by the Department for Transport in July 1998 (Department for Transport, 1998). The overall aims of the study are to make recommendations for:

- an integrated and sustainable plan for the strategic road, rail and water networks in the study area; and
- a plan of specific interventions to address the most urgent key strategic problems in the study area through to 2021.

Applied GIS and Spatial Analysis. Edited by J. Stillwell and G. Clarke
 ISBN: 0-470-84409-4

The study is required to address the Government's aims for transport, which refer to economy, accessibility, safety, integration and environment. It is also required to address a set of study-specific objectives, which are

- to reduce congestion on the motorways and A1;
- to re-establish the primary role of the trunk network for strategic traffic flows;
- to facilitate sustainable economic regeneration of depressed areas, especially the Objective 1 area of South Yorkshire and the Objective 2 area of West Yorkshire; and
- to sustain economic growth in other parts of the study area.

At the time of writing,[1] the study, which is being carried out for the Government Office for Yorkshire and the Humber, is still ongoing. Further details of the team, the work and the findings of the study can be found at www.swymms.org.

At an early stage in the development of the study, it was decided that it would be necessary to use both a strategic model and a detailed model in the analysis of transport options. The need to use both reflects the fact that with present technology (in terms of modelling methods and computing power) transport planners are forced to choose between:

- models which allow detailed representation of the origins and destinations of travel (zone system) and of the transport network, but represent only a few of the choices which households and businesses make in response to changing travel conditions; and
- models which represent more of the choices and linkages between choices, and hence are more appropriate to forecasting reactions to change, but which incorporate much less details about the spatial system and the transport network.

The study-specific requirements imply that it is necessary to study not only the impacts of transport change on transport use, but also the interactions between transport change and local economic performance. It therefore needs to examine how demand for transport is affected by land-use changes and how changes in transport may affect land use – with particular reference to impacts on economic performance. The consideration of transport must take account of both planned interventions and other changes such as worsening congestion.

In order to examine these changes and their linkages systematically, particularly to compare the consequences of different strategies, it was recognised that a strategic land-use/transport model was required, i.e. a model considering a wide range of responses to change with relatively little spatial detail, and that this should be used as the main tool in developing the proposed strategy. At the same time, it was also necessary to use more conventional, detailed models to examine alternative changes to the transport infrastructure in the process of developing the recommended *plan* of interventions within the preferred strategy. To ensure maximum consistency

[1] July 2002.

between the two levels, the outputs from the strategic model are used as a way of generating the levels of travel demand which are assumed in the detailed model.

In the remainder of this chapter, we explain the components and structure of the strategic transport and land-use/economic models in Sections 11.3 and 11.4 respectively and briefly recognise the interaction between them in Section 11.5. Section 11.6 demonstrates how the model is used for testing particular policies. We begin, however, with an explanation of the overall design of the model.

11.2 Model Design and Application

11.2.1 Scope of the Model

The strategic model draws heavily on the study team's previous experience with a number of similar models developed since 1995 (see, for example, Copley *et al.*, 2000; Simmonds and Skinner, 2001). The approach adopted is that of a linked model system which brings together distinct transport and land-use/economic models. The basis of the approach is that the land-use/economic model generates the activities (and in some cases the interactions between activities) which generate the demand for transport; and that the transport model represents the interaction between demand and supply in transport, and generates information about the ease or difficulty of travel which influences subsequent land-use and economic change.

A significant advantage of this approach is that because the land-use/economic and transport models remain distinct entities, it is very easy to run each one on its own. This helps in the development stage, where to a considerable extent the two models can be developed in parallel. It is also useful in the application stage: the transport model can be used on its own to test the transport impacts of different alternatives under common land-use forecasts, and the land-use/economic model can be run to test (for example) the land-use impacts of different economic scenarios assuming no change in transport conditions.

The transport model was implemented using MVA's START package (Roberts and Simmonds, 1997). The land-use/economic model was implemented using DSC's DELTA package (see Simmonds and Still, 1999; Simmonds, 2001). The MapInfo GIS package is used extensively in mapping the inputs and outputs of the model. In the following sections we describe the design of the South and West Yorkshire Strategic Model (SWYSM) model and illustrate its application.

11.2.2 Scope of the Model – Space and Time

The focus of the study is on the 'motorway box' formed by the M62, A1(M), the M18 and the M1. It was recognised that the forecasting process needs to consider not only what happens within this box, but also to consider in equal detail what happens in a significant area around it, and in less detail what happens further away. Accordingly the model covers the whole of South and West Yorkshire plus a ring of adjoining areas such as Skipton and Harrogate; this is the fully modelled area (Figure

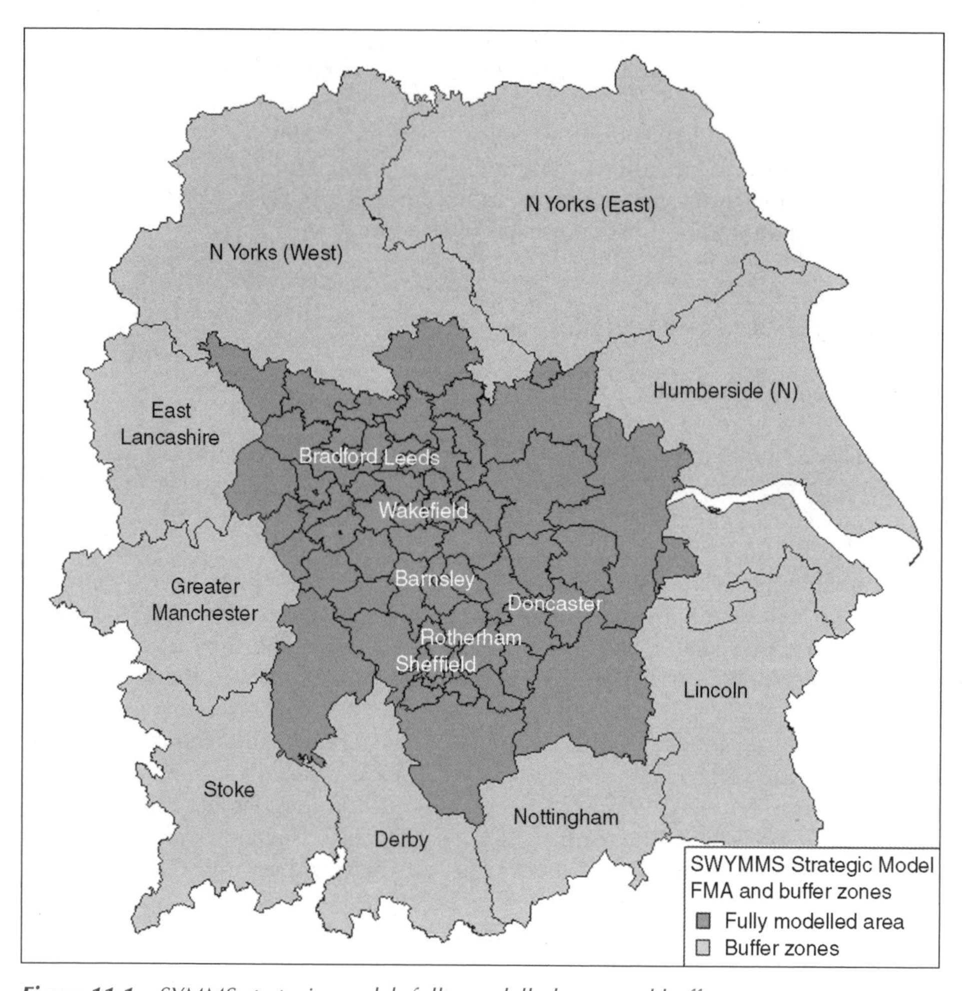

Figure 11.1 *SYMMS strategic model: fully modelled area and buffer area*

11.1). Around this, a wider ring of adjoining areas (the 'buffer area', extending westwards to Manchester and eastwards to Hull) is covered in less detail, with larger zones and fewer model processes working. Beyond these is a set of 'external zones' used to represent movement to and from the rest of Britain.

Within the fully modelled area, the zones were defined initially as 1991 wards, or groups of wards, within district boundaries; these conditions facilitate the use of data both from the 1991 Census of Population and from other sources. The grouping of wards into zones within districts was intended to distinguish the main urban centre of each district (in the event, some subdivision of wards was required), and then as far as possible to create zones of reasonable size and shape in relation to the transport networks. The process was constrained by the number of zones which is compatible with practical computing times and memory requirements, especially in the transport model where both time and memory increase as the square of the number of zones. In total, there are 92 zones in SWYSM – in contrast with the 570 zones in the more detailed highway and public transport models. The model was set up so that the base year is 2000 and the forecasting process works forward to 2020; this involved an element of forecasting to create the 2000 base.

11.3 Transport Modelling: The START Package

Unlike many conventional transport models which have to be calibrated to reproduce the base year situation, START is incremental in approach: matrices of trips are estimated outside the model to describe the base year pattern of travel, making use of all available data. The model then modifies the pattern of travel in response to factors 'external' to transport – mainly changes in land-using activities such as the number, mixture and car-ownership of residents, the number and type of jobs, etc. and to changes within the transport system itself. The following sections summarise the preparation of the data for the base situation, the external forecasting model and the transport model proper.

11.3.1 Transport Model Base Data

Data on base year travel by motorised modes were obtained from the SWYMMS detailed highway and public transport models, which were in turn developed mainly from the extensive surveys carried out during 2000. Walking and cycling, which are not represented in the more detailed models, were estimated separately. Most of the START supply representation was obtained by aggregating data already assembled in the detailed models. The primary sources of data for the detailed models were surveys of road traffic and public transport trip making undertaken in spring 2000. Approximately 140 road traffic interview sites were employed, covering all entry points to the motorways and the A1 within the study area. Public transport passenger surveys were undertaken in the centres of the 10 largest study area towns, with details of non-central movements provided via operator ticketing systems. Cartographic network data for the highway system were created using GIS techniques, and public transport networks were coded from timetables.

11.3.2 External Forecasting Model

The external forecasting model (EFM) predicts changed travel patterns due to zonal changes in land use, including population, employment and car ownership. The model requires considerable detail about present and future land-use by zone (age groups, household types, etc.), since such factors are known to influence the amount and the type of travel. In SWYSM, all of these are forecast by the land-use model (see below).

The EFM assumes that future residents of a given age, employment status, car-ownership level, etc. will, if other factors remain equal, wish to make the same trips as comparable present residents. It works by factoring the different trip matrices in proportion to the numbers of residents by type and zone, i.e., in proportion to changes in the land uses that produce and attract trips. The output of the EFM is therefore an artificial travel pattern – what would happen if the land-use changes occurred but the time and money costs of travel were unchanged? The impact of time and money changes is considered in the transport model proper.

11.3.3 The Transport Model

The transport model (Figure 11.2) contains the demand model, which adjusts travel in response to changes in transport conditions; and the supply model, which adjusts transport conditions in response to changes in the pattern of travel. These are

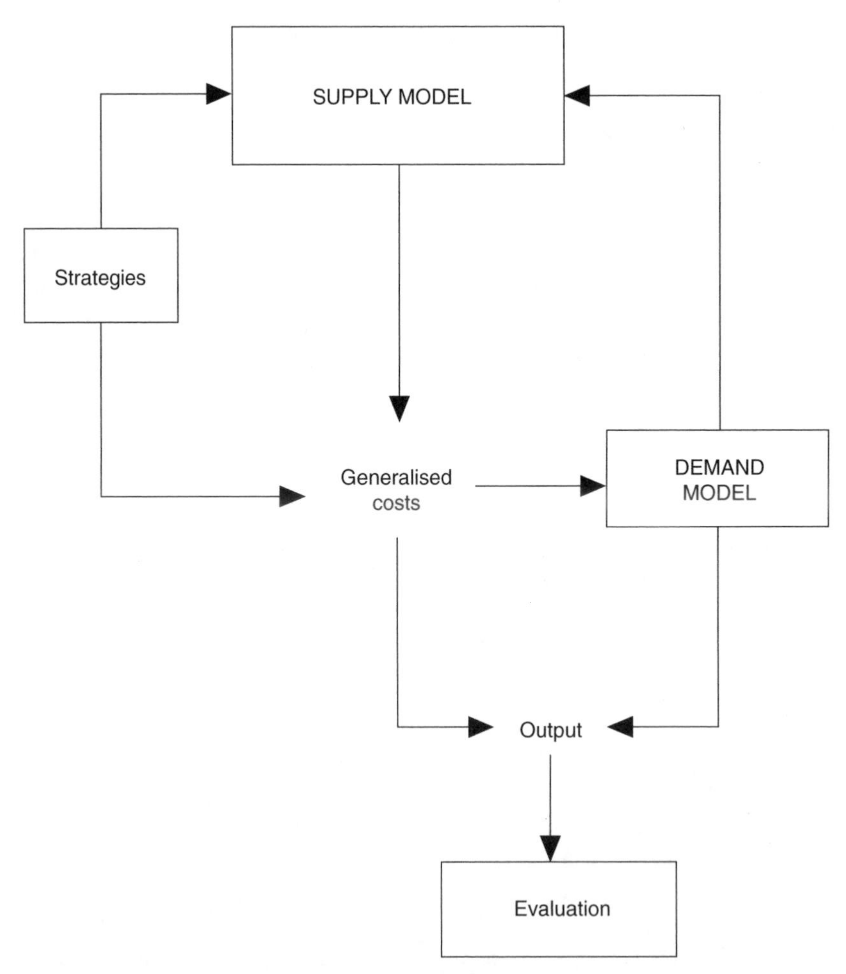

Figure 11.2 The structure of the START transport model

automatically run in turn until the travel pattern and transport conditions are consistent with each other. The key variable linking them is generalised cost, which is used to describe every journey within the modelled area by a single value bringing together money, time, inconvenience or discomfort, etc.

The *Demand* component of the transport model deals with all of the key choices available to travellers. These are:

- frequency – how often to travel;
- destination – where to travel to;
- mode – car, bus, rail, tram, walk, cycle, taxi;
- time of day – when to travel;
- route – highway or public transport route alternatives; and
- parking – type, location and duration.

These choices are organised into a hierarchy. The most sensitive choices are dealt with first, at the bottom of the structure, and the least sensitive choices at the top of the structure. The relative sensitivities, and hence the hierarchy, differ between trip purposes.

Changes in the pattern of freight movement are forecast by the land-use/economic model. Only road freight is currently represented in SWYSM; this is equivalent to assuming that the road share of freight movement for each commodity and each origin-destination (OD) pair will remain constant. Growth in road goods vehicle movements is forecast by DELTA as a result of changing land use and economic conditions. The prospects for modal shift in freight are being considered within SWYMMS, but outside SWYSM.

The *Supply* component of the transport model ensures that the level of service of the transport system, as defined by monetary costs and the various components of travel time, is compatible with the flows on the various transport networks. The highway supply definition is based upon the following key features:

- links that represent the speed/flow relationships for flow groups in each zone (typically defined in terms of inbound and outbound radial, clockwise and anticlockwise orbital);
- motorway and other 'limited access' links defined as specific roads; and
- a set of fixed car routes for each OD pair, including intra-zonal movements, defined in terms of distance travelled upon each link (with allocation of trip proportions to routes carried out via the demand model).

A similar arrangement is applied to public transport, with the addition of specific rail network links. The public transport definition allows for mixed mode trips. Crowding relationships are included so that generalised costs increase as demand approaches capacity. Fares and frequencies can be made responsive to changes in patronage. A number of indicators of the environmental impact of transport are calculated within the transport model, and some of these influence subsequent land-use change. In addition, for SWYMMS, more extensive analysis of environmental impacts was carried out by Environmental Resources Management as an input to the strategy appraisal process.

11.4 Land-Use/Economic Modelling: The DELTA Package

The DELTA land-use/economic modelling package has been developed by David Simmonds Consultancy since the mid-1990s. It was designed as a practical tool to represent urban and regional change and to interact with an appropriate transport model. The approach draws upon previous modelling experience and upon the wide range of relevant research carried out in geography, urban economics, etc. (for a fuller description of the model, see Simmonds, 2001).

Since land uses take a long time to respond to transport changes, the land-use model needs to represent change over time, in contrast with the transport modelling which describes transport supply and demand at particular points in time. Another requirement is to recognise that different processes operate at the urban and the regional levels: for example, different factors affect the total economic activity *in* an area and the location of employment *within* it. The model accordingly contains urban processes which change in or between the 92 zones, and regional processes in which the units are (approximate) travel-to-work areas – typically districts, in SWYSM.

The urban processes represent both changes in buildings and changes in activities. The processes of physical change are:

- development: the amount of floorspace by zone and type (residential, retail, office, industrial); this is driven by the economic scenario and the modelled property market, and controlled by inputs measuring what is allowed by the planning system; and
- the housing quality model, which models the way in which an area may decline to slum status, or be revived from slum to high-quality.

The urban activity sub-models are:

- the transition model, which represents household (and hence population) changes in terms of movements through a simplified lifecycle;
- the car-ownership model, which predicts the proportions of households by type and zone owning 0, 1 or 2+ cars, mainly in response to increasing incomes;
- the location model, which locates or relocates a proportion of households and of employment in each year, and also models the property market within which location occurs; and
- the employment status model, which updates the work status of residents and the commuting pattern in response to the spatial changes in households and employment.

At the regional level, there are three models, all of activities. These are:

- the migration model moves households between areas;
- the investment model allocates investment to areas (taking account of changes in accessibility and property costs); and
- the production/trade model (a spatial input–output model) estimates production by sector and area and the patterns of trade between areas.

The processes are considered in a fixed sequence within each one-year step, as shown in Figure 11.3. However, there are also numerous time lags between the different

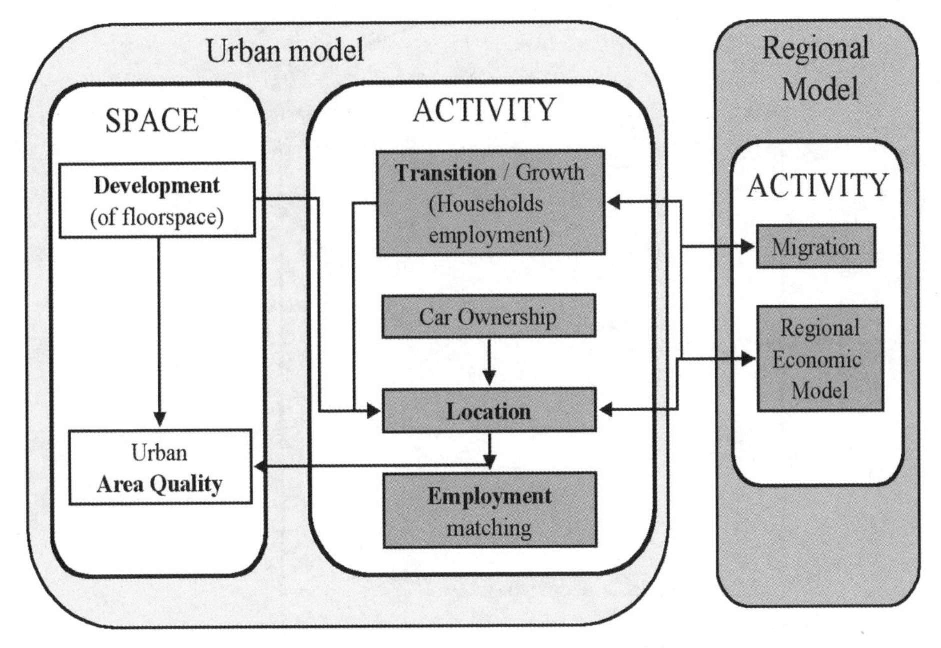

Figure 11.3 The DELTA sub-models: land-use/economic model structure

processes which are equally or more important to the overall performance of the model.

11.5 Land-Use/Transport Interaction

The model operation over time is a sequence of five one-year steps through DELTA, followed by a run of START to represent the resulting state of the transport system (Figure 11.4). Land-use changes have immediate impacts on transport, through the EFM described earlier. At the area level, changes in the costs of trade (goods and passengers) affect the location of production and of new investments. At the urban level, generalised costs are used to calculate various accessibility measures which affect the location of households and businesses within each area. In addition, the environmental impacts of transport influence the locational preferences of households.

11.6 Use of the Strategic Model

11.6.1 The Strategy Testing Process

As with most models, the strategy testing process using SWYSM is comparative. A strategy option is examined by appropriate changes to the model inputs; running the

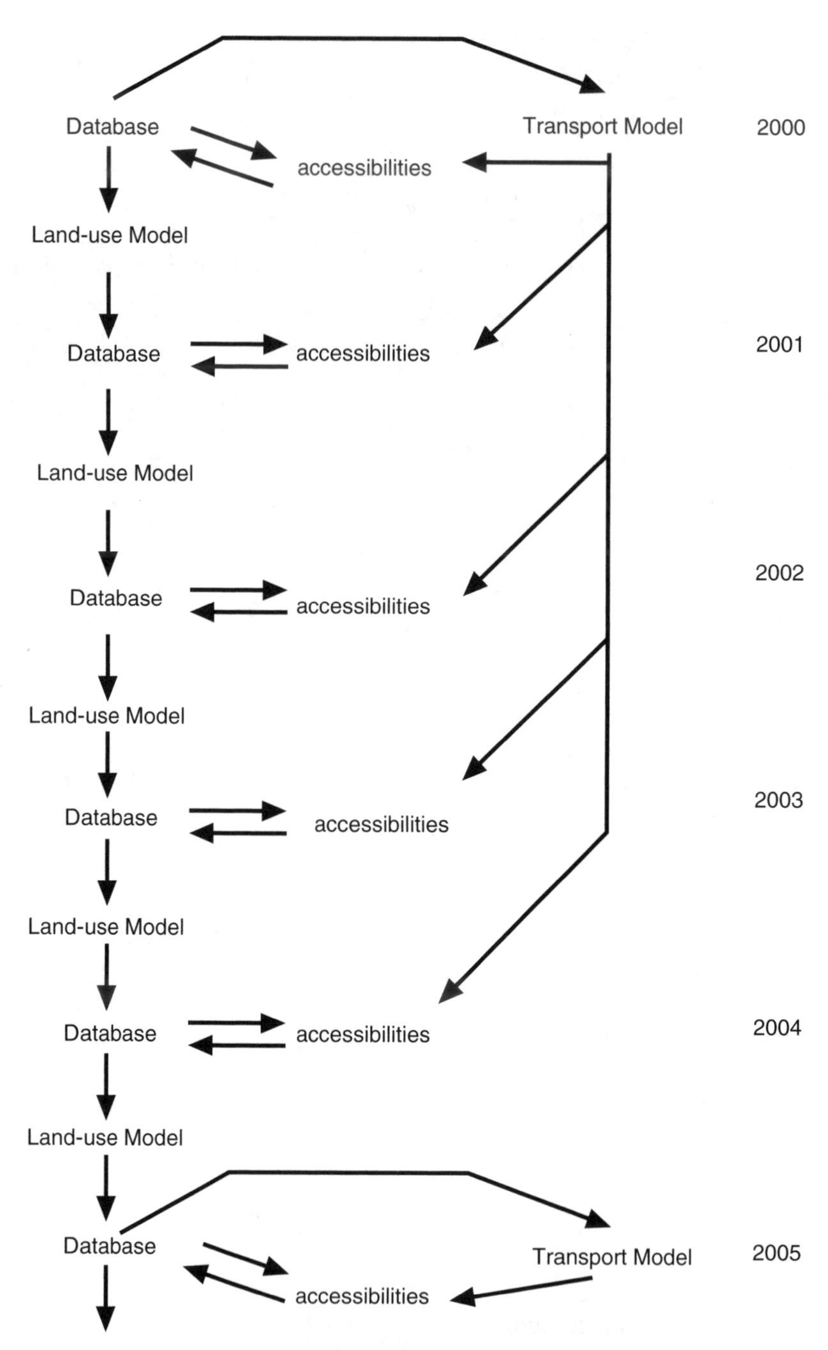

Figure 11.4 *START/DELTA time series arrangements for SWYSM*

model over time; and comparing the resulting outputs with those from a previously run reference case. This process applies whether transport options, land-use options or combinations of the two are to be tested.

Transport policy changes are introduced into START either as direct adjustments to costs or as changes in the characteristics of the transport supply. They influence land-use and economic change through the accessibility and environmental values that are passed from START to DELTA. Land-use policy changes are introduced into DELTA via a special policy file, which is used to specify changes in the amount of land available for each type of development in each zone in each year. Policies may not have immediate effects: for example, an additional allocation of land may not be used immediately or at all.

An important characteristic of using a land-use model is that land-use policies are tested in the same way as transport policies. This is not always the case. Whilst it is conventional for transport policies to be tested by forecasting their effect taking account of user preferences and interactions, land-use planning often assumes simple linkages between policy changes and outcomes. Formal land-use modelling allows a more rigorous approach to be taken.

11.6.2 Strategy Testing in SWYMM

The strategic model was used extensively in the Strategy Development phase of SWYMMS. This was divided into three main themes: travel reduction options; economic development options; and combined transport/economic development options. For the first theme the transport model was frequently used on a stand-alone basis, as direct transport effects were dominant. The full model was used for investigation of options for the second and third themes. The following paragraphs illustrate the results from just one part of this exercise.

11.6.3 Example Strategy Results

One of the tests carried out under the economic development options theme (identified as test ED3 within SWYMMS) was to look at the combined impact of two packages of road building and widening schemes. The packages were a group of schemes aimed at economic regeneration, and a group of schemes aimed at the relief of traffic congestion in areas where economic regeneration is a priority concern.

These schemes were concentrated in the South Yorkshire area between Sheffield and Doncaster, plus improvement to the A1 north of Doncaster; they are illustrated in Figure 11.5. They were introduced into the transport supply in 2005, and the model system was run to 2020. The following discussion of their impacts is based on comparison with the reference case; except for the road schemes, there were no other differences in transport or land-use policy between the ED3 test and the reference case.

The changes in the road networks are introduced into the model through the supply side of the transport model in 2005 (and in all the subsequent transport model years). These changes have the effect of allowing faster travel by road for some parts of

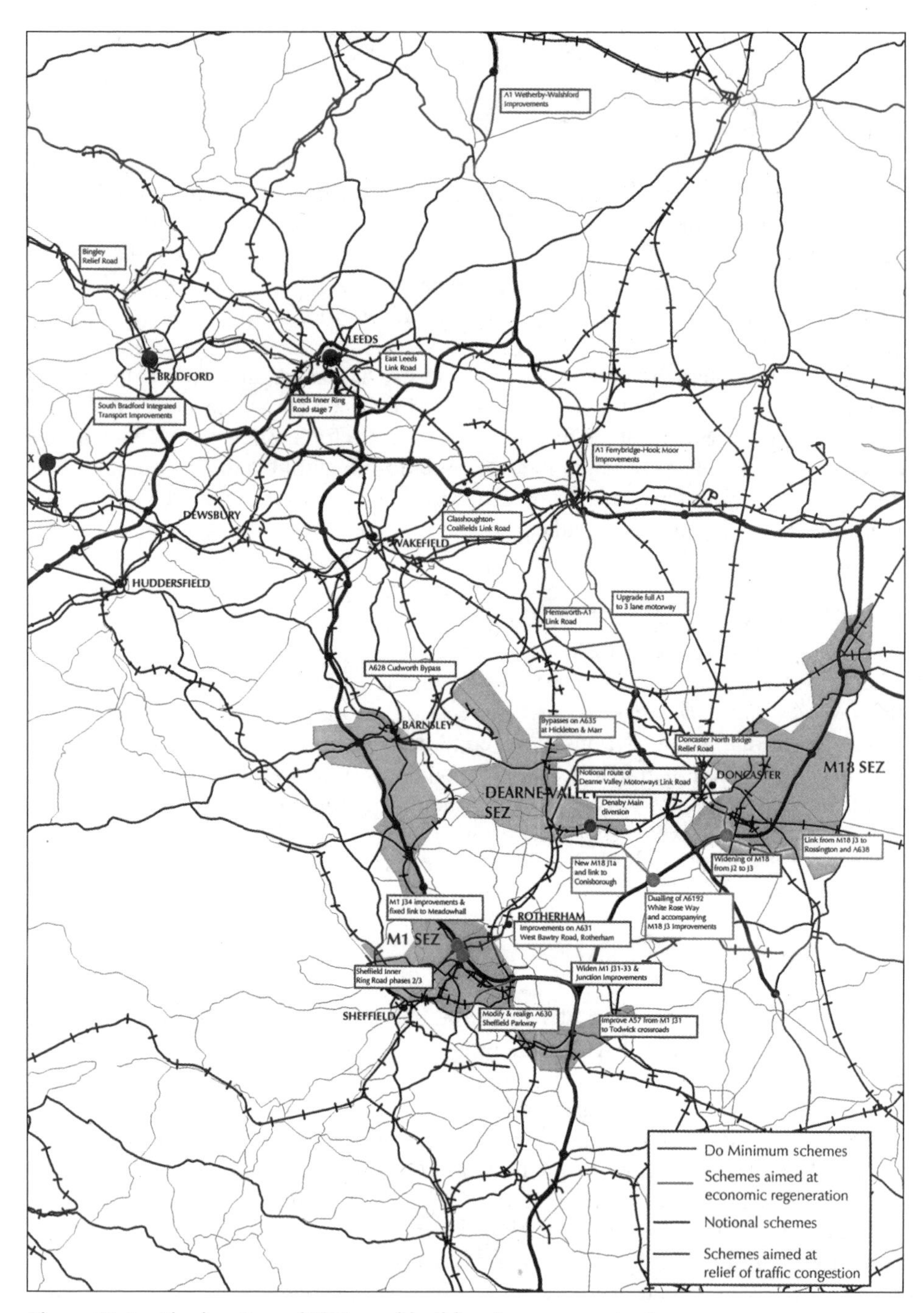

Figure 11.5 *The location of ED3 road building/improvement schemes*

journeys in South Yorkshire (including some journeys between origins and destinations outside South Yorkshire). The journey time reductions come about partly from higher road design speeds, and partly from increases in capacity which reduce the effects of congestion.

As a result of the journey time reductions, the generalised cost matrices output from the transport model show very slight reductions for affected zone pairs relative to the equivalent matrices for the reference case. As a result, there are some very slight changes in trip patterns, including a slight increase in commuting between Doncaster and Sheffield/Rotherham.

All of these effects are very slight – mostly measured in fractions of 1% change from the reference case. The average cost of delivering goods from the Doncaster area to the Sheffield/Rotherham area, for example, falls by slightly less than 1%. These effects are apparent in the data passed to the land-use model at the end of the 2005 run of the transport model (that is, they are slightly different from the equivalent data in the reference case). The differences from the reference case will affect the land-use/economic model in a number of ways. At the area level:

- the changes in cost or generalised cost of travel will have some impact on the pattern of trade, and this will have an immediate impact on the level of production of each sector in the affected areas; and
- the same changes will also affect the accessibility of each area and hence its attractiveness to investment: the change in the level of investment (or reinvestment) will affect the output of each local economy in the longer term.

The effect on investment is, as one would expect, slight, and markedly different across sectors. Figure 11.6 shows the percentage changes in capacity by sector in the Doncaster area in 2020, i.e., the accumulated effect on investment (relative to the

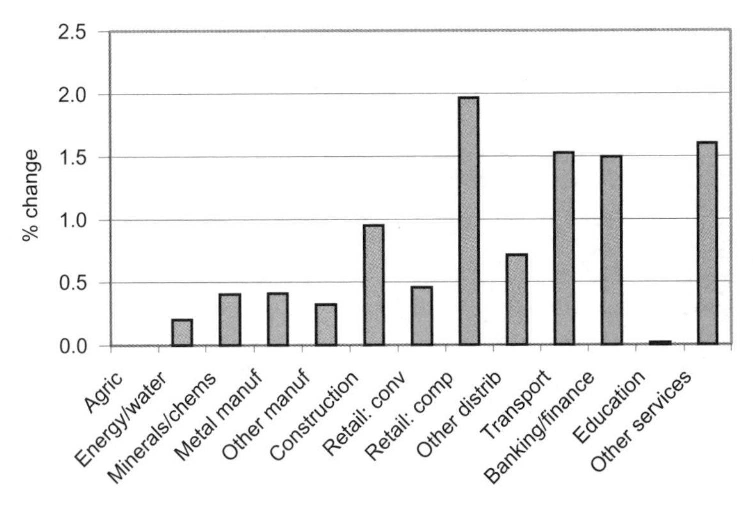

Figure 11.6 *Changes in productive capacity, Doncaster area, 2020, test E3 compared with reference case*

reference case) of slightly better accessibility over 15 years. It can be seen that the greatest impact on capacity is a 2% increase for comparison retailing, followed by increases of around 1.5% for transport and communications, banking/finance and other services.

Most of this increase occurs over the period 2006–2015, and the increased investment is largely responsible for the resulting slight increase in economic activity in the Doncaster area. As is shown in Figure 11.7, total (all-sectors) value added in the area gradually increases very slightly compared with the reference case, reaching a gain of slightly more than 1% after 10 years.

Figure 11.7 also shows the much smaller positive impact on the Barnsley and Sheffield/Rotherham areas. It should be noted at this point that the model system reflects what has become known as the 'two-way road effect', that is, that transport improvements intended to benefit a specific area may assist the economic regeneration of that area, or they may increase the problems of the area by making it easier to serve that particular market from other locations. In the present case, the concentration of schemes in South Yorkshire does have a (very slight) positive effect in that sub-region, though it will be seen below that not all of the indicators are positive for all the districts affected. This comes about mostly through the improvement in accessibilities making South Yorkshire in general, and Doncaster in particular, a marginally more attractive location for investment in various sectors.

The gain in value-added in Doncaster in 2006, the first year after the opening of the road schemes, is rather larger than that in any subsequent year. This is due to the initial impact on the pattern of trade, as Doncaster businesses find it marginally easier to win custom elsewhere as a result of the decreased transport costs. Note that within the results for later years there are multiplier effects:

- the additional production due either to the initial impact on the pattern of trade or to the additional investment creates additional intermediate demand for goods and services; and

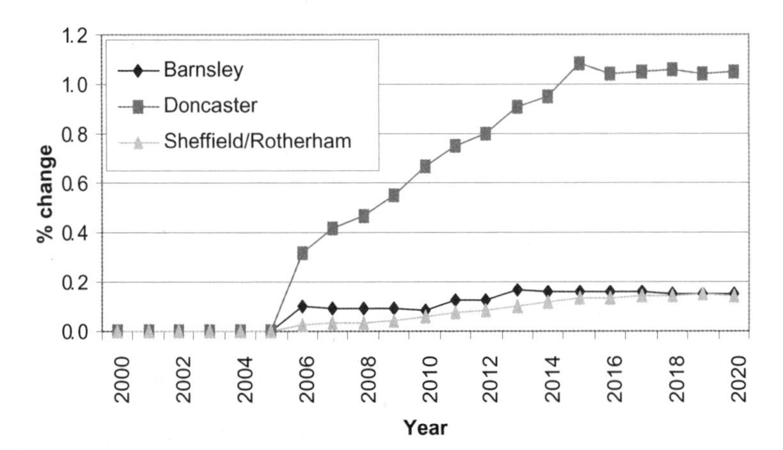

Figure 11.7 *Impact of ED3 on value added in South Yorkshire areas, 2006–2021*

- the increases both in capacity and in production create additional employment which leads to increases in consumer final demand.

The additional employment opportunities affect the pattern of migration, leading to a slight increase in population of the Doncaster area relative to the reference case, which in turn helps to retain expenditure within the area and to reduce the 'leakage' of multiplier effects which would otherwise occur.

Returning to the aspects of the model which are directly affected by the outputs from the transport model, at the urban (zonal) level:

- the local pattern of location of households and businesses will be affected by the changes in accessibility resulting from the changes in generalised cost (with further consequent impacts through competition in the housing and commercial floor-space markets); the pattern of household location will also be affected to some extent by the environmental changes resulting from the changed traffic patterns; and
- the likelihood of working-age individuals being in or out of work will be slightly modified, with consequent impacts on residential location.

These effects and their further consequences (for example, in the development model) work through over a number of years.

The increases in economic activity generate increases in employment. These are shown by district for the South and West Yorkshire authorities, in Figure 11.8. Note

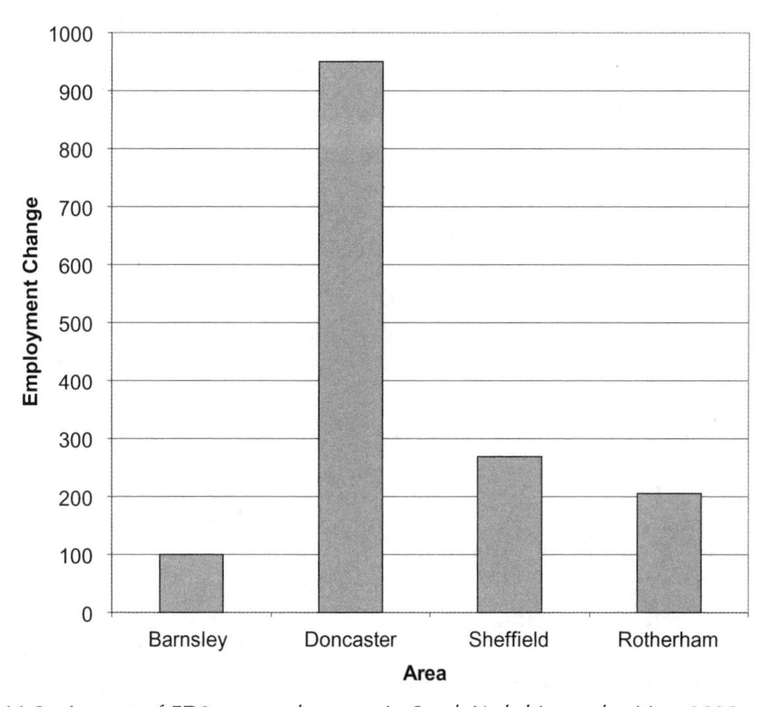

Figure 11.8 *Impact of ED3 on employment in South Yorkshire authorities, 2020*

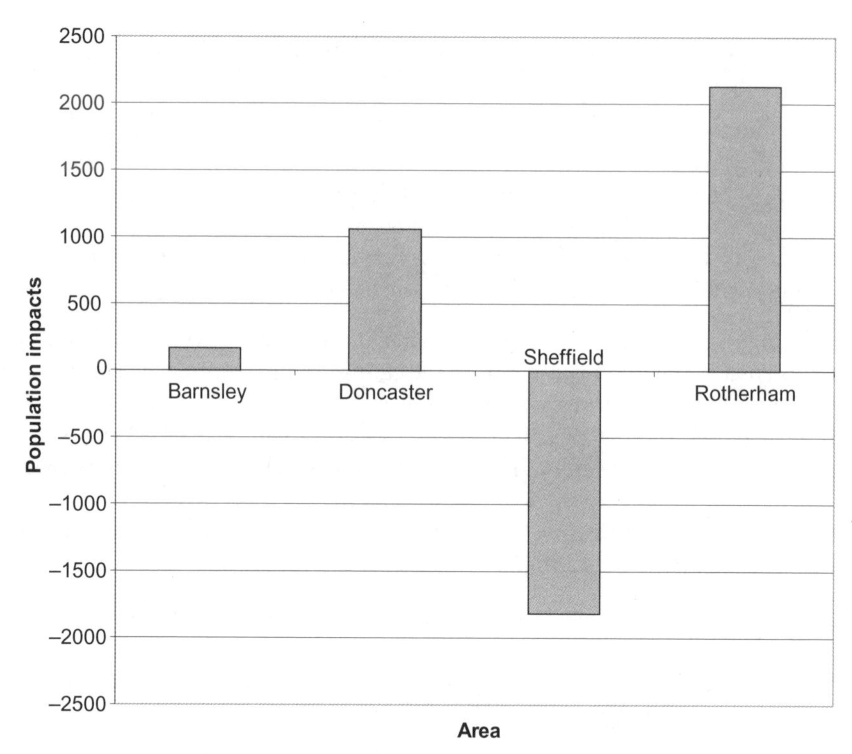

Figure 11.9 *Impact of ED3 on population in South Yorkshire authorities, 2020*

that whilst there are some very slight negative impacts on parts of West Yorkshire, the gain to South Yorkshire is greater, i.e., the gains to South Yorkshire are largely at the expense of other areas rather than of West Yorkshire.

The growth in employment in Doncaster makes a slight improvement to the relative employment prospects there, and this, as already noted, leads to some changes in migration patterns. The overall population impacts by 2021, inclusive of migration effects, are shown in Figure 11.9. Whereas the largest change in employment was in Doncaster, followed by Sheffield and Rotherham, the largest change in population is in Rotherham, with half as much growth in Doncaster and a negative effect on Sheffield.

The localised differences between the employment and population impacts illustrate the scope for transport improvement to modify commuting patterns as well as economic activity. One can see that there is a risk that measures which help to capture additional employment may also promote an exodus of population, and in a small way that is happening to Sheffield in this particular case.

It has to be kept in mind that the impacts considered here are very slight when considered as percentage changes though, as one would expect, there is greater variation between individual zones than between local authorities. This is illustrated in Figures 11.10 and 11.11, which show the percentage impacts of the ED3 test in 2021 on employment and population. The impacts of this particular test are small enough

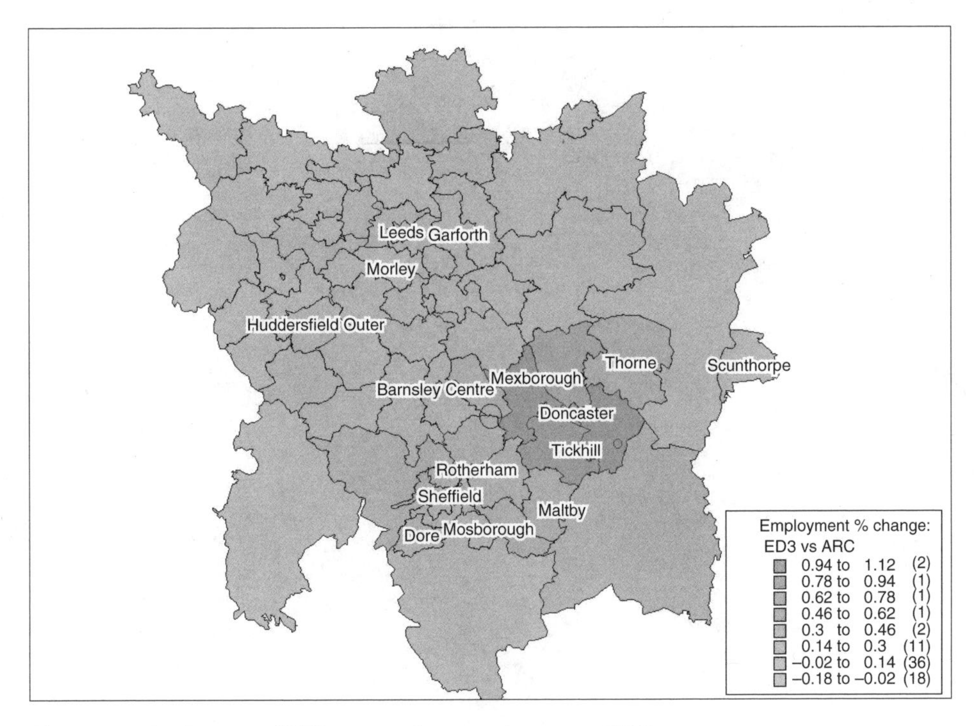

Figure 11.10 *Impact of ED3 on employment by zone, 2020*

that there are no significant impacts either in the land-use impact (little change in residential or commercial rents) or in transport (little additional congestion due to the growth in economic activity and population). In other tests, however, we found that these induced effects could be significant. Indeed we found that land-use policies aimed at encouraging growth in certain locations without complementary transport measures to provide the necessary capacity could actually have negative effects, that is, could actually have the effect of discouraging growth in their target locations.

11.7 Discussion

11.7.1 The Role of the Model in the Study

One of the objectives of using the strategic land-use/transport interaction model in SWYMMS was to provide a formal, quantified analysis of the likely impact of transport and other schemes intended to promote economic regeneration. As we have seen in the ED3 results discussed above, the strategies designed to improve access to regeneration areas and to relieve congestion in their vicinity would have very slight but generally positive impacts in South Yorkshire (particularly the Doncaster area), whilst being neutral for West Yorkshire. Whilst the impacts identified are very small percentage changes from the underlying reference case, it is worth remembering both

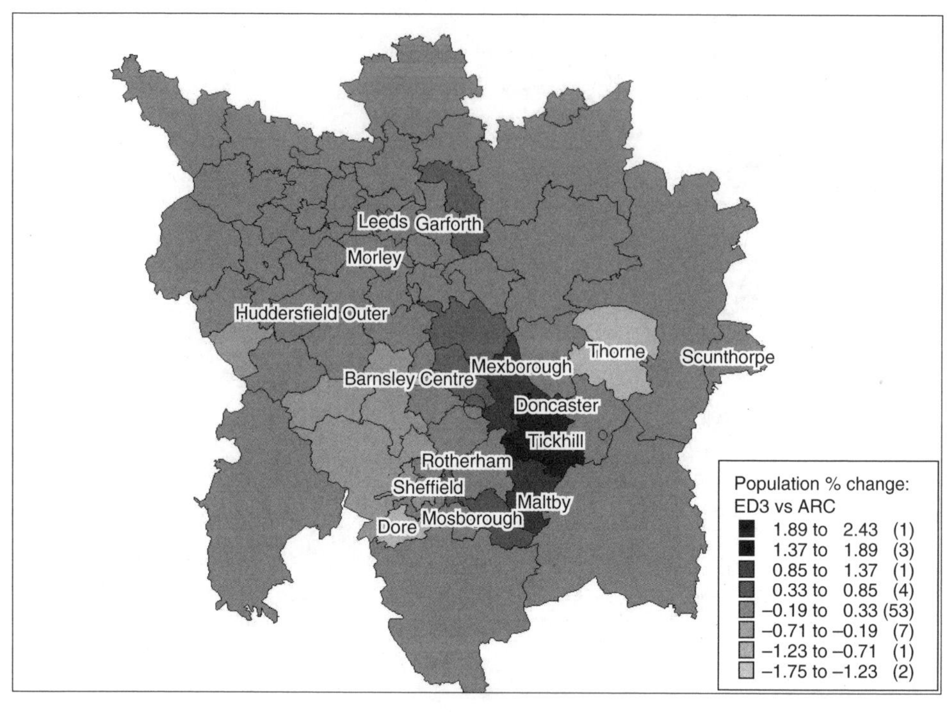

Figure 11.11 *Impact of ED3 on population by zone, 2020*

- that the formal analysis of such impacts may serve as much to guard against exaggerated claims about the economic consequences of transport strategies as to support reasonable claims (whether the consequences are expected to be positive, as in the ED3 case, or negative as might be expected for some demand management measures); and
- that in many cases the small percentages nevertheless represent politically significant numbers of people and jobs.

It is also appropriate to note that the impacts predicted by the land-use/transport interaction model do not always go in the expected direction. An example of this arose in testing another strategy within SYWMMS, which involved substantial (and entirely hypothetical) incentives to the development of retail, office and industrial floorspace in town/city centres and selected Strategic Economic Zones of South Yorkshire. (Details can be found in the SWYMMS Scenarios and Strategies Report, as test ED6.) No other changes were introduced. The development incentives did lead to significant extra construction in the zones affected, but, to the initial surprise of the modelling team, the impact on total employment in South Yorkshire was negative. Detailed examination of the model outputs showed that this was due to

- much of the additional development in South Yorkshire being in congested locations;

- businesses relocating into the additional floorspace thus moving into congested locations and contributing to making congestion even worse (with no provision in the strategy for transport improvements); and
- the resulting increases in congestion, and the increases in the proportion of activity in the most congested locations, making South Yorkshire slightly less competitive as a location for production and slightly less attractive as a location for investment.

Other planners with whom we have discussed these results have described them as an entirely unsurprising and intuitively obvious example of what is likely to happen if land-use and transport plans are not properly integrated!

11.7.2 Model Development

Without embarking on a general review of land-use/transport interaction models, or of the SWYMMS project, it is appropriate to note some of the technical characteristics of the modelling process adopted, and to consider how these are being developed in subsequent projects and may develop further in future.

As noted earlier, the strategic transport model was supported by more detailed highway and public transport models, which were used both to prepare the aggregated descriptions of transport supply used in the strategic model and as the modelling tools for more detailed analysis of highway and public transport impacts in the subsequent development of the plans through which the preferred strategy may be implemented. The possible integration of the detailed and strategic levels of modelling into a single model system, combining the land-use and time-of-day responses currently unique to the strategic model with the more detailed zone system and network representation of the detailed models, is one of the possibilities for future application. This would of course be more demanding of computer resources, but might be justified in terms of avoiding the need for complex interfaces between the two levels, and in terms of the benefits of having a finer zoning system in the land-use/transport interaction modelling. To some extent this approach is already being adopted in work by the authors' firms for the Central Scotland Transport Corridor Studies, where a land-use/transport model with over 700 zones is being used.

Another important aspect is the treatment of land-use planning in the modelling system. The treatment of planning policy and of the planning process in land-use/transport interaction models is typically one of extreme simplification, requiring inputs which tend to imply a very simple 'zoning' system of planning rather than the more subtle approach which operates in the UK. Current developments of the DELTA model are seeking to improve on this, particularly to represent the ways in which the planning process can influence the type of development within broad categories such as 'housing', and in particular to represent the possibility that planning activities in regeneration areas can introduce developments radically different from the existing stock (such as new luxury housing in decaying inner-city industrial areas). More broadly, attention is also being given to the possibility of modelling the planning process as a responsive system working within a quantified policy context, rather than simply quantifying the expected outputs of the planning system – the amounts of development which will be permitted – as is done at present.

11.8 Concluding Comment

We believe that the SWYMMS strategic model has made a valuable contribution to the study for which it was commissioned, which may well continue to be used on behalf of the various regional and sub-regional planning bodies in relation to other strategic issues. It is representative of the state-of-the-art in forecasting land-use/transport interaction for major UK studies, but further developments in this field can be expected and indeed are already emerging.

Acknowledgements

The authors are grateful to the numerous organisations which contributed to the study and to the work described here; to the editors for the invitation to publish this paper, and to the Government Office for Yorkshire and the Humber for permission to do so; and to their respective colleagues under the overall direction of Dr Denvil Coombe for their strenuous efforts in data collection and in the development and operation of the model. The authors remain responsible for this description of the model and of the use made of it.

The views expressed are not necessarily those of the Government Office or of the Secretary of State for Transport, Local Government and the Regions.

References

Copley, G., Skinner, A., Simmonds, D. and Laidler, J. (2000) *Development and Application of the Greater Manchester Strategy Planning Model*, European Transport Conference, 2000, PTRC, London.

Department for Transport (1998) *A New Deal for Trunk Roads in England*, http://www.dft.gov.uk/itwp/trunkroads/.

Roberts, M. and Simmonds, D. (1997) A strategic modelling approach for urban transport policy development, *Traffic Engineering & Control*, **38**: 377–84.

Simmonds, D. (2001) The objectives and design of a new land-use modelling package: DELTA, in Clarke, G. and Madden, M. (eds) *Regional Science in Business,* Springer, Berlin.

Simmonds, D. and Skinner, A. (2001) Modelling land-use/transport interaction on a regional scale: the Trans-Pennine corridor model, *Proceedings of the European Transport Conference 2001*, PTRC, London.

Simmonds, D. and Still, B. (1999) DELTA/START: adding land use analysis to integrated transport models, in Meersman, H., van de Voorde, E. and Winkelmans, W. (eds) *Proceedings of the 8th World Conference on Transport Research, Volume 4: Transport Policy*, Pergamon, Amsterdam.

12

The Relocation of Ambulance Facilities in Central Rotterdam

Stan Geertman, Tom de Jong, Coen Wessels and Jan Bleeker

Abstract

This chapter considers an important strategic planning question – the relocation of emergency service facilities – as the focus for applied research and, in so doing, exemplifies the reciprocal relationship between academic (or fundamental) and applied research in the field of geo-information and communication technology (geo-ICT). The underlying indicators of accessibility used in the application are introduced initially and a set of optimal locations of ambulance facilities in the study area of central Rotterdam are determined for the client using a planning support tool known as *Flowmap* in conjunction with a proprietary GIS. Some further experiments to improve the solution are also presented.

12.1 Introduction

Since the early 1990s, interest in information systems for storing, managing, processing, analysing and displaying geographical data has boomed. The last decade has witnessed some new areas of application for this 'geo-information and communication technology' (geo-ICT). The expansion started in sectors such as defence, forestry and mining and has extended into geo-marketing, insurance, car navigation and telecommunications. Moreover, geo-ICT expanded in size, functionality and diversity (Tomlinson, 1987; Burrough, 1986; Ottens, 1993; Steinitz, 1993). Increasing numbers of commercialised information systems were put on the market and new systems, or updated versions of old systems, have been developed, offering novel

Applied GIS and Spatial Analysis. Edited by J. Stillwell and G. Clarke
 ISBN: 0-470-84409-4

capabilities (e.g., Internet-GIS) as well as traditional functions in a way not previously available (e.g., three-dimensional mapping/animation) (Geertman, 1999).

Research activities within the field of geo-ICT can be broadly divided into those that are fundamental and those that are applied. Fundamental research is frequently referred to as 'blue skies' research and is often undertaken in an academic context. Examples of fundamental research include studies of ontological and/or semantic issues (e.g., Winter, 2001), advances in the use of open technologies (e.g., OCG, 2002), the creation of new error propagation methods (e.g., Burrough and McDonnell, 1998), and the development of geographical modelling methodologies such as those based on artificial intelligence (Openshaw, 1997; Fischer, 1999). Applied research involves the application of these new technologies in real-world contexts, frequently in public sector planning at local or regional level. Whilst the distinction between fundamental and applied research is drawn for purposes of clarification, most institutions involved with geo-ICT related research undertake a mixture of both types of activity. This has led to the critical view amongst commentators in the higher education sector that there is a growing amount of research, which is money-driven, and that fast, clear-structured and short-term applied research is taking priority over more fundamental longer-term research, which is suffering as a consequence. However, there are some advantages of applied work, not least of which is the necessity to develop policy-relevant tools for understanding problems and providing solutions. Consequently, an appropriate combination of more theoretical fundamental research activity and more applied developmental work seems desirable. At Utrecht University in the Netherlands, for almost two decades now, various research groups have been working with geo-ICT. Although some of these are more oriented towards fundamental research and others towards applied work, in general all groups show elements of both and, in certain cases, their research activities have resulted in the commercialisation of geo-ICT products (such as *Flowmap* and PCRaster) and in the creation of (quasi-) private organisations (such as Nexpri and Carthago).

Flowmap, for example, is a stand-alone software package dedicated to the analysis of spatial interaction data, data that describes movement of people, goods, energy or information through space on a regular or irregular basis usually via some form of transport network. An example of regularly occurring interaction is commuting whereas irregular interaction might be residential migration. *Flowmap* has capabilities for the selection and display of this kind of data, together with various methods and models to analyse historical flows and predict potential interaction. These latter capabilities make *Flowmap* a very suitable tool to incorporate in planning support systems, especially in the field of facility/service localisation (Geertman *et al.*, 2002). Although *Flowmap* has many GIS functions, it is not designed as a full GIS package, but rather as an extension for specific spatial analyses that are difficult or impossible to realise within mainstream GIS packages.

Nexpri, the Netherlands Expertise Centre for Geographical Information, can be regarded as an example of a spin-off organisation. It began its existence in 1989 as a subsidy-based institution related to five Dutch universities, but in 1996 it became an entirely market-oriented organisation related to Utrecht University and is soon to change its status to that of a private company. Its activities primarily concentrate on two fields of applied geo-ICT research: one concerns the organisational issues related to the implementation and functioning of geo-ICT in organisations; the other concerns different kinds of location-oriented research questions such as the identification of the best location for a new dealer in the branch network. There is a strong interrelationship and active interaction between the research groups at the University and the applied research undertaken by Nexpri.

In this chapter we discuss one particular example of the benefits of this reciprocal interrelationship between fundamental and applied Geo-ICT research and endeavour to demonstrate how the two-way interaction between the two research communities can be of considerable advantage to both. The example involves an important strategic planning question – the relocation of ambulance facilities. Nexpri was contracted by the Rotterdam Municipal Health Authorities to undertake a study of the provision of ambulance services in the Rotterdam Harbour area and to identify how the location of ambulance stations might be altered to improve the efficiency and effectiveness of the service.

Initially, attention is focused on some of the underlying concepts related to the research question in general terms, i.e., how can facility siting be performed and how to measure accessibility. Thereafter, the real-world planning question will be elaborated in more detail, followed by an explanation of the way in which the *Flowmap* software has been applied to solve this planning question. Subsequently, reflections on the application will be presented and recommendations made about future applications of this instrument to these kinds of research questions. Finally, some remarks are provided on the experience of working in this reciprocal interrelationship between fundamental and applied geo-ICT research.

12.2 Facility Siting with Accessibility Analysis

12.2.1 Introduction

Facility siting is a form of applied research aiming at finding the absolute or relative optimal location or set of locations for a predefined facility or group of facilities. Facility location is closely related to the concept of geographical accessibility that can be defined as the effort taken to get to some place and the level of acceptability to the target group. So the concept of accessibility incorporates the transport link between origin and destination and the ability to travel by the target group, as well as the characteristics of the destination and the objective of the trip. For example, what is accessible by private car may be inaccessible by public transport; what is within an accessible walking distance for younger people may be effectively inaccessible for senior citizens; what is an acceptable effort to travel to buy antique furniture is likely to be unacceptable when shopping for daily groceries.

Moseley (1979) has put the concept of accessibility into a schematic framework in which activities require transport and transport involves people; from this it follows that actual accessibility depends on the properties of each of the components in this scheme. However, accessibility can be measured in many different ways. Robinson (1990) distinguishes between composite measures, comparative measures and time–space approaches. The time–space approaches are most useful for studies based on the perspective of the individual: 'Is service C accessible for a person living in place P with mode of transport T?' The other approaches are more useful for measuring aggregate accessibility. Composite measures are a combination of several different individual measures like 'How many buses arrive before 8.00 a.m.?' or 'Is there a Sunday bus service?'. Comparative measures also take differences into account between social groups like car users or public transport users. In addition the comparative measures can be further subdivided into potential and actual accessibility measures (Hilberts and Verroen, 1992). Moreover, accessibility can be measured for many individual locations or these measures can be combined into an aggregate measure for a whole region. The latter is relevant, for instance, when one wants to measure the effect of infrastructure changes on the accessibility of (or inside) a larger region. Finally, accessibility measures can be based upon a single-purpose trip as presented in the scheme above or on multipurpose trips (also known as combination trips) where for instance the accessibility of a shopping centre is based on its location relative to the home–work trip of the customer (Dijst and Vidacovic, 1995).

Despite its geographical dimension, facility siting is not a field of study restricted to human geography. Other disciplines, such as operations research, have made contributions too and still do. Therefore, before discussing the methodology of accessibility analysis in depth, we discuss some alternative approaches for research on facility siting: the optimisation approach from operations research; the buffer/overlay approach from geographical information science; and the values of accessibility potential that have their roots in quantitative geography.

12.2.2 The Optimisation Approach

Whenever a facility-siting project is considered, there is always the question of whether or not it really involves an optimisation problem and requires an optimisation approach. This is because there are several reasons why such an approach is not appropriate (Floor *et al.*, 2001):

- not all factors of relevance in a public planning or policy environment can be included in an optimisation approach;
- the typical 'hard' and 'singular' result of an optimisation approach often is very difficult to incorporate in a public planning or policy environment;
- optimal results can be highly sensitive to the choice of spatial aggregation and also to temporal changes; a good example of this is the sensitivity to the modifiable area unit problem (MAUP) as described by Fotheringham *et al.* (1995);
- there is not necessarily one optimal solution, but many different optimal solutions depending on the chosen target functions (like minisum or minimax);
- although exact solution methods are known for these target functions, for most practical applications they require a prohibitive amount of computer memory or time, in which case usually some approximation method has to be used;
- it is very difficult to optimise using transport network-based impedance functions instead of airline distances. The traditional optimisation approach is exemplified by the ALA and NORLOC algorithms described by Cooper (1964) and Törnqvist

et al. (1971) for locating services in a plane and Maranzana (1964) for locating services on a transport network. There is no known published example of locating services in a plane using distances via a transport network.

In a recent publication (de Jong *et al.*, 2001) we argue that the only useful part optimisation techniques can play in a public planning or policy environment is to provide a yardstick against which the outcome of a project can be measured in terms of accessibility.

12.2.3 The Buffer/Overlay Approach

In an earlier study (de Jong and Ritsema van Eck, 1997), the traditional GIS approach was outlined. It mainly involves the use of buffer techniques to determine a straightforward 'accessible' area followed by overlay techniques to clip out this buffer area and to present the remainder as the 'inaccessible' area. This approach can be criticised for a number of reasons. Firstly, transport networks are not taken into account. Not only does travel speed differ depending on mode of travel and network quality but also natural (waterways, steep slopes) or man-made (railway tracks, airfields) barricades will have a diminishing effect on accessibility. Secondly, all sites within the buffer zones have similar weight, so a location with all opportunities at the maximum allowed distance scores just the same as a location with all opportunities at minimum distance. Depending on the type of facility siting that is being undertaken, other shortcomings of the buffer/overlay approach may also be of relevance: competition with already established service outlets is not taken into account; and depending on the subject of analysis, sometimes the limited capacity of service outlets cannot be ignored.

12.2.4 Potential Values

To overcome some of the shortcomings of the GIS approach, Geertman and Bosveld (1990) used the potential value technique based on a transport network instead of using buffers. This can be seen as a first attempt to incorporate some accessibility analysis into GIS. Potential values are the best-known measure for the nearness of clients/customers/potential users and were first introduced by Hansen (1959) to describe accessibility. Potential values can be calculated as follows:

$$Pot(i) = \sum_j D_j \cdot C_{ij}^{-\beta} \tag{12.1}$$

in which:

$Pot(i)$ is the potential value for client site i;
D_j is the weight (size or number) of services provided for service location j;
C_{ij} is the distance between origin site i and destination site j; and
β is the distance decay parameter.

This equation can be used to find good sites for new residential areas, for example, based on the number of jobs offered at the destination locations. Sites relatively close to a large D_j score a higher potential value than sites further away. However, being relative measures, potential values are difficult to interpret. This problem can be overcome by calculating two sets of potential values where the distance decay parameter differs by exactly one. After dividing the outcome of the one set of values by the other the result will be an accessibility measure in the form of a weighted average distance for each origin site to all destination sites (Geertman and Ritsema van Eck, 1995).

Also in practice we see that, because of the ease with which potential values can be calculated within a raster-based GIS, the measure is often calculated by using just the geometric distances between cell centroids. As these distances are comparable to the simplest form of airline distances, the first criticism mentioned above regarding the buffer/overlay approach is in this case valid too. There is no satisfactory way in raster-based GIS to handle the effect of the discontinuous aspects of travel via a network, like multilevel overpasses and limited access to highways. Therefore, in our opinion, raster-based GIS are unsuitable for any realistic analysis involving calculations of effects of travel via transport networks.

There are some other criticisms of the potential value approach that are harder to overcome. In gravity modelling the distance decay function can be calibrated using the (log) mean trip length to represent distance behaviour correctly. A similar mechanism is not available when it comes to the calculation of potential values (often simply a power function is selected with distance decay parameter 2 in an unfounded analogy with Newton's Law of Gravity). Moreover, the results are highly sensitive to opportunities at extremely short distances (Ritsema van Eck and de Jong, 1996). As these particular distances are often arbitrary (intrazonal distance estimation or depending on the location of the centre point that represents a whole zone), this renders any outcome potentially unreliable.

12.2.5 The New Methodology of Accessibility Analysis

A comparative potential accessibility measure that overcomes most of the criticisms mentioned above can be developed. Moseley's schematic framework involving population linked to activities by a transport network can be redefined in GIS terms as origin locations linked to destination locations by transport network distances. Based on this adapted scheme, we can use GIS to calculate, for each destination location, which of the origin locations are within reach (and sum up the selected attribute values of those locations) and vice versa. This creates an accessibility measure called the proximity count or cumulative opportunity index. The proximity count is a simple count of locations that are within a predefined distance range. Instead of counting the number of locations, a weight variable for each location (such as the number of jobs) can be summed. For each site, all locations are counted that meet the distance requirement:

$$PCount(i) = \sum_{j^*} D_j \quad (j^* = \text{over all } j \text{ such that } C_{ij} < \text{Range}) \qquad (12.2)$$

where $PCount(i)$ is the proximity count for origin site i.

A location or location related value is not uniquely allocated to any site. In fact, locations can be counted for more than one site. The proximity count therefore gives an indication of the potential (job) market not taking any competition effects into account. It can be used to determine which sites reach above certain (minimum) threshold values. Of course the choice of distance is critical: if it is set too large then all origins will be counted for all destinations and all proximity counts will end up equal. As an alternative, the threshold value can be preset and the threshold distance can be calculated as the minimum distance at which enough locations are to be reached (de Jong and Ritsema van Eck, 1997). Proximity counts and threshold distances are especially useful for determining whether or not the potential market size of a proposed site (service or retail) is large enough (de Jong *et al.*, 1991). In the next section, we exemplify this with the help of a real-world strategic planning question, the relocation of ambulance facilities.

12.3 The Relocation of Ambulance Facilities

In the Netherlands, as in many other countries in the western world, increasing the efficiency of public services is one of the main solutions for handling decreasing funds for public facilities. One of the areas in which the cutback in public funds is translated into a quest for improved efficiency is in the number and the location of ambulance service stations. The Dutch national government is striving to unify smaller service areas into larger service regions. This forces different ambulance services to collaborate, which in turn leads to a more efficient and effective use of ambulance provision, and a reduction in costs. One of the newly formed service regions is the ambulance service region of the Rotterdam Harbour area (Figure 12.1).

Ambulance provision is divided into three categories. The first category consists of the urgent journeys (so-called A1 journeys) where, according to the law, an ambulance has to arrive at its destination within 15 minutes of the Central Post for Ambulance (CPA) journeys receiving a call for help. The second category involves less urgent journeys (A2 journeys) where there is no strict time limit other than to arrive 'as quickly as possible'. The last category consists of so-called 'ordered' journeys (B journeys) consisting mainly of pre-arranged journeys between hospitals and pre-arranged journeys of patients from home to hospital and vice versa (e.g., for radiation therapy). In 1998, 74478 ambulance journeys occurred in the study area of which 45% belonged to the A1 category and 41% to the B category.

The existing ambulance services and the CPA asked the Municipal Health Services of the Rotterdam Harbour area to investigate which ambulance stations in the Rotterdam Harbour area should be established within the new configuration and to determine how many ambulances would be needed in the area to fulfil the requested ambulance provision. To answer this question the Municipal Health Services contracted Nexpri to undertake an analysis of the effectiveness of current service provision and to recommend how ambulance services within the Rotterdam Harbour area might be improved.

The locations of the 16 existing ambulance stations and the classified total demand for A1 journeys in 1998 for each four-digit postal area in the Rotterdam Harbour

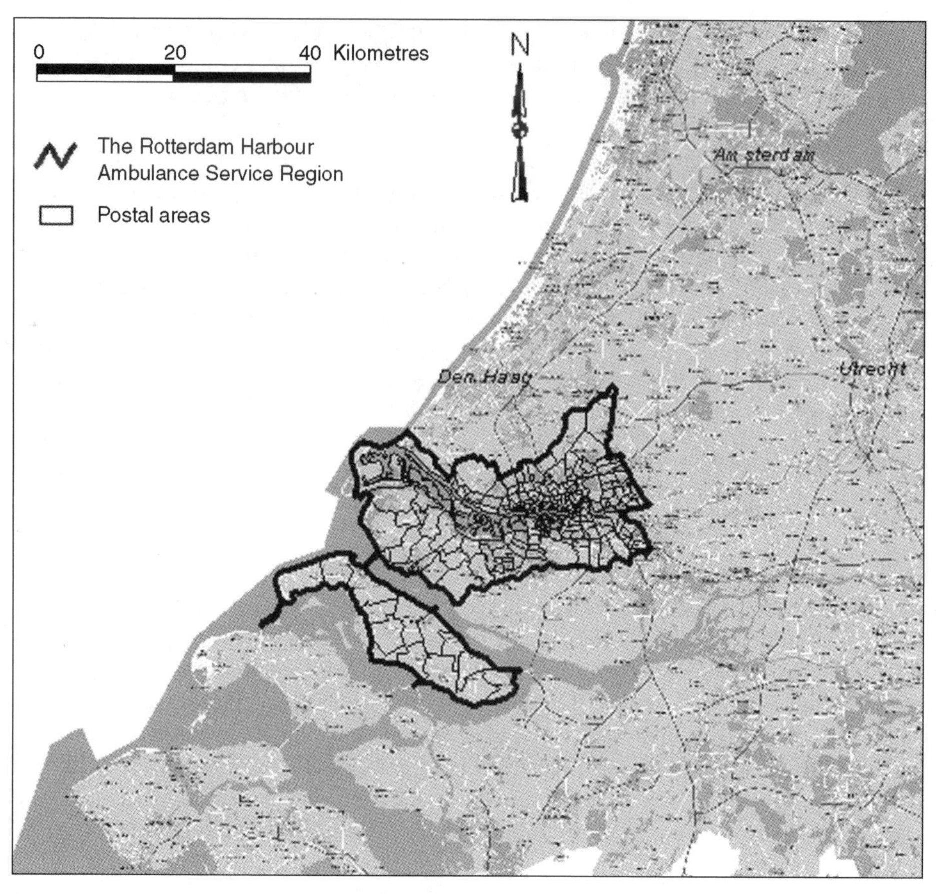

Figure 12.1 Rotterdam Harbour ambulance service region

area are shown in Figure 12.2, indicating the high demand intensity (counted number of calls for A1 journeys) in the centre of Rotterdam (the eastern part of the Rotterdam Harbour area) and decreasing towards the perimeter of the area.

Initially, a measure of the effectiveness of the existing ambulance services provided from the 16 stations was calculated with the help of ArcView and ArcView Network Analyst. The service area associated with each ambulance station was constructed by calculating an 8.5 minute travel time buffer based on the road network. To determine the required locations of stations to be able to reach patients within 15 minutes in response to an A1 call for assistance, a 95% confidence interval was calculated based on the Poisson distribution of A1 arrival times. The upper 95% confidence limit was assigned at the legal norm of 15 minutes. This resulted in a mean arrival time of 8.5 minutes within the confidence interval. The time period of 8.5 minutes takes into account the problem of congestion at peak hours as well as the average delay time (2 minutes) between the incoming call for assistance at the ambulance station and the moment the ambulance begins its journey.

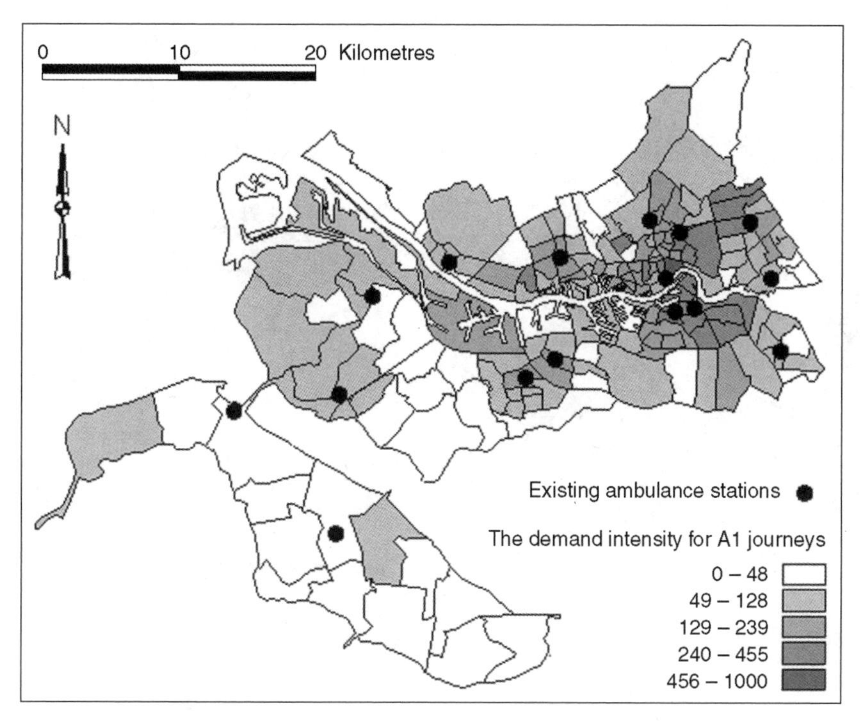

Figure 12.2 *Location of existing ambulance stations and the counted demand intensity for A1 journeys in 1998 per four-digit postal area*

Within the city centre of Rotterdam, the ambulance stations have overlapping service areas. Moreover, the travel time zones indicate that some areas in the more peripheral areas are not covered at all (Figure 12.3). The conclusion of this is that, when taking the threshold value of 15 minutes as a maximum for A1 journeys (8.5 minutes as an average), some parts of the Rotterdam Harbour area are in fact over-served by existing ambulance services and some are under-served. This implies that a more efficient distribution of ambulance services over the Rotterdam Harbour area might be identified.

As a starting point for this redistribution of ambulance station locations, four pre-requisites must be taken into account:

- each part of the Rotterdam Harbour area is potentially suitable as the location of a new ambulance station, so no areas are excluded initially;
- the new locations for the ambulance stations should be situated as close to demand as possible;
- the overlap of service areas of the ambulance stations should be minimised; and
- the total range of the new ambulance stations should cover at least all the urban areas within the Rotterdam Harbour area and preferably also the remaining rural areas within this region.

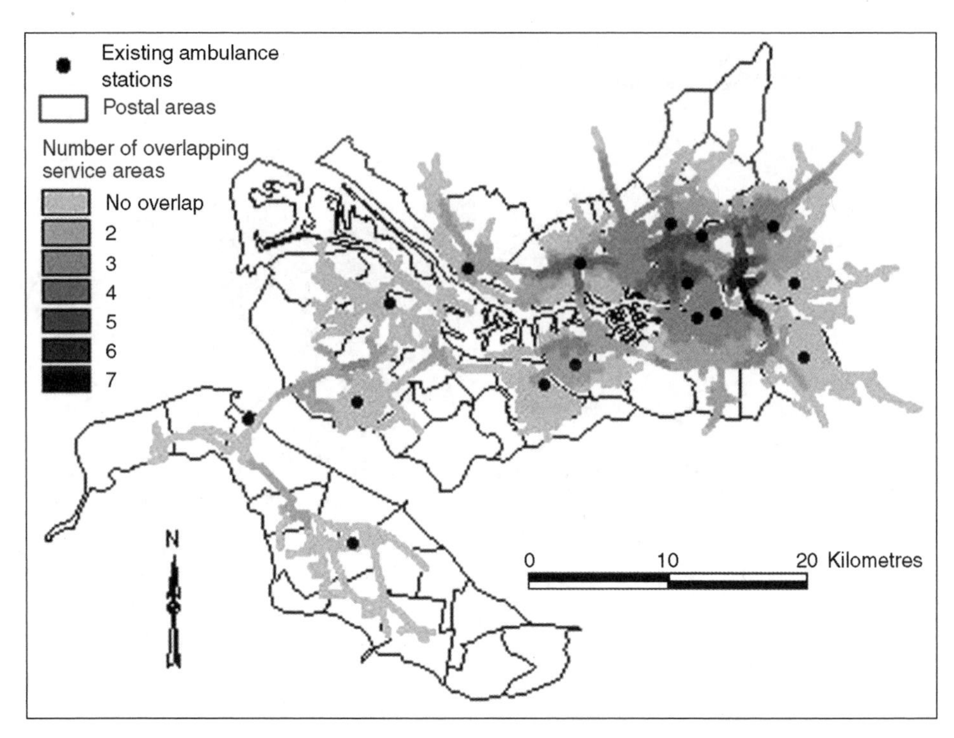

Figure 12.3 *Service areas for current ambulance stations*

Taking these prerequisites into account, network-based analyses have been performed with the help of ArcView and *Flowmap*. ArcView has been used primarily for database management, map-making, and the requisite analytical GIS functions of buffering and network analysis. *Flowmap*, on the other hand, has performed all the additional network calculations and measures that cannot be executed within the proprietary GIS (for additional information concerning *Flowmap*, see http://flowmap.geog.uu.nl). The demand for ambulance facilities is represented by the intensity of requests for ambulance service for each individual postal area. Moreover, use has been made of a hexagonal grid, corresponding in size with the average four-digit postal area. The grid cells represent potential new ambulance locations in the Rotterdam Harbour area. The choice for a hexagonal grid was made because the distance from the centroid of a square grid to the centroids of its neighbouring square grids is different (the diagonal is longer than the vertical or the horizontal), while with a hexagonal grid this distance is equal. In addition, a detailed road network has been used that links all the four-digit postal areas and hexagonal grid cells. Each of the component parts of the road network possesses a travel time, which also takes into consideration the delay time as a consequence of congestion.

Two accessibility measures, the proximity count and the proximity coefficient, have been computed for each of the hexagonal grid cells using *Flowmap*. The proximity count gives an indication of the demand for ambulance services within a user-defined

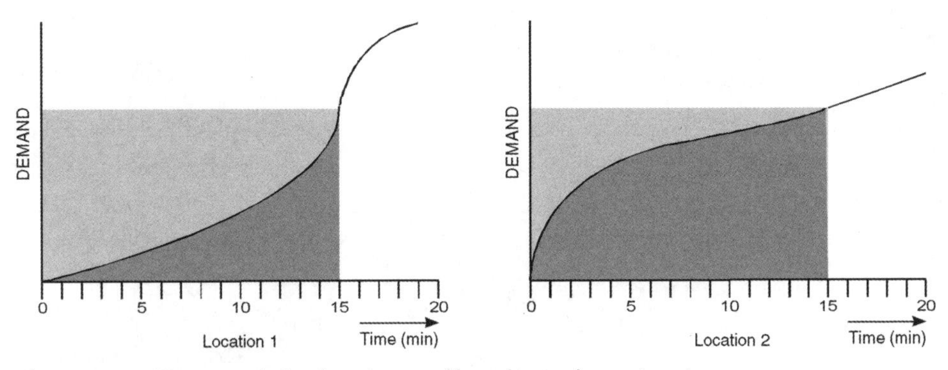

Figure 12.4 *The cumulative location profiles of two alternative sites*

travel time from each hexagonal cell. In this case, the proximity count indicates the intensity of requests for ambulance services within an average travel time of 8.5 minutes. This proximity count can be presented in a map, but also in a so-called location profile (Figure 12.4). A location profile for a potential ambulance station indicates the distance (in this case expressed in travel time) between that location (hexagonal cell) and the cumulative requests for the ambulance service. In fact, the shape of the location profile provides some information on the accessibility of each specific location. A convex profile (as shown in location 2 in Figure 12.4) indicates that the site is relatively close to the demand for ambulance services; a concave location profile (as shown in location 1) indicates that the site is relatively far away from the location of demand for the service.

This relationship can also be expressed in one measure, a so-called proximity coefficient, calculated by expressing the area under the graph (dark shaded area) as a percentage of the total area (dark and light shaded areas). Hexagonal grid cells with high proximity coefficients possess high accessibility to demand and are therefore very suitable sites to locate ambulance stations. The inverse is true for grid cells with low proximity coefficients. Figure 12.5 shows the proximity coefficients – classified on natural breaks – for each of the hexagonal grid cells within the Rotterdam Harbour area (in fact it appears that other classification methods in ArcView showed similar results).

Based on the proximity count maps and location profiles, the optimal future locations of ambulance stations were assigned by trial and error. The proximity coefficients were used to locate the ambulance stations so as to maximise accessibility and minimise the overlap between service areas. It appears that in and near the city centre of Rotterdam these coefficients possess a high value, so represent sites with good accessibility, as we would expect. The siting of ambulance stations in the (south) western part of the Rotterdam Harbour area is mainly driven by the achievement of the fourth prerequisite: that the range of the new ambulance stations should cover at least all the urban areas within the study area and preferably also the remaining rural areas within this region. Figure 12.6 illustrates the proposed sites for the ambulance stations.

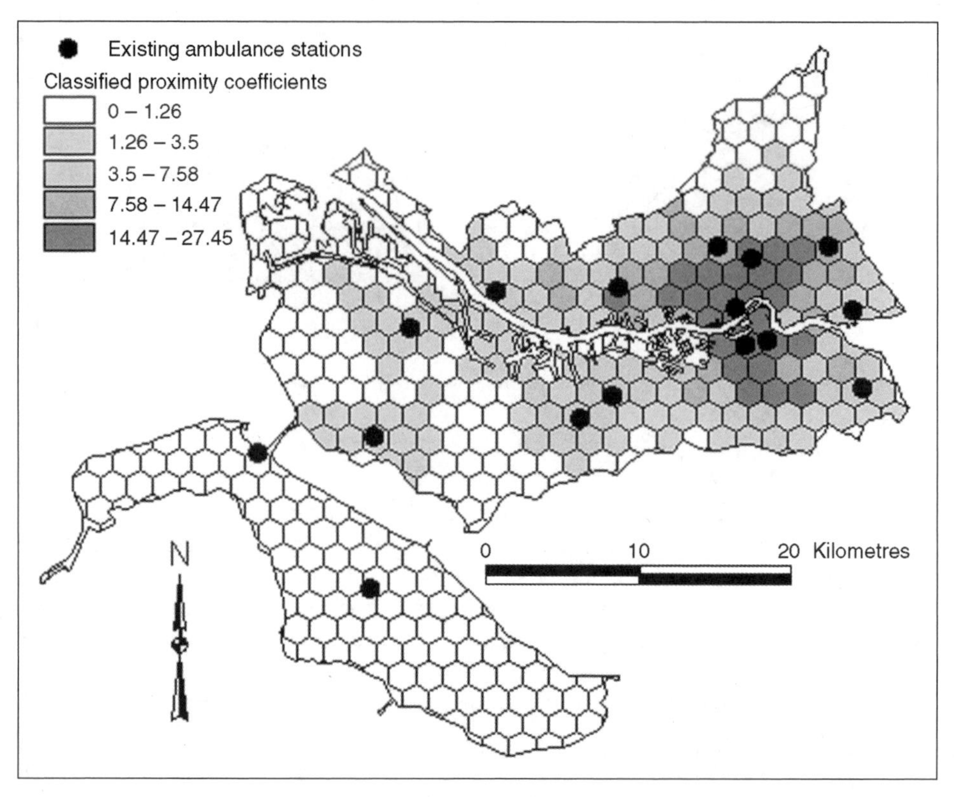

Figure 12.5 *Classified proximity coefficients for hexagonal grid cells in the study area*

This allocation process resulted in a total of 13 newly located stations (three stations less than the initial situation), with a reduction in the level of over-provision (less overlapping service areas) and an improvement in under-provision (fewer areas that do not fulfil the threshold of 15 minute A1 journeys) (Figure 12.7).

The results of this analysis were delivered to and accepted by the Municipal Health Services at the end of 2000. The existing ambulance services, the CPA and the Municipal Health Services of the Rotterdam Harbour accepted the results of the study. The political debate about the reconstruction of the ambulance service network is still in progress.

12.4 Further Developments

The analysis conducted for the client shows the solution of a real-world planning problem by applied geo-ICT research with the help of more fundamental research than had been required to create *Flowmap* in the first instance. Although acceptable to the Rotterdam Municipal Health Services, the research undertaken was not particularly innovative and begs the question of whether it is possible to improve on the proposed facility locations.

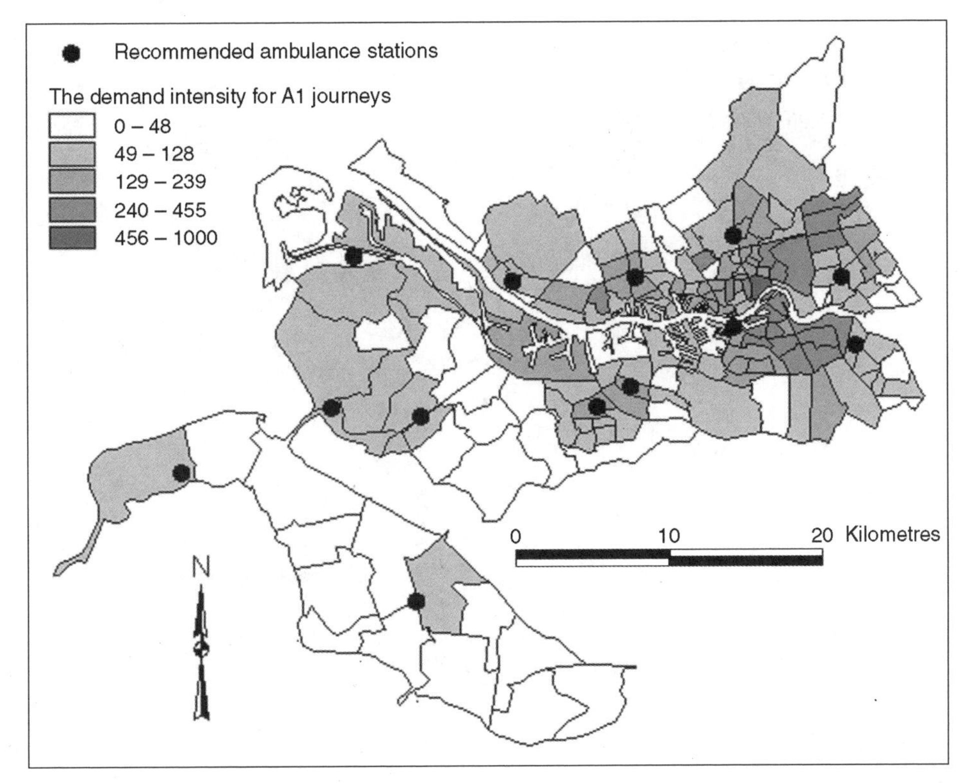

Figure 12.6 *The proposed sites for ambulance stations*

One development is to make use of the so-called 'alternate' algorithm to examine the distribution of ambulance stations. This alternate algorithm is an approach to optimisation that can be adapted to a GIS environment when dealing with the impedance associated with transport networks. As such it can be considered a combination of different approaches to the problem of facility siting. The basic principle of this adaptation is to 'tessellate' space into a large number of cells and to calculate the optimisation criterion for all the cells. The most extreme value found in a catchment area is the best location for a new site. Where the best site differs from the current site it replaces the current site. Thus, new catchment areas and best sites should be calculated until no more substitution takes place. This procedure does not guarantee a true optimal solution; there is some sensitivity to the MAUP problem and the procedure may end in a suboptimal solution. However, in general, it gives a good indication of what can be achieved through optimisation.

When performing an optimisation, it is necessary to be aware that there is not just one optimal solution, but there may be several solutions, depending on the 'target' function that is to be minimised. In the case of ambulance stations, for example, one can handle at least two different target functions. On the one hand, there is a good case for minimising the mean unweighted travel cost, ensuring the best accessibility for the whole area. On the other hand, one can minimise the mean travel cost

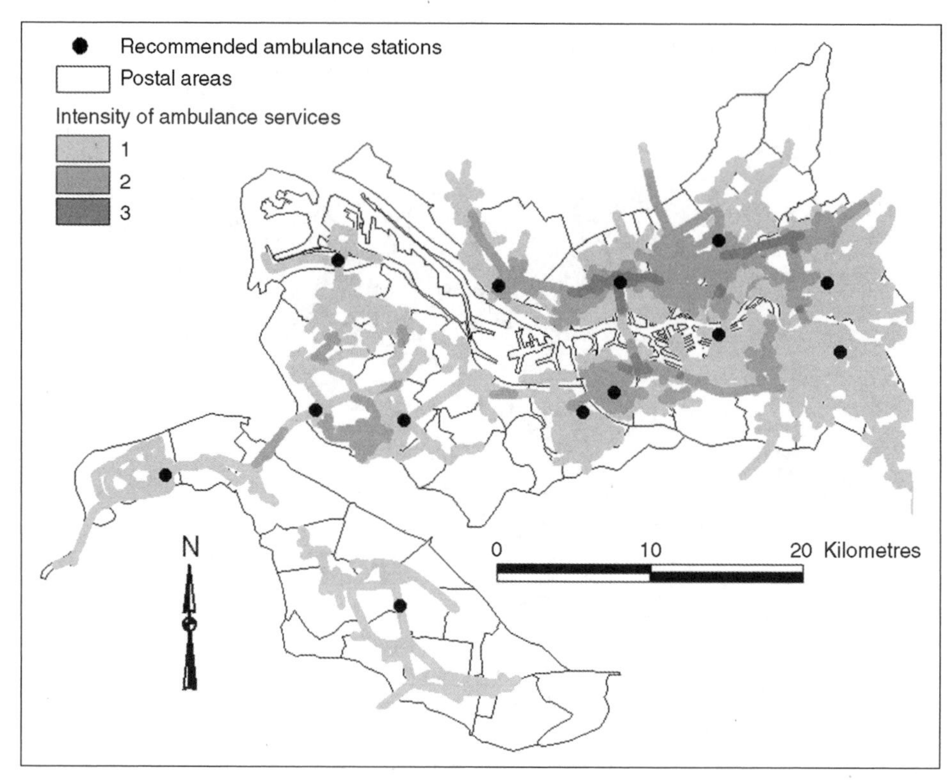

Figure 12.7 Service areas of proposed ambulance stations

weighted by the distribution of demand, ensuring the best accessibility where demand is highest. In the first case, sites tend to be found towards the centre of their catchment areas. In the second case, sites are more oriented towards the centre of gravity of demand within the catchment area. These two points may very well not overlap and as a consequence the choice of target functions can result in very different solutions.

The consequences of handling different target functions can be exemplified using the problem of siting ambulance stations. We have divided the study area into almost 1400 hexagonal cells each with a radius of 50 metres. Each hexagon has been allocated to the nearest of the proposed 13 ambulance stations. The study area is thus divided into 13 catchment areas as shown in Figure 12.8a.

Based on these initially proposed 13 catchment areas, some access time statistics have been calculated (Table 12.1). These statistics represent the base position shown in the second column. If the target criterion is to minimise the mean area access time without weighting, the catchment areas in Figure 12.8b are derived and the access times change accordingly. The mean area access time is reduced from 9.76 to 8.26 minutes by relocating 12 of the 13 proposed ambulance stations. This optimisation procedure can be repeated again but on this occasion under the criterion of minimising the mean travel cost weighted with the distribution of demand (Figure 12.8c).

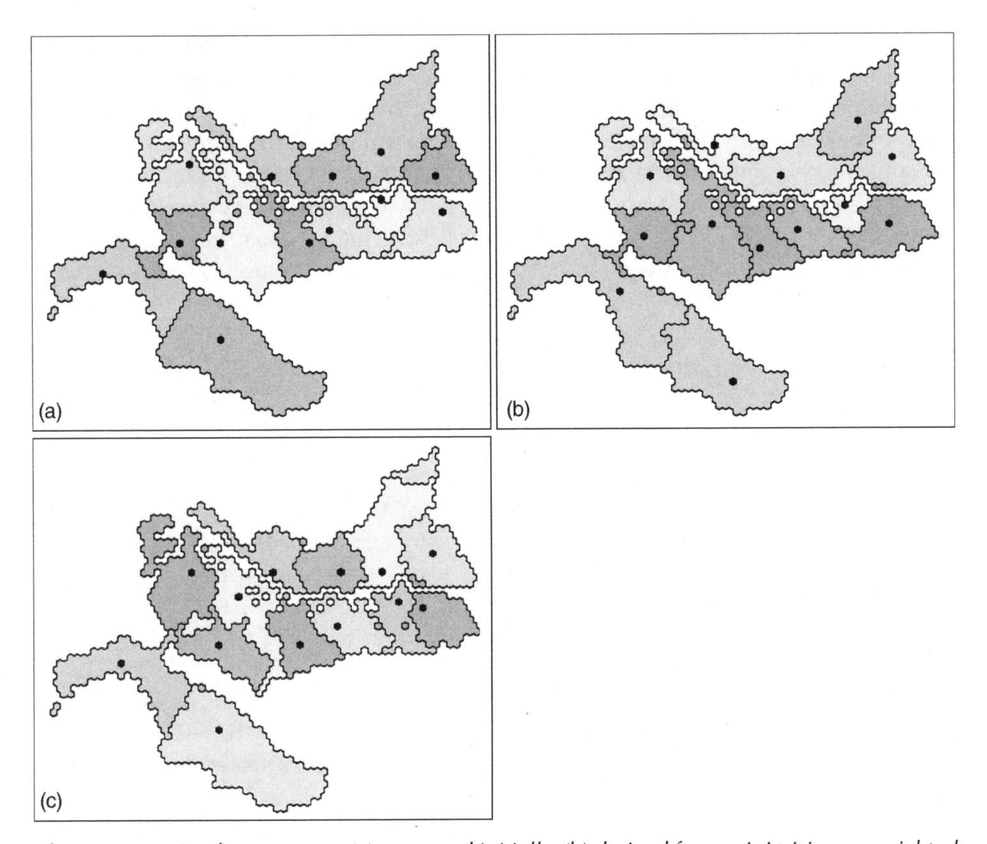

Figure 12.8 *Catchment areas: (a) proposed initially, (b) derived from minimising unweighted mean travel costs and (c) derived from minimising weighted mean travel costs*

Table 12.1 *Access times under different target criteria*

	Initial proposals (minutes)	Minimise mean area access time (minutes)	Minimise mean demand access time (minutes)
Mean area access time	9.76	8.26	9.29
Mean demand access time	7.09	6.70	5.62
Worst-case access time	43.22	38.24	34.59

This ensures the best accessibility where demand is highest. Based on the resulting catchment areas, the mean demand access time is reduced from 7.09 to 5.62 minutes by moving 12 of the proposed 13 ambulance stations.

This approach, using the alternate algorithm, can be applied in all cases where the relocation of service facilities, possibly in combination with an addition or reduction, is at stake. The added value of this kind of analysis is not so much the maps shown in Figure 12.8, but primarily the statistics that go with these maps. Similar statistics can be calculated for all sorts of scenarios based on the relocation of fewer ambulance stations, in which case the optimal statistics provide a yardstick to measure the effect of each of the scenarios. Although of academic interest to continue such analysis until the optimal location has been found, it should be stressed that in practical situations accessibility rarely is the only factor playing a part in the decision-making process. As a consequence, the trend surface that can be constructed around the optimal location is of far greater importance as it can be combined in a multi-criteria fashion with other criteria to arrive at a decision space that provides the decision-maker with some direct insights.

12.5 Conclusions

This chapter has provided a particular example of the benefits of a reciprocal relationship between fundamental and applied research in the context of public service planning. An analytical tool (*Flowmap*) and a proprietary GIS have been used to generate a set of proposals for the relocation of ambulance stations in central Rotterdam. Some additional developments have been reported to show more state-of-the-art capabilities and possibilities.

In relation to the specific policy problem, efficiency has been improved by reducing the number of ambulance facilities from 16 to 13. Moreover, the effectiveness has been improved by a reduction of the areas experiencing either over-provision or under-provision of service. In this way the goals of the applied research have been achieved and have become input into the policy process. The outcome of the political process is still unclear, although the upcoming pressing financial situation undoubtedly will push in the direction of an efficiency operation in which the outcomes of this research will be of help.

As far as the interrelationship between fundamental and applied geo-ICT research is concerned, two final conclusions can be drawn. First, without the input of the former and its outcome (*Flowmap*), the research question would have had to be answered in a far more rudimentary way. In fact, *Flowmap* offered different tools that substantially improved our insight into the planning problem and the generation of a solution. Second, the research performed can only be attained by individuals working closely together on fundamental and applied research. It is the direct contact that is of additional value to both parties and which will result in novel solutions. This collaboration, the solution that has been produced and the insights generated through the research process, give impetus to undertake further applied work to support public service planning.

Acknowledgement

The authors would like to thank the editors of this book for their fruitful comments on an earlier version of this chapter.

References

Burrough, P.A. (1986) *Principles of Geographical Information Systems for Land Resources Assessment*, Monographs on soil and resources survey 12, Clarendon Press, Oxford.

Burrough, P.A. and McDonnell, R.A. (1998) *Principles of Geographical Information Systems*, Oxford University Press, Oxford.

Cooper, L. (1964) Heuristic methods for location-allocation problems, *SIAM Review*, **6**: 37–52.

Dijst, M. and Vidacovic, V. (1995) Individual action space in the city: an integrated approach to accessibility, Paper presented at the conference 'Activity-based Approaches: Activity Scheduling and the Analysis of Activity Patterns', May 25–28 1995, Eindhoven.

Fischer, M.M. (1999) Intelligent GI analysis, in Stillwell, J., Geertman, S. and Openshaw, S. (eds) *Geographical Information and Planning*, Springer, Heidelberg, pp. 349–368.

Floor, H., de Jong, T. and Ritsema van Eck, J.R. (2001) The adaptation of optimization techniques for usage in combination with Planning Support System; including a case study of siting an Islamic secondary school in the city of Utrecht, The Netherlands, Paper presented at the Cupum 2001 Conference, Honolulu.

Fotheringham, A.S., Densham, P. and Curtis, A. (1995) The zone definition problem in location-allocation modelling, *Geographical Analysis*, **27**: 60–77.

Geertman, S. (1999) Geographical information technology and physical planning, in Stillwell, J., Geertman, S. and Openshaw, S. (eds) *Geographical Information and Planning*, Springer Verlag, Berlin, pp. 69–86.

Geertman, S.C.M. and Bosveld, W. (1990) A GIS-based model for goals achievement analysis and design, *Proceedings of the First European Conference on Geographical Information Systems*, 356–65, EGIS, Utrecht.

Geertman, S., Jong, T. de and Wessels, C. (2002) Flowmap: a support tool for strategic network analysis, in Geertman, S. and Stillwell, J. (eds) *Planning Support Systems in Practice*, Springer, Heidelberg, pp. 155–76.

Geertman, S.C.M. and Ritsema van Eck, J.R. (1995) GIS and models of accessibility potential: an application in planning, *International Journal of Geographical Information Systems*, **9**: 67–80.

Hansen, W.G. (1959) How accessibility shapes land use, *Journal of American Institute of Planners*, **25**(1): 73–6.

Hilberts, H.D. and Verroen, E.J. (1992) Het beoordelen van de bereikbaarheid van lokaties. *INRO-TNO Rapport INRO-VVG 1992-21*, Delft.

Jong, T. de and Ritsema van Eck, J. (1997) Threshold surfaces as an alternative for locating service centers with GIS, in *JEC-GI '97 Proceedings. Vol 1*, IOS Press, Amsterdam, pp. 799–808.

Jong, T. de, Ritsema van Eck J. R. and Toppen F. (1991) GIS as a tool for locating service centers, *Proceedings of the Second European Conference on Geographical Information Systems*, EGIS, Utrecht, pp. 509–19.

Jong, T. de, Maritz J. and Ritsema van Eck, J. (2001) Using optimisation techniques for comparison of the accessibility criteria of facility siting scenarios; a case study of siting police stations in South Africa's Bushbuckridge area, in *Proceedings of the 4th Agile Conference on Geographic Information Science*, Masaryk University Brno, Czech Republic.

Maranzana, F. (1964) On the location of supply points to minimize transport costs, *Operational Research Quarterly*, **15**: 261–70.

Moseley, M.J. (1979) *Accessibility: The Rural Challenge,* Methuen, London.

OCG (2002) Implementations of OCG specifications. http://www.opengis.org/cgi-bin/implement.pl.

Openshaw, S. (1997) Building fuzzy spatial interaction models, in Fischer, M.M. and Getis, A. (eds) *Recent Developments in Spatial Analysis. Spatial Statistics, Behavioural Modelling and Computational Intelligence*, Springer, Heidelberg, pp. 360–83.

Ottens, H. (1993) The application of GIS technology in Europe, in Harts, J., Ottens, H. and Scholten, H. (eds) *EGIS '93 Proceedings*, EGIS Foundation, Utrecht, pp. 1–8.

Ritsema van Eck, J.R. and de Jong, T. (1996) Data points in GIS-based network analysis, *Stepro Working Paper*, Faculty of Geographical Sciences, University of Utrecht, Utrecht.
Robinson, G.M. (1990) *Conflict and Change in the Countryside*, Belhaven Press, London.
Steinitz, C. (1993) Geographical information systems: a personal historical perspective, the framework for a recent project, and some questions for the future, in Harts, J., Ottens, H. and Scholten, H. (eds) *EGIS '93 Proceedings*, EGIS Foundation, Utrecht.
Tomlinson, R.F. (1987) Current and potential uses of geographical information systems: the North American experience, *International Journal of Geographical Information Systems*, **1**: 203–18.
Törnqvist, G., Nordbeck, S., Rystedt, B., and Gould, P. (1971) *Multiple Location Analysis*, Lund Studies in Geography Series C, Number 12, Lund University, Lund.
Winter, S. (2001) Ontology – buzzword or paradigm shift in GI science? *International Journal for Geographic Information Science*, **7**: 587–90.

13

A Probability-based GIS Model for Identifying Focal Species Linkage Zones across Highways in the Canadian Rocky Mountains

Shelley M. Alexander, Nigel M. Waters and Paul C. Paquet

Abstract

Road fragmentation is a concern for the survival of sensitive species within and adjacent to protected areas in the Rocky Mountains of North America. The Idrisi Decision Support Module and Dempster–Shafer theory are used to establish potential linkage zones for lynx and wolves across the unmitigated sections of major highways in Alberta. Wolves and lynx acted as focal species for large and medium carnivore guilds, respectively. Linkage zones, derived from probability surfaces, showed minimal overlap between these two species. This suggests that the spatial requisites of crossing structures for wildlife are guild-specific at a minimum and mitigation should be designed to capture multiple guilds. Spatial probability modelling is a powerful method for representing habitat associations, determining ecologically important linkage zones and generating testable hypotheses. The uncertainty inherent in resource management and planning is reduced and managed by using such innovative spatial modelling tools.

13.1 Introduction

Road, rail, power and telecommunication networks create a grid upon landscapes commensurate with the extent of human endeavour. These networks are vital to our economy and provide a means to expand the human experience and improve travel

Applied GIS and Spatial Analysis. Edited by J. Stillwell and G. Clarke
 ISBN: 0-470-84409-4

convenience. Associated with these 'benefits' is the ecological cost of habitat fragmentation, which occurs when contiguous patches of habitat are divided into smaller, isolated units (Andren, 1994). Most wildlife species have evolved resilience, to the extent that they can endure ecological change caused by natural fragmentation (e.g., floods, rockslides, rivers) (Weaver *et al.*, 1996). However, human fragmentation tends to occur with an intensity and extent unprecedented in natural systems, and some species fail to persist when confronted with this human disturbance.

Of the previously listed networks, roads carrying traffic may have the greatest cumulative disturbance effect. In addition to habitat loss and isolation, vehicular traffic can kill animals that attempt to cross between isolated habitats, and may cause others to avoid crossing roads for fear of detection. In ecology this phenomenon is termed 'the barrier effect' (Noss and Csuti, 1997). The barrier effect can alter animal community composition, create meta-populations, reduce biological diversity and increase the threat of extinction (Trombulak and Frissell, 2000; Spellerberg, 1998; Forman and Alexander, 1998; Bascompte and Solé, 1996).

In recent years, 'road ecology' has become more prominent in ecological investigations, as shown by conference series such as the International Conference on Wildlife Ecology and Transportation (Evink *et al.*, 1998a,b, 1999) and the 'Roads and Rails' workshop series (Columbia Mountain Institute of Applied Ecology, Revelstoke, British Columbia 1997–2001). Reducing wildlife mortality on roads has been one focus of previous studies, as roadkill can pose a serious threat to species with small populations or low resilience. Low resilience results from life history traits, which include, among others, late age of first reproduction, low reproductive rate, and low fecundity (Weaver *et al.*, 1996). For example, a species that cannot reproduce rapidly enough to offset road mortality will suffer population decline. Over time and across space, such a decline could force a population into extinction. Road fragmentation has been identified as an extinction concern for carnivores in and adjacent to protected areas in the Rocky Mountains (Noss *et al.*, 1996). Specifically, roadkill has been noted as one of the largest known sources of mortality for wolves within and outside of protected areas of the Canadian Rockies (Noss *et al.*, 1996; Banff-Bow Valley Study, 1996).

The creation of linkages between isolated patches of habitat is one method for improving biotic exchange across barriers (Forman and Alexander, 1998; Swart and Laws, 1996; Fahrig and Merriam, 1995). In the present study region (described below) few sources of empirical data document the effects of traffic on wildlife movement or the spatial design requisites for placing linkages (highway mitigation) across highways. The present research was designed to address these deficiencies in knowledge and understanding of landscape connectivity for multiple species in the Canadian Rocky Mountains.

In this study, wildlife movement was documented in four different road fragmented environments in the Rocky Mountains of Alberta. Wildlife movement was measured, using snow tracking surveys, across and adjacent to the Trans-Canada Highway (TCH) (14 000 vehicles: Annual Average Daily Traffic, AADT) and the 1A Highway (4000 AADT) in Banff National Park, and the Highway 40 (5000 vehicles: Summer Average Daily Traffic, SADT) and Smith Dorrien/Spray Trail (2500 SADT) in Kananaksis Country. In the study area (described subsequently), little empirical

Table 13.1 *Summary of objectives*

Objectives	General hypotheses/methods
Identify landscape attributes predicting wildlife movement for individual species	Species preferences are defined by a probability of relationship to habitat attributes
Test for species interactions	Species interactions will be analysed using GIS
Create a linkage zone model using evidential reasoning	Application of IDRISI GIS and D–S weight-of-evidence modelling

research exists on multiple-species crossing behaviour before road expansion, on the effects of traffic on wildlife movement or on the spatial requisites for crossing structures.

The modelling effort presented in this chapter is unique. This work provides an original application of Bayesian (Dempster–Shafer, D–S) inference to determine optimal road crossing sites (i.e., linkage zones across highways), generates additional working hypotheses about the efficacy of wolves and lynx as focal species, and creates a platform for the integration of spatial wildlife data in highway planning. The matrix in Table 13.1 specifies the objectives of the research relative to developing the linkage zone model, and presents the general hypotheses or methods by which objectives were met.

The following section explores the literature pertaining to road fragmentation effects, wildlife habitat preference, modelling habitat selection using GIS and the integration of Bayesian analysis in defining wildlife linkage zones across highways.

13.2 Background Literature

13.2.1 When Roads and Wildlife Corridors Collide

This research focused on the interface between wildlife movement corridors and highways, and the identification of optimal sites for placing wildlife crossing structures (i.e., mitigation). A number of key literature reviews have summarised research on road-related, ecological disturbance effects, including Trombulak and Frissell (2000), Jackson (1999), Spellerberg (1998) and Forman and Alexander (1998). *The Proceedings of the International Conferences on Wildlife Ecology and Transportation*, ICOWET (e.g., Evink *et al.*, 1998a,b, 1999) offer examples of recent advances. Research trends have focused on mortality, pollution/contamination effects, and mitigation.

Highway mitigation, which refers to the implementation of measures that reduce wildlife mortality and/or facilitate movement across highways, and spatial trends in mortality remain at the forefront of road fragmentation studies. Mitigation effectiveness, the barrier effect and placement requisites remain little studied (Spellerberg, 1998; Forman and Alexander, 1998). Community studies that address movement requisites of multi-species are rare or non-existent, with the exception of the present study. The use of GIS has featured prominently in the recent *Proceedings of ICOWET* (Evink, 1999) and non-spatial decision support systems have become focal (Alexander and Waters, 1999; Smith, 1999; Maurer, 1999).

13.2.2 GIS Applications in Mitigating Road Effects

Habitat modelling has been advanced by the advent of geographic information systems (GIS), which greatly increase the scale, complexity and precision of spatial analysis. GIS is an excellent medium for predicting patterns of biological diversity, identifying areas of conservation significance and assessing the habitat potential of unstudied sites (Lenton *et al.*, 2000). In the region, studies have applied GIS in defining habitat selection by wildlife (e.g. Alexander and Waters, 2000a; Paquet, 1993). Relatively few have examined the spatial attributes of road crossing sites (Alexander and Waters, 2000b; Singleton and Lehmkuhl, 1999). Beyond the present research, no studies have examined multi-species interactions or predicted spatial linkages at a community level. GIS analysis has been used to identify mortality hotspots in determining placement of highway mitigation, although mortality hotspots have never been demonstrated to reflect crossing preferences. They simply measure mortality, not crossing attempts. Intuitively, these sites may reflect road alignment or curvature effects, rather than animal preferences. These road features may increase the likelihood of vehicle–wildlife collision and give the illusion that more animals cross at these sites.

Modelling the spatial requisites of movement corridors can be used to ameliorate the barrier effect, by confirming mitigation sites that have a high probability of selection for use by wildlife. Jackson (1999) summarises the types of mitigation popularly employed including modified drainage culverts, wildlife/drainage culverts, upland culverts, oversize stream culverts, expanded bridges, viaducts, wildlife underpasses, wildlife overpasses and fencing. Efficacy studies have demonstrated that species tolerances to structural design are highly variable (Jackson, 1999). Expanded bridges (open spans), viaducts and wildlife overpasses appear to be the most effective across communities in the Rocky Mountains of Banff National Park (Alexander and Waters, 1999). Placement of mitigation is argued as the most important component of mitigation success, followed by mitigation design. Jackson (1999 and all included references) lists the following efficacy criteria: size and openness, placement, human disturbance, substrate, vegetation cover, moisture, hydrology, temperature and light.

Barrier effects and mitigation efficacy vary by species (Trombulak and Frissell, 2000). For terrestrial mammals, GIS analysis helps identify the spatial attributes (habitat selection) likely to be suitable for movement corridors for individual species, guilds or community assemblages of species. Habitat selection is detailed in the next section and the use of GIS to infer habitat selection is presented in Section 13.3.3.

13.2.3 Habitat Selection: Spatial Attributes that Define Movement Corridors

Habitat selection modelling is predicated on the assumption that wildlife 'select', 'prefer' or 'avoid' specific habitat attributes (e.g., they may select closed vegetation for security or avoid roads). The selection of habitat may be inferred by examining use versus the availability of habitat. Peek (1986) identified 'ultimate factors' and 'proximate factors' as two facets of habitat selection. Ultimate factors include food, cover or shelter from enemies and adverse weather, nesting or denning sites and interspecific competition. Proximate factors may include stimuli of landscape, terrain and competitors.

To exemplify, habitat selection by wolves was described as 'a complex interaction of physiography, security from harassment, positive reinforcement (e.g., easily obtained food), population density, available choice, and distribution theory' (Carroll *et al.*, 2000). In a multi-species environment, such selection must be explored for individual species. Alternatively, the interactions between species may be investigated to determine the degree to which a given focal species may act as a surrogate for others. In this study, relationships to habitat metrics and the spatial interaction of species were investigated for each study species. These results are the topic of a different paper (Alexander *et al.*, 2003, in prep). The latter analysis allowed inference on which species is the 'best surrogate species' and identified three focal species, representing three functional ecological scales: marten – micro-scale; lynx – meso-scale; and wolf – macro-scale.

13.2.4 Bayesian Approaches in Ecological Modelling

Ecologists are revisiting the use of classical statistics in the evaluation of ecological phenomena (Ellison, 1996). Ellison (1996) argues that this introspection is due to the fact that the central assumptions of classical statistics are violated in ecology: true randomisation is difficult, replication is often small, misidentified or nonexistent and ecological experiments rarely are repeated independently. Alternatives to classical statistical tests have been presented, and include variants on probability theory (e.g., Bayesian probability, Dempster–Shafer theory) and information theory (e.g., Bayesian information criterion, Akaike's information criterion) (Wade, 2000; Anderson *et al.*, 2000).

The application of information and Bayesian theory is neither new, nor confined to conservation biology, but is popular in current ecological research (Anderson *et al.*, 2000). For examples of relevant ecological applications, see Bartel (2000), Anderson *et al.* (2000), Dragicevic and Marceau (2000), Urbanski (1999), Ellison (1996) and Pearl (1987). The integration of Bayesian theory into GIS for use in wildlife management and ecosystem planning is far less developed. One notable example is the work of Lenton *et al.* (2000), which predicted species biodiversity in Africa.

A central assumption of these alternative approaches is that testing null hypotheses adds little to scientific understanding (Anderson *et al.*, 2000; Ellison, 1996). Ellison (1996) argued that 'we normally expect to accept a single ecological alternative hypothesis by rejecting a statistical Ho: otherwise we would not have done the experiment in the first place'. Anderson *et al.* (2000) add that nearly all null-hypotheses are false on *a priori* grounds and that *p*-values are dependent on sample size (i.e., with a large enough sample one can always reject a null-hypothesis), even though such statements might be controversial. They argue that the former points to the difference between statistical significance and biological importance. Testing prior hypotheses (Ellison, 1996) or multiple working hypotheses (Anderson *et al.*, 2000) has been suggested as an alternative to classical statistical approaches. Prior hypotheses are based on the current state of knowledge about underlying distributions of the data. These priors are compared with new information to see if the new information adds to the current knowledge. Multiple working hypotheses (Anderson *et al.*, 2000) exhaust all possible combinations of explanatory variables for a given phenomena, such as species distribution. In this case, the range of hypotheses (i.e., combination of predictive variables) is compared using an information criterion, to provide a rank that indicates the optimal combination of variables.

13.2.5 Integrating Bayesian Theory and GIS in Habitat Assessment

Research on wolves in Banff National Park (BNP) indicated that landscape features, such as slope, aspect, elevation and vegetation type can influence movement (Paquet, 1993). For example, steeper slopes require greater energetic expenditure for an animal to cross and may result in the displacement of movement around a steep feature. Alternatively, dense vegetation may provide escape or hiding cover and may change a movement path when an animal fears detection or predation. A disproportionate use of physiographic attribute classes is assumed to reflect a 'preference' for those landscape elements. A 'preference' suggests that the animal is selecting a specific attribute class over another. Attribute classes are aggregated values within an attribute. For example, for the variable 'aspect', which ranges from 0 to 360 degrees, the 'aspect classes' are groupings of aspect values, such as those labelled north, which includes all aspect values between 337.5 and 22.5 degrees.

The statistical evaluation of preference is often handled with techniques, such as logistic regression, which treat species occurrence as a binary value representing presence as (1) or absence as (0) (Pearce and Ferrier, 2000; Bartel, 2000). Although there are variations in analytical procedures, often a site's presence and absence have been defined by stating a threshold count of 'wildlife sign' that indicates presence. Sites that have a number of observations of a species that exceed a threshold are considered 'presence' sites (1), those below the threshold are 'absence' sites (0). For instance, if the greatest number of species observed for all sites surveyed is 10, the analyst may specify that any sites with greater than five observations indicate presence (value 1) and any sites where there were fewer than five are discarded from the sample (value 0).

Binary data (1,0) are then subjected to techniques such as logistic regression. In this approach, the imposed threshold that defines the difference between presence and absence is arbitrarily defined: a value of 1 might be five tracks, 20 tracks or any number in between; a value of 0 might be assigned to any sites with fewer than five tracks. This assumption that sites with fewer than five tracks are not 'used' may be an ecologically incorrect assumption. Moreover, Conroy and Noon (1996) argued that although presence provides affirmation of occurrence, absence does not provide any information – it could be an artefact of the sampling effort.

Binary approaches do not represent adequately the spectrum of preferences observed within landscape attributes, may needlessly discard information and are inherently indefensible because, in most cases, habitat attractiveness and preference for movement corridors exists as a continuum. For example, species may use sub-optimal habitat during regular foraging or movement (Tischendorf and Fahrig, 2000) because habitat is most often heterogeneous. In this study, highly suitable lynx habitat was spatially discontinuous and lynx traversed areas of low habitat suitability when moving between highly suitable patches. As an alternative to binary approaches, detailed above, we used a Bayesian classification approach that treats habitat use as a continuous probability. This approach considered relative track density by attribute classes and the probability of use was indicated by the weight-of-evidence of a trend in habitat use, based on the difference between observed and expected track counts for each attribute class. Thus, instead of specifying a threshold value, all track density values are considered important in predicting the presence of a species.

Fuzzy set theory is well suited to problems that deal with habitat analysis, where there is 'no clear separation between areas that are suitable and those that are not' (Bartel, 2000; Eastman, 1999; Zadeh, 1965). Although typically applied to spatial boundaries, such as problems related to soil classification (Dragicevic and Marceau, 2000), fuzzy set theory is extended easily to describe the fuzziness in species associations with habitat types. Fuzziness, as a measure of inconclusiveness (i.e., lack of exact separation in selection), is handled well with Bayesian probability or its variant D–S theory (Eastman, 1999). D–S theory was developed by Bayes' researchers to handle absence of information (Gordon and Shortliffe, 1985). The distinction between Bayesian probability theory and D–S theory is elaborated in the methods section below.

13.3 Study Area and Methodology

13.3.1 Introduction

Research was conducted in Banff National Park (BNP) and Kananaskis Country (KC), Alberta. Figure 13.1 shows the study area. BNP and KC are approximately 110 kilometres west of Calgary. BNP is 6640 square kilometres in area and is the most heavily visited national park in Canada with over 5 million visitors per year (Banff Bow Valley Study, 1996). KC is a 4250 square kilometres Forest Landuse Zone that is reserved for hunting, ranching, resource development, recreation and tourism activities. Peter Lougheed Provincial Park (PLPP) is a 514 square kilometres park in the northwestern corner of KC, where hunting and resource development are not permitted, with the exception of ongoing maintenance of existing hydro-electric projects. The study region is characterised by rugged mountainous terrain, steep valleys and narrow (2–5 kilometres), flat valley bottoms.

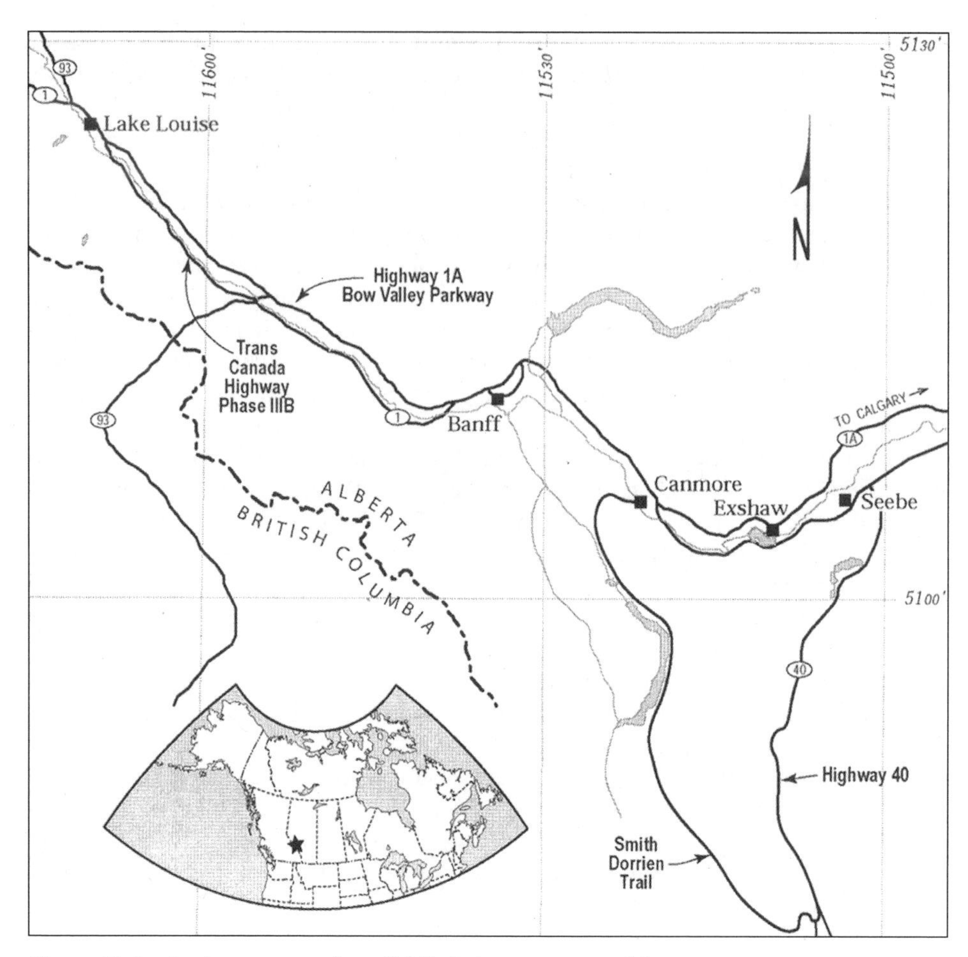

Figure 13.1 Study area map for wildlife linkage zone model

In BNP, track data were collected on the non-twinned (two lane), unfenced section of the Trans-Canada Highway (TCH) from Castle Junction to the British Columbia border, and along the entire length of the Bow Valley Parkway (1A). This section is termed Phase IIIB of the TCH 'twinning project' in BNP. Apart from the TCH and 1A the study area transportation corridor contains the Canadian Pacific Railway (CPR) main line, Trans-Alta power line, and a few seasonal campsites and tourism facilities. The BNP study roads do not have structures that facilitate the movement of wildlife over or under the highway or to reduce wildlife–vehicular collisions (i.e., it is not fenced). The previous structures are termed 'highway mitigation'.

In KC, research was conducted on Highway 40 (Hwy 40) and the Smith Dorrien/Spray Trail (Figure 13.1). Highway 40 is a paved two-lane highway with moderate traffic volumes, and the Spray Trail is a gravel two-lane road primarily offering access to primitive backcountry activities. The service areas of both transportation routes are under pressure for recreation and tourism development, which will result in increased traffic volumes. With the exception of wildlife crossing signs, neither route is mitigated for wildlife.

13.3.2 Methodological Framework

Figure 13.2 shows the methodological framework for production of the final linkage zone model. This image is useful for understanding each of the following steps of model creation described below.

Research was conducted during the winter months (from November to April) over three years (1997–2000). Roads were surveyed between 18 and 48 hours after the end of a snowfall. Tracks in snow were observed from a field vehicle, while driving slowly (15–20 km/h) and verified on foot (Van Dyke *et al.*, 1986). Tracks entering or exiting the road right of way were recorded for 13 terrestrial species, including the two focal species presented in this paper, wolf (*Canis lupus*) and lynx (*Felis lynx*). Tracks of other species were observed but not reported herein, for coyote (*Canis latrans*), fox (*Vulpes vulpes*), cougar (*Puma concolor*), bobcat (*Felis rufus*), marten (*Martes americana*), fisher (*Martes pennanti*), wolverine (*Gulo gulo*), elk (*Cervus elaphus*), moose (*Alces alces*), sheep (*Ovis canadensis*) and deer (*Odocoileus virginianus* and *Odocoileus hemionus*). Tracks were only observed in the right of way, as vegetation obscured tracks beyond this area. It was assumed that tracks entering this area were of animals attempting to cross the road.

Data collected at crossing sites included an Easting/Northing (UTM), which was collected with a handheld GPS (Garmin II, non-differentially corrected). Other data collected included species type, number of individuals and a range of behavioural parameters that are not analysed here. Repeat road surveys were conducted, as conditions permitted, approximately 3–4 days after initial surveys until the next new snowfall. Transects of one-kilometre length were fixed perpendicular to each road treatment and surveyed for tracks of the above study animals and in addition for hare, weasel and squirrel. Again, only the results of wolf and lynx are used for the probability mapping section in this chapter. Transects are straight survey lines across a study area, in this case marked every few metres with flagging tape on trees. Flagging tape was placed every 50 metres, which allowed researchers to repeat survey exactly the same area. Forty transects were surveyed randomly in Banff National Park and twenty in Kananaskis Country, between 18 and 108 hours after snow and immediately following each road survey (Thomson *et al.*, 1988). Transects were surveyed on foot and required an extended survey period relative to the road survey. A one-metre resolution differential GPS (Trimble Pathfinder) was used to collect UTM coordinates for every 50 metre interval start point, for every transect. These data were used to geo-reference track counts for import into the Idrisi GIS (Eastman, 1999) for spatial modelling. Data were standardised by sampling effort, to minimise the effects of sampling bias. For example, if one site was surveyed five times and another three times, the total track count for each site was divided by the number of samples per site (i.e., five and three) to standardise between the transect sites.

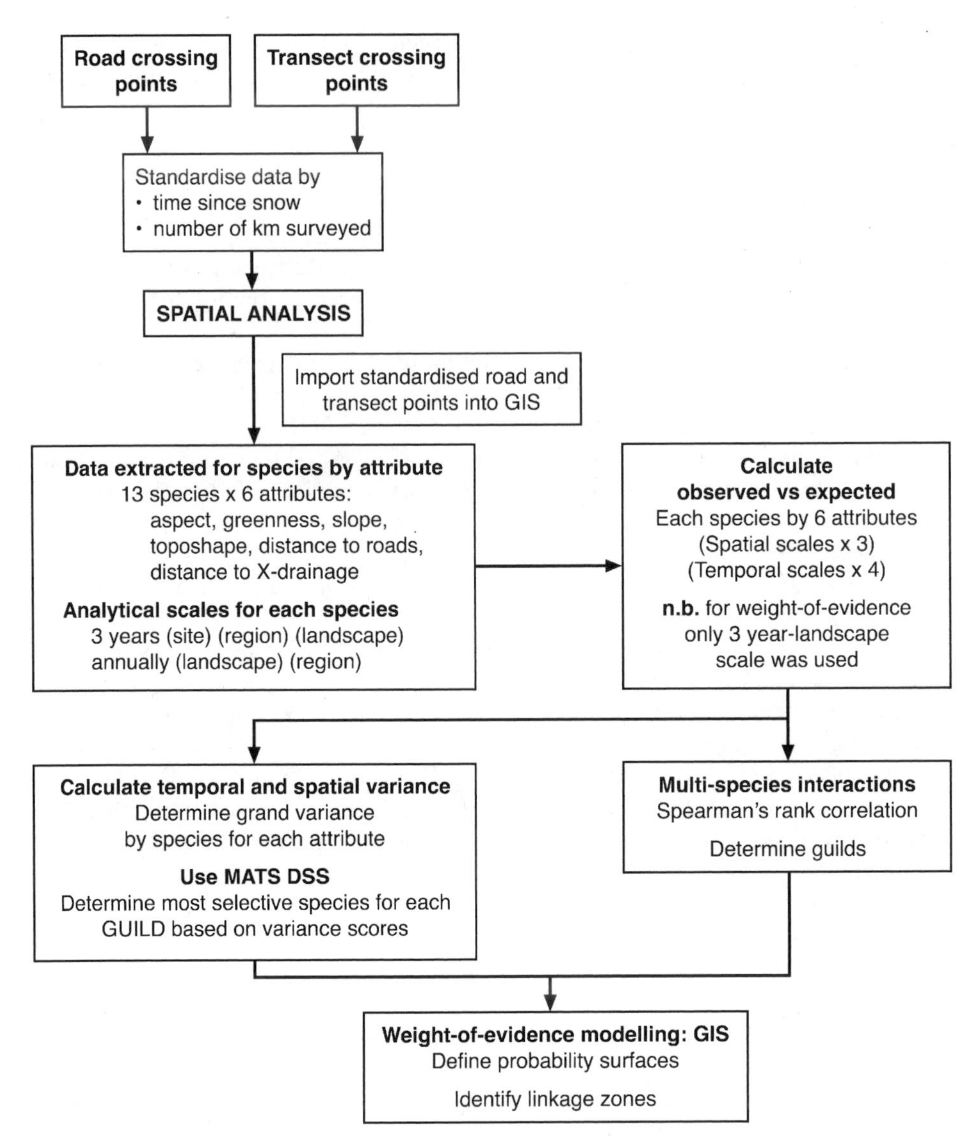

Figure 13.2 *Flowchart of model development*

13.3.3 Inferring Selection from Use Density

Density of tracks by attribute classes (i.e., track density in each aspect class) was determined by using the EXTRACT function in Idrisi (Eastman, 1999). With the EXTRACT function, the total track density by each attribute class is summarised in table format. These data tables were saved as text files and imported into Excel (MS Office 97).

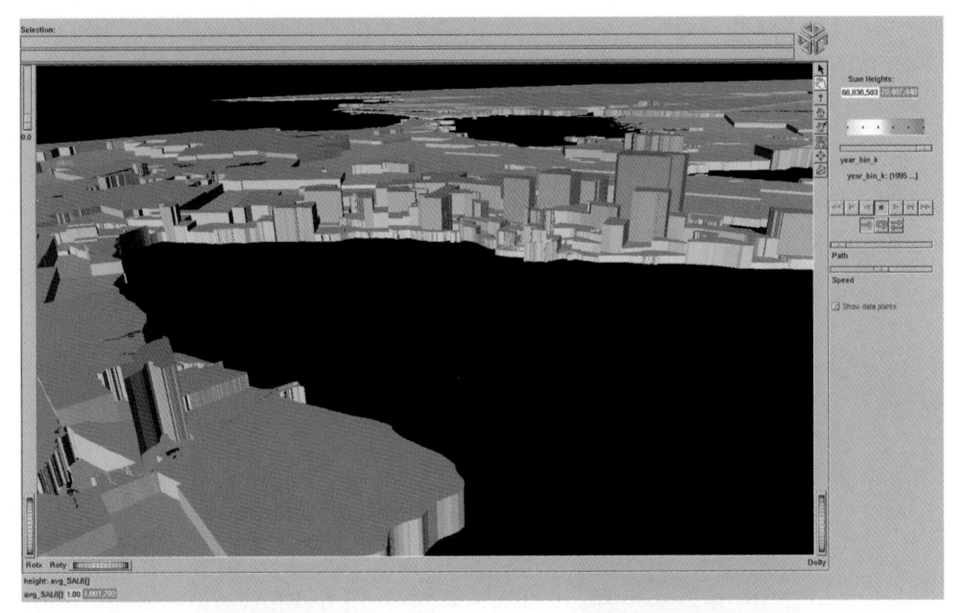

Plate 1 Retail sales volumes in the Southern Ontario market (see Chapter 2)

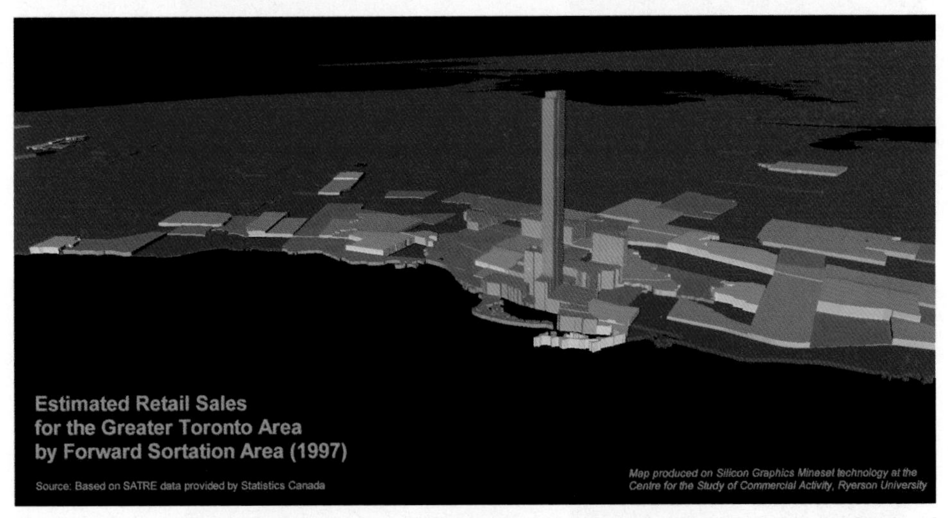

Plate 2 Estimated retail sales for the Greater Toronto Area. From Centre for the Study of Commercial Activity, Ryerson University, Canada (see Chapter 2)

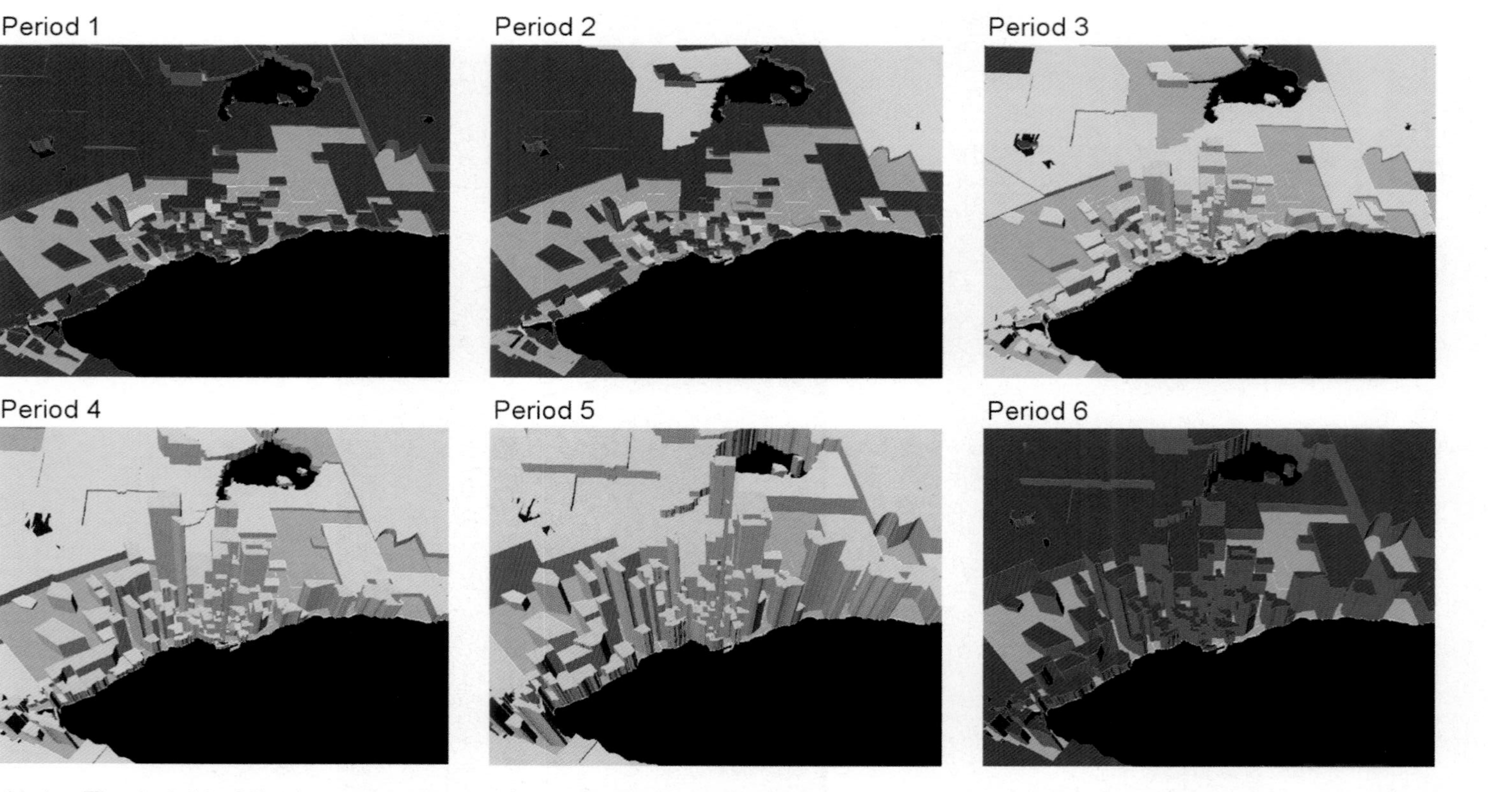

Note: *The height of the bars represent sales volume. The color represents year-on-year sales change, blue: sales loss, green: unchanged, orange: sales growth. Periods 3 to 5 show the significant growth associated with Y2K celebrations, and sharp year-over decline for the first week of January 2001 (Period 6). The data are reported at the FSA level.*

Plate 3 Spatial–Temporal animation of weekly sales data for a retailer operating in Ontario, Canada (see Chapter 2)

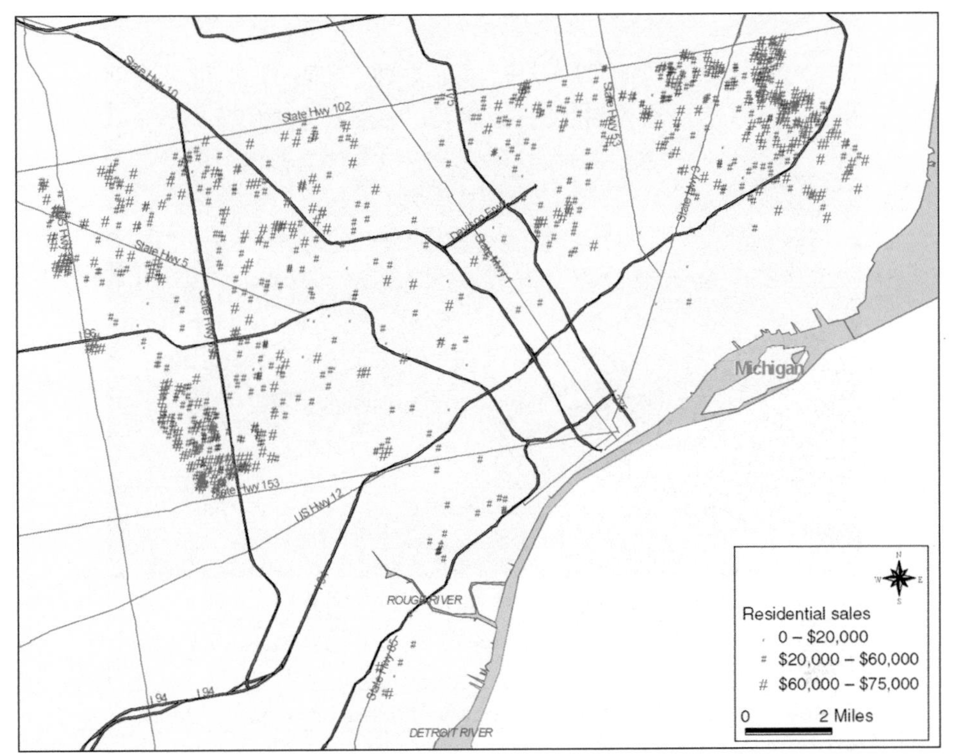

Plate 4 MLS home sales (see Chapter 5)

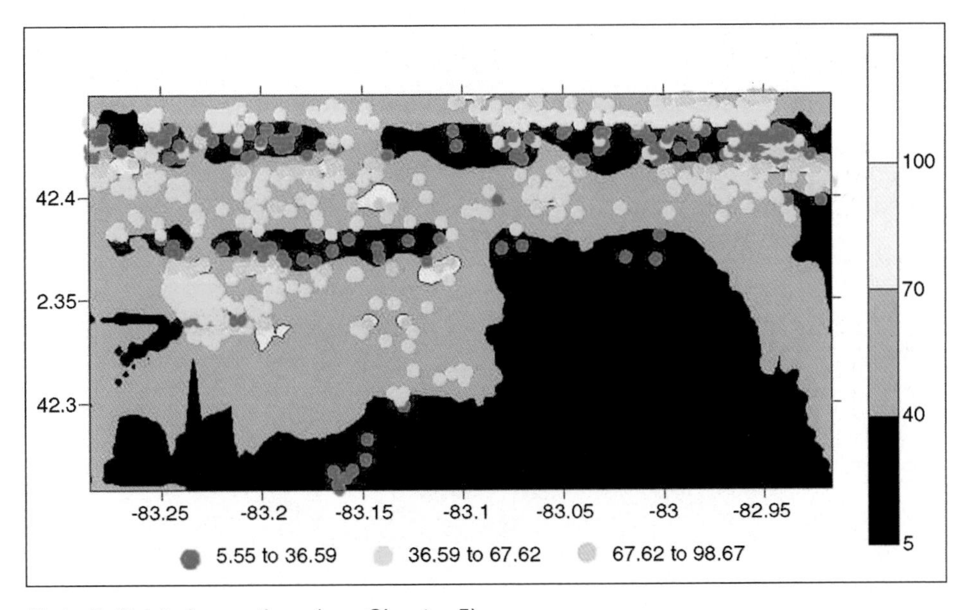

Plate 5 Point observations (see Chapter 5)

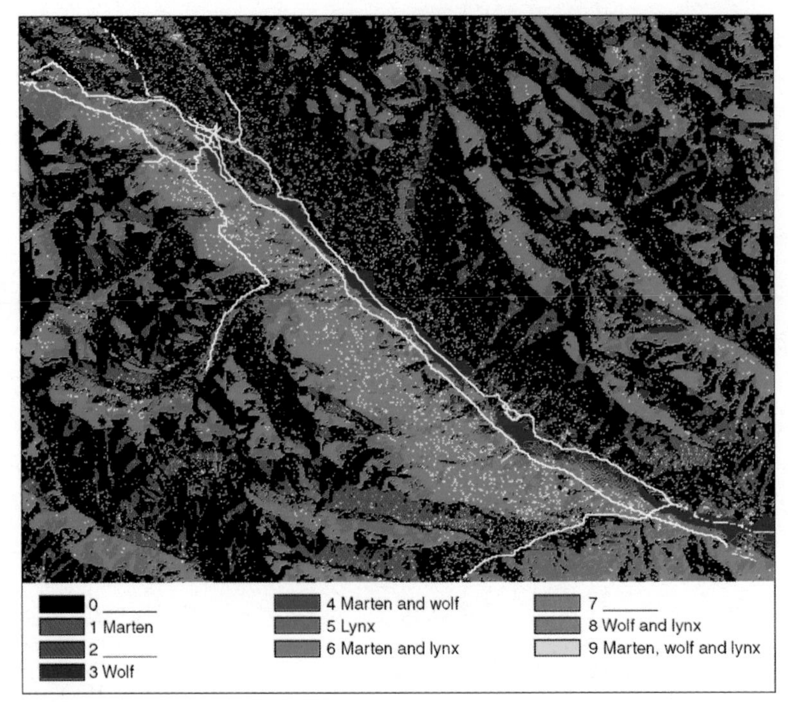

Plate 6 Intersection of >90% probability of detection of guilds, Banff (TCH III) (see Chapter 13)

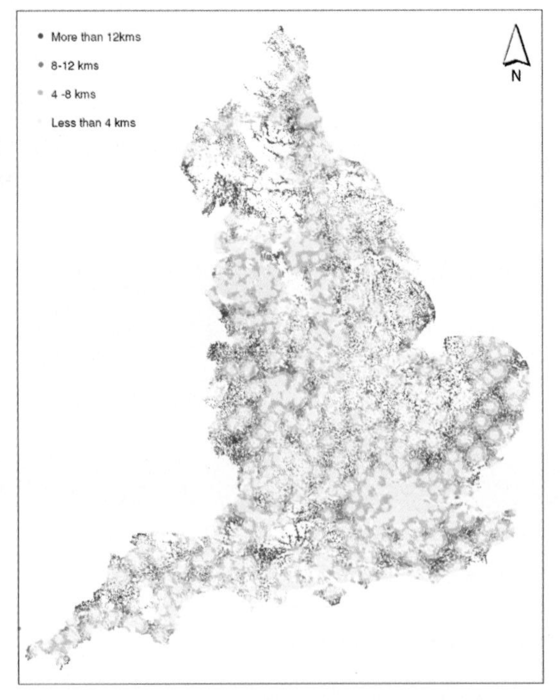

Plate 7 Household access to hospitals in England. From RSS2000 Technical Report (see Chapter 17)

A study frame (i.e., the spatial extent of area actually surveyed) was used to extract the distribution of class types within each attribute (e.g., north class or aspect). The study frame was a Boolean image, in which any site with a value of one (1) was an area that was sampled, any site with a zero (0) value was not sampled. The use of the frame is essential to avoid issues of 'framing bias' and 'ecological fallacy', which can yield incorrect results about habitat selection (Verbyla and Chang, 1994). The output from this operation provided a summary of total cell counts by attribute class. The extracted study frame values (i.e., the distribution of attributes samples) were used to determine the proportional representation of each attribute class in the sampling frame, which was then used to determine the expected number of tracks for each attribute class, as a proportion of total tracks observed. A value was then derived to indicate over- and under-use of attribute classes. This value was determined by subtracting the expected from the observed track density and dividing by the expected density (*N*) for each sub-class. Table 13.2 shows the distribution of marten points by aspect class for the three years. Column numbers are indicated at the top.

The first column of Table 13.2 indicates the attribute classes (here using the example of aspect classes). In this case, the aspect classes represent groupings of aspect values that are similar. For example see the row North (column one) in Table 13.2, where north aspect is defined as all slopes that face the directions 337.5 to 22.5 degrees around north (0 degrees); northeast is 22.5–47.5 degrees, and so on. Column two shows the total number of tracks observed by aspect class, as tabulated from the EXTRACT function in Idrisi. Column three is the expected track count, the derivation of which is discussed below. The fourth column shows the total sampled raster cells for each aspect class, followed by the proportional representation of the total number of cells (column five). The cells represent locations on the landscape that were sampled. The attribute value on any cell is an average over the cell (or pixel) resolution, which in this case is 30 metres. The proportion in column five was used to derive the expected track count (column three) by multiplying the proportion by the total number of marten tracks observed. Column five sums to 1.00 when the error

Table 13.2 *Species, track counts by aspect class*

1 Aspect Class	2 Observed	3 Expected	4 Total Aspect	5 Prop	6 Diff(o-e)	7 Over/Under use	8 RANK
Background	67	0	0		67	0	
North (337.5–22.5)	219	263.5402	8031	0.076234	−44.5402	−0.16901	7
Northeast	556	335.1761	10214	0.096956	220.8239	0.658829	1
East	285	205.7523	6270	0.059518	79.24768	0.385161	3
Southeast	184	145.3721	4430	0.042052	38.62794	0.265718	4
South	466	281.1948	8569	0.081341	184.8052	0.657214	2
Southwest	730	642.4592	19578	0.185840	87.54083	0.136259	5
West	305	662.7062	20195	0.191699	−357.706	−0.53977	8
Northwest	92	286.9047	8743	0.082992	−194.905	−0.67934	9
Flat – No aspect	553	633.8944	19317	0.183365	−80.8944	−0.12761	6
Total	3457	3457	105347				

point (67 points in the Background class 0) are considered. The difference between observed and expected values appears in column six, and finally, the observed minus expected value is divided by the total expected (column three) in each class, to produce the values in the seventh column. This provides a relative index of observed versus expected use for comparison between scales of analysis and between species.

In Table 13.2, over-use is indicated in column seven by a positive value and under-use by a negative value. It was assumed that over or under-representation of tracks by attribute class has biological significance and represents a continuum of use by species of attribute classes. Column seven values were ranked by their relative level of over- and under-use. Over-use defines a greater amount of use of a habitat class than expected, based on its availability. Alternatively, under-use implies less use than expected, based on availability. To exemplify, in column eight, the highest rate of under-use (rank 9) is found in the class northwest, and the highest rate of over-use (rank 1) is found in class northeast. The rank scores in column eight provide a means to treat the nominal data as a linear variable, when determining the probability of detection for each attribute class and creating a probability surface in the GIS, for the aspect variable. This procedure was carried out for all 13 species sampled (refer to Section 13.3.2).

13.3.4 Dempster–Shafer Weight-of-Evidence Modelling

A univariate analysis of track occurrence by landscape attribute indicated that the probability of detecting carnivores is a function of slope, aspect, greenness and toposhape. Two other variables, distance to cross drainages and distance to roads were evaluated, but yielded no trend in selection or avoidance. Slope and aspect are surfaces derived from a DEM (30 metre resolution). Greenness was derived from a Landsat TM image using a Tasselled Cap (TC) analysis. TC is a simple linear index produced by taking the weighted average of the input bands from a TM Image (Richards and Jia, 1999). TC Greenness is proportional to vegetation productivity, where higher greenness values are assumed to represent better suitability. Toposhape is a measure of surface curvature that defines the landscape as a series of critical features, including peaks, ridges, saddles, flat areas, ravines, pits and five types of hillsides – convex, saddle, slope, concave and inflection (Eastman, 1999; Warntz and Waters, 1975).

Probability functions describing optimal conditions for habitat linkage zones were derived from the index of selection and avoidance. These are represented by the over-use and under-use observed for certain habitat classes, such as those shown in Table 13.2, previously. The decision rules and probability functions are described below.

Probability surfaces were constructed for wolves and lynx using the FUZZY module in Idrisi (Eastman, 1999). Although 13 species were studied in this research, only the probability surfaces for wolves and lynx are presented in this chapter. For habitat relationships and probability maps relating to other species, refer to Alexander (2001). An example of how decision rules are translated to probability curves is shown in Figure 13.3, which shows the aspect classes and ranks, which were identified previously in Table 13.2 (see also the discussion in Srinivasan and Richards,

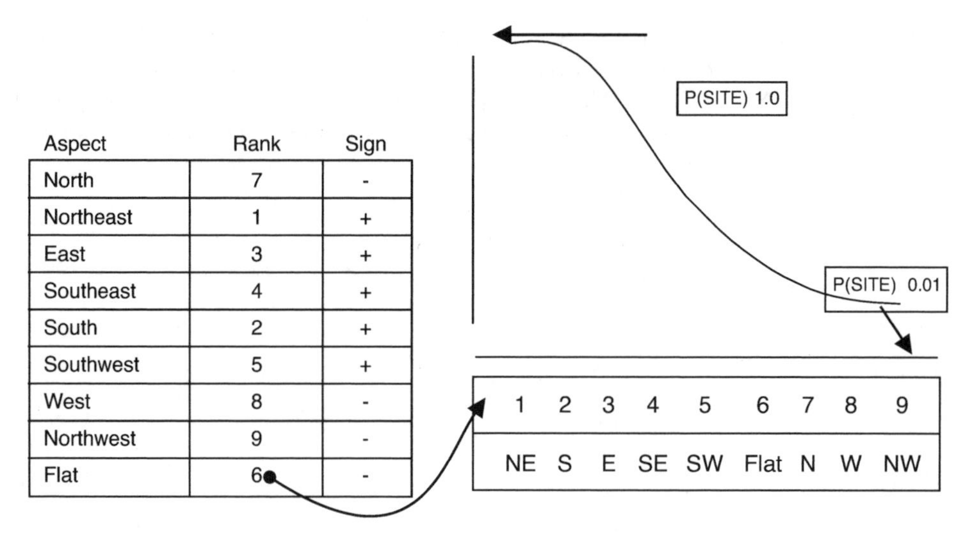

Figure 13.3 *Species use by aspect*

1990). The sign indicates the selection value (see description of Table 13.2), and the rank at which the sign changes may be used as an inflection point on the probability curve.

In the Idrisi GIS software, fuzzy sets can be expressed by sigmoidal, j-shaped, linear and user-defined probability curves. The curve indicated in Figure 13.3 is a monotonically decreasing sigmoidal function. For each probability curve, an associated hypothesis must be stated. From the curve, a hypothesis may be stated that the probability of detecting a marten site *p*(site) decreases with increase in rank.

The BELIEF module in Idrisi was used to aggregate all the 'lines of evidence' (slope, aspect, greenness and toposhape – as represented by their various hypotheses, e.g., species are more likely to be found on slopes with certain aspects as described in Figure 13.3). BELIEF aggregates these various 'lines of evidence by applying rules of combination based on Dempster–Shafer (D–S) Weight-of-Evidence modeling' (Eastman, 1999). D–S theory is a variant of Bayesian probability theory (Srinivasan and Richards, 1990), which allows us to decompose a large body of evidence into its component parts. The importance or influence of these component parts can then be assessed as probability statements that subsequently can be recombined to determine the probability of finding a particular species at a given location. If this species is then found in association with others, such as a guild of wildlife species, these locations, when adjacent to roads, may be especially attractive as sites for the mitigation of highway induced wildlife mortality. In this case, the evidence was aggregated using BELIEF, to create a linkage zone model, which indicates regions with the highest probability of detecting a given species. These high probability sites should indicate optimal locations for placing crossing structures.

D–S theory defines hypotheses in a hierarchical structure. For example, if the frame of discernment (Srinivasan and Richards, 1990), or decision frame using Eastman's

(1999) terminology, contains two hypotheses [A,B], the structure of the hypotheses for which D–S incorporates evidence includes all possible combinations of these hypotheses i.e. [A], [B], [A,B]. All three hypotheses are recognised because often evidence supports one or other of this combination of hypotheses. D–S theory employs the following rule of combination:

$$\begin{aligned} m(Z) &= \sum m_1(X) \times m_2(Y) \text{ when } X \cap Y = Z \\ &= 1 - \sum m_1(X) \times m_2(Y) \text{ when } X \cap Y = \Phi \end{aligned} \tag{13.1}$$

If

$$\sum m_1(X) \times m_2(Y) = 0 \text{ for } X \cap Y = \Phi \tag{13.2}$$

Then

$$m(Z) = \sum m_1(X) \times m_2(Y) \text{ when } X \cap Y = Z \tag{13.3}$$

where $m(Z)$ = basic probability assignment (the mass of support) for the hypothesis [Z] and X and Y are independent variables (Eastman, 1999; Gordon and Shortliffe, 1985).

D–S theory varies from Bayesian probability theory, in that it recognises the existence of ignorance due to incomplete information – as indicated by the previous non-site hypotheses. In Idrisi, D–S theory is ‘able to determine the degree to which we, with our current state of knowledge (evidence), cannot distinguish among a subset of hypotheses (a condition known as ambiguity) or among all hypotheses (a condition known as ignorance)’ (Eastman, 1999). For example, the mass of support for the probability of detecting wolves by slope is based on the empirical evidence that wolves prefer lower slopes. The weight of this evidence tends to support the non-use of steep slopes (>30 degrees), given the paucity of use in sites with higher slope steepness. Low slope alone, however, may not be enough to explain preference for shallow slopes, thus the hypothesis is expressed as a non-site (i.e., high slopes lack wolves) rather than as a site hypothesis that would state directly that low slopes have a high probability of being wolf sites. By contrast, all other lines of evidence, toposhape, greenness and aspect, were used to support the site hypothesis.

The BELIEF module is used to construct a knowledge base from the input data (i.e., all four predictive attributes for each species) and the hypothesis that each data layer supports (i.e., site or non-site, as discussed above). All four lines of evidence are integrated into a knowledge base by combining the various lines of evidence and producing basic probability assignments for each of the hypotheses. The BELIEF module will produce maps of the probability that an animal will be found at a given location based on the three lines of evidence that support the habitat hypothesis. The line of evidence, in this case slope, which supports the non-habitat hypothesis, produces a ‘plausibility’ map. The difference between the belief and plausibility maps produces the belief interval that may be useful for determining where to look for new sources of support for the habitat hypothesis. In the present chapter we are primarily concerned with our existing evidence for species existence with respect to their habitat properties.

A first iteration of probability surfaces was constructed using four attributes (aspect, greenness, slope and toposhape). A lack of clear linkages was evident, and maps were re-analysed. A second iteration of probability maps was then developed without greenness (i.e. with three attributes: aspect, slope and toposhape). Greenness was removed because it is highly correlated for all species, and the initial probability image showed little variation in greenness within one kilometre of roads. Finally, data assimilation techniques were used to improve the identification of corridors (linkage zones) in a third iteration of maps. Data assimilation techniques involve using regional data to 'fine-tune' global models. Climatologists frequently use this method to enhance global weather models (Khattatov *et al.*, 2000). In the third iteration of the linkage zone model, known species sites were 'assimilated' to calibrate the final model geographically. The assimilation was done by adding a layer depicting distance from known sites of the focal species, represented by a decay function where suitability of habitat decreases with distance from observed sites. This calibration was used because the first probability surface was generated using habitat associations from the three-year landscape scale analysis, a global model.

13.4 Results and Discussion

13.4.1 Introduction

In this section the results of three iterations of the linkage zone model are presented. The initial belief surface combined the four variables that showed non-uniform use by the focal species (i.e., slope, aspect, greenness and toposhape). A second surface was developed in response to an examination of the first surface. Specifically, conclusions regarding the lack of usefulness of 'Greenness' resulted in the development of iteration two, which included only slope, aspect and toposhape. In the final iteration (3), the model was enhanced by including known species occurrences. This is an assimilation approach that was investigated because there is a general understanding that, in movement, wildlife may be more inclined to use sites they have previously used more than those they have not. This was an exploratory iteration to determine the effect on the final model.

13.4.2 Iteration 1: Slope, Aspect, Greenness, Toposhape

In the first iteration of BELIEF maps, the valley bottoms showed uniformly high probability of detection for lynx and wolves. This probability was high for any areas below rock or cliff faces. Although a few sites of high probability could be identified crossing the valley (i.e., linkage zones), there was little visible separation between high and low classes in the valley bottom. This lack of separation was more pronounced in the wolf probability surface. Consequently, we reviewed the probability surfaces and found that greenness was the only surface with extremely low geographic variability within 1 km of roads. Thus, the effect of adding greenness to the probability model appeared to be that of enhancing suitability of the entire image and reducing separation on the valley floor. This effect was not expected, but is reasonable given that all valley areas are highly productive, relative to the surrounding peaks. However, as the objective of the research was to identify suitable linkages across roads (i.e., the valley floor) the lack of separation in greenness at lower elevations is problematic. Greenness was removed from the calculation in the second set of map iterations and linkage zone separation was improved.

13.4.3 Iteration 2: Slope, Aspect, Toposhape

In this second iteration of the BELIEF module, non-suitable lynx habitat linkages were evident in the valley bottoms, illustrating the constrained movement opportunities for lynx at the interface with roads. Areas that exhibit low probability of detecting lynx were areas of flat slope and aspect. The region adjacent to the TCH Phase IIIA (southern side) showed enhanced suitability relative to other regions in the Bow Valley. Ground validation showed a combination of old-growth spruce and young, dense pine stands. In this unique situation, there is a spatial convergence of two habitat niche requirements for lynx, namely, denning and hunting. This type of spatial arrangement in forest types is not common in the Bow Valley study frame.

The second iteration of the wolf probability image reconfirms the topographic effect of the landscape on wolf habitat suitability (Weaver *et al.*, 1996). The highest probability of detection is limited to extreme valley bottoms. Lakes appear in the images as high probability sites for wolves, which is appropriate in winter only, when lakes are frozen and used as travel routes (Paquet, 1993). In some cases, glaciers were identified as high probability wolf sites, because they have low slope angle. This is a consideration for future research that may look at linkages outside the valley floor. The identification of glaciers, lakes and rivers as highly probable linkages between valleys is misleading, as they may be inhospitable for travel by wolves. A glacial and river hydrology overlay could be used to remove these areas, but is not necessary for the present research.

The limitation of wolves to the valley floor, as depicted in the probability image provides insight into the reasons why wolves have a difficult time persisting in road fragmented mountain terrain (Weaver *et al.*, 1996). In addition to finding the valley floors easier for movement, wolves are obligate predators of ungulates, which as a rule are human tolerant, valley bottom dwellers. These constraints placed on wolves increase their interaction with road surfaces in the valley and, consequently, increase their probability of being killed.

13.4.4 Iteration 3: Slope, Aspect, Toposhape, Observed Species-sites

In the third iteration of probability surfaces, shown in Figures 13.4 (lynx) and 13.5 (wolves), the data assimilation method was used. The probability of detection was

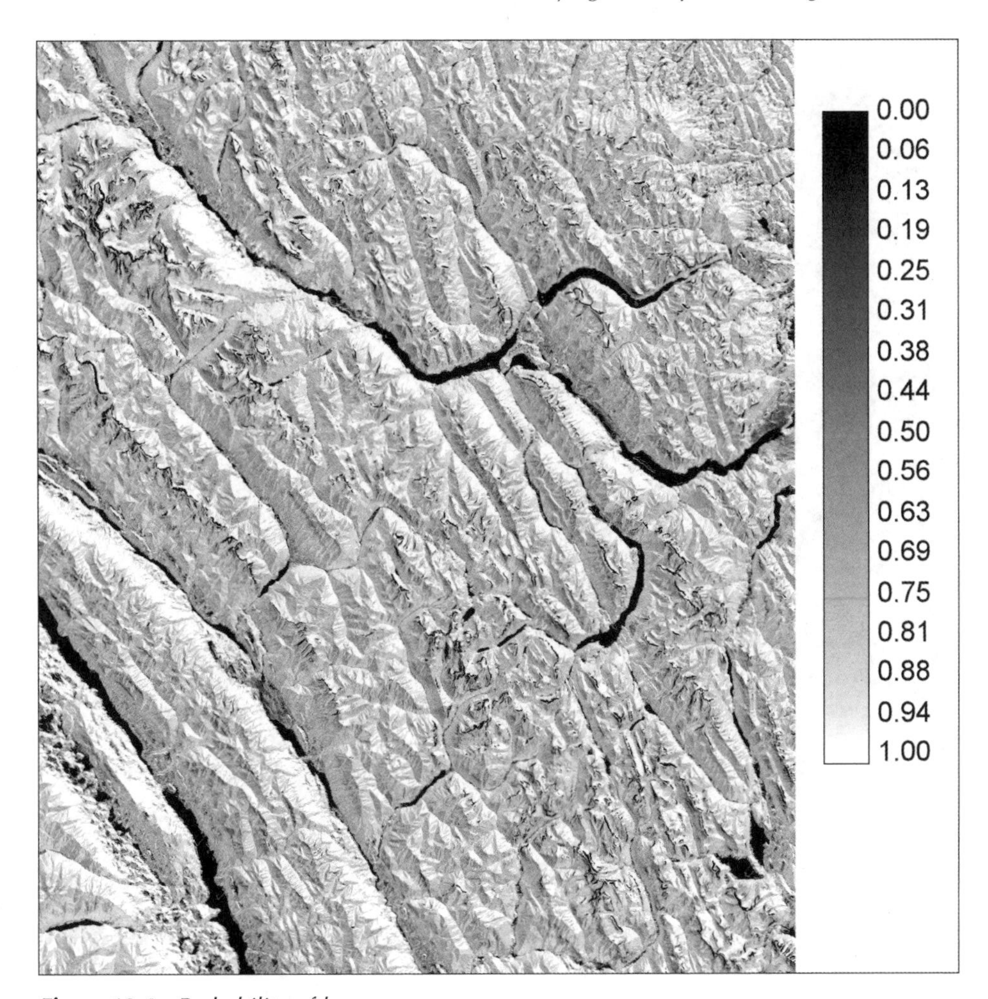

Figure 13.4 *Probability of lynx*

assumed to decline rapidly with distance from known (observed) points. Hence, the probability decayed over a distance from zero to 300 metres. The data assimilation created a more regionally specific (species-precise) predictive surface. However, the inclusion of this data enhanced sites within the existing linkages, but did not change the overall linkage zones.

Cross-valley linkages with higher than 95% probability (from Figures 13.4 and 13.5) were more frequent for lynx than for wolves, and crossing sites for wolves have wider spatial extent than lynx. By comparison Figure 13.4 (lynx probability) shows less contiguous highly suitable habitat for lynx than that for wolves, shown in Figure 13.5. The highest potential habitat for lynx is dispersed and is not confined to the valley floors. This offers lynx the benefit of not interacting with human disturbance on a continuous basis. However, lynx may be required to traverse very unsuitable patches of habitat (i.e., valley floors), in order to access other high quality habitat

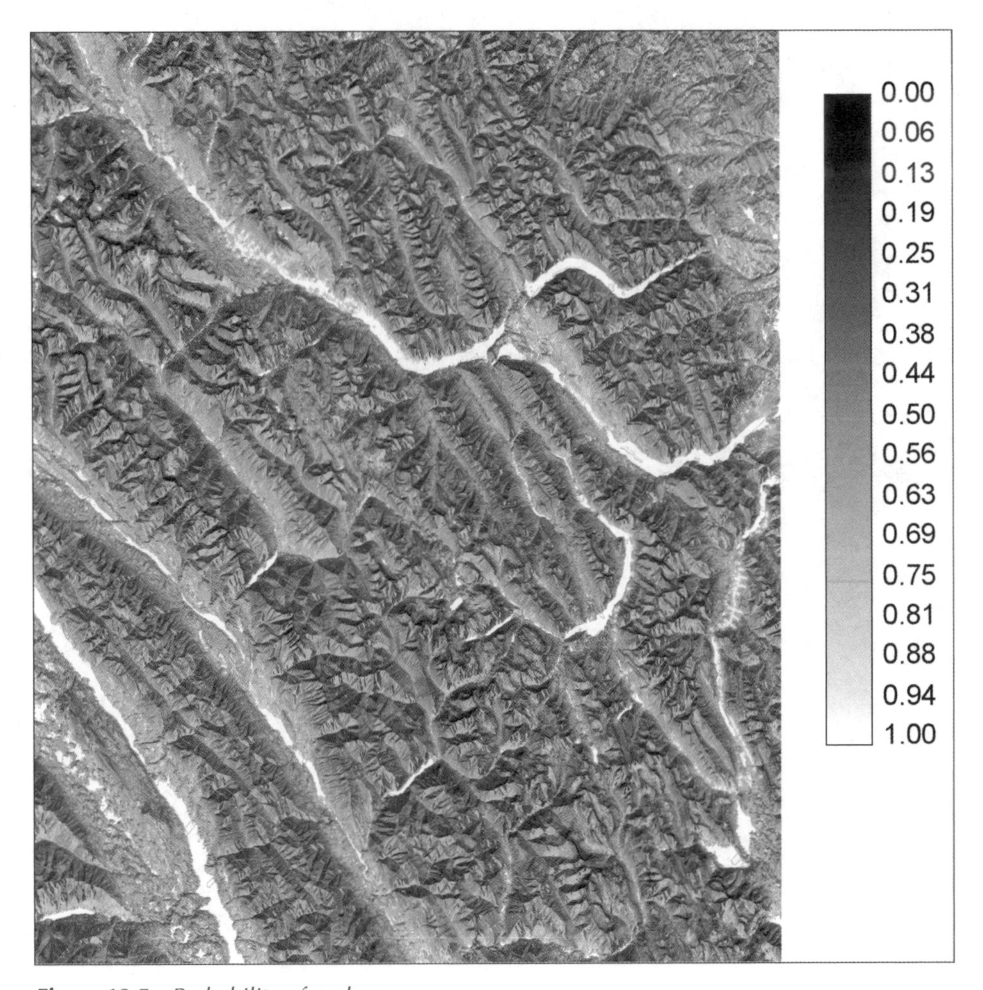

Figure 13.5 *Probability of wolves*

during dispersal or times when local prey abundance is low. Figure 13.5 also shows that wolves are highly constrained to the valley floors.

A comparison of Figures 13.4 and 13.5 also yields evidence that fewer connections are needed across highways for wolves, relative to lynx. The limited number of crossing opportunities suggests that the placement of mitigation will be critical to wolves. Lynx movement may be accommodated by placement of mitigation at a greater frequency. The suggestion to place lynx structures at higher frequency than wolves is grounded in the arguments of body size–home range and functional scale requirements (Holling, 1992). Lynx have smaller home ranges, thus, more may be accommodated in the same space as that required by wolves (assuming equal suitability for both species).

Plate 6 shows regions where there is >90% probability of detecting three focal species. Marten are included in this figure for demonstration purposes (i.e., how the methodology may be extended to multiple species). Plate 6 was created by combining three Boolean images – one for each species. Marten were coded in Boolean as [1,0], wolves as [3,0] and lynx as [5,0]. This allows the images to be combined, such that the output may be used to detect sites for individual species (e.g., marten at zones with value 1), or to identify sites of various combinations of species. For example, if the user would like to identify a site where marten and wolves intersect in the landscape, then he/she would look for sites with the value 4 (i.e., 1 – marten + 3 – wolf). Likewise, sites where all three species may be detected have a value of 9 (1 + 3 + 5). This figure shows only the study region in Banff, west of Castle Junction. Evident in Plate 6 is the limited frequency and high dispersion of sites where all focal species have greater than 90% probability of crossing (i.e., the yellow sites). This suggests linkage zones must be species- or guild-specific, supporting previous arguments about connectivity requisites (Tischendorf and Fahrig, 2000; Conroy and Noon, 1996).

In major linkage zones where there exist fragmented regions of overlap, mitigation design might enhance the chances of both species crossing. For example, in the region directly west of Castle Junction there is some overlap between species. Mitigation design in this region should address the species-specific habitat needs that improve movement probability, such as maintaining a wide swath of land. Corridor width is one of the over-riding indicators of corridor success (Harrison and Bruna, 1999). Such design would increase the functional connectivity of the landscape (Tischendorf and Fahrig, 2000).

13.5 Policy Implications

As a result of the research findings, a number of policy implications have been presented to management agencies in the region, which are summarised in Table 13.3. These policy recommendations should serve as general guidelines for the spatial requisites of highway mitigation in the Rocky Mountain ecosystem.

Table 13.3 *Policy implications of the research*

1	Placement of mitigation should occur in high probability crossing areas, as defined on the GIS predictive model, herein.
2	Mitigation should be between 250 and 1000 metres in diameter (Alexander and Waters, 2000).
3	Using the above, mitigation should use elevated and buried highway sections, as often as possible. Exclusion fencing will be important in non-corridor areas.
4	Mitigation should be species- or guild-specific to maintain functional connectivity and ecosystem integrity.
5	Lynx and wolves may be used as focal species for mitigation (Alexander, 2001).
6	Higher resolution GIS images (e.g. <30 metres) are needed to manage appropriately for smaller functional levels (i.e. marten).
7	The adoption of GIS probability modelling and decision support is a cost-effective means of identifying crossing sites and provides a traceable process for decision makers.

13.6 Conclusion

This chapter examined the use of probability modelling in determining sites with the highest probability of detecting wildlife along four road systems in the Rocky Mountains of Alberta, Canada. The principal goal of the research was to locate optimal sites for placing mitigation structures (overpasses, underpasses or elevated highways) in order to restore structural and functional habitat connectivity for wildlife. Structural connectivity was addressed in this research by using spatial modelling and probability theory to find the optimal linkages in the landscape. Functional connectivity was addressed by identifying which species were good representatives of two levels of ecological organisation (i.e., meso- and macro-scales). Maintaining structural connectivity for two levels of ecological organisation alone will not achieve functional connectivity, if the design attributes of crossing structures are not appealing to the species involved (e.g., too narrow or too sparsely vegetated). This requires further examination of individual species' tolerances to habitat structure and geometry.

Biophysical attributes that predict movement were tested for non-uniform use by each of 13 study species. Attributes included aspect, greenness, slope, distance to drainages, distance to roads and toposhape (curvature). Strong evidence was found that aspect, greenness, slope and toposhape were useful predictors of habitat selection. It was observed that these preferences were well defined using probability functions. Moreover, in the analysis of track density estimates, this technique was argued to be superior to classical statistical approaches such as logistic regression or compositional analysis.

It was expected that greenness would be a useful predictor, based on work with grizzly bears (Mace *et al.*, 1999). However, in this case greenness was not suitable as a predictor because of the frame of analysis. There was little separation in greenness classes within the region of 1 kilometre from roads. Consequently, the inclusion of greenness obscured most linkage zones by applying an overwhelming 'greenness' suitability to habitat near roads. Therefore, greenness was removed from the analysis.

Structural connections were evident along the valley in BNP and these varied by species. Zones of overlap for both species that had high probability of use, were not abundant in the valley. This confirms the hypothesis that mitigation must be species-specific, or at least, guild-specific. Further tests should examine the periodicity of placement that may best suit each species/guild. In addition, design constraints should be examined. Other metrics should be tested for use in predicting species occurrence. For example, some species are known to associate with forests that have high structural complexity at the ground. Expert knowledge on relationships between metrics and finer scale predictors could be gleaned and incorporated into the weight-of-evidence model.

The strength of this modelling technique lies in its flexibility, its explicitness and its lack of assumptions associated with the more traditional statistical methods. Future data could be used to assess the predictive power of the model. Future analysis or evidence may suggest that other metrics are important, and these could be incorporated into the model. Moreover, if relationships within attributes (by class) are found to change, the associated probability functions can be modified. Human preferences could also be included in this model, at a later date. For example, highway structural engineers may want to evaluate the costs of mitigating each identified linkage zone, based on the site criteria. The flexibility of this tool enhances its suitability for use in resource management, because it makes the tool practical. The model is explicit in its assumptions and rules. Explicitness is a desirable quality for resource management agencies, such as Parks Canada, who are subject to public scrutiny and who must be accountable for decisions, and for replication and adaptation of the scientific research.

Acknowledgements

This project was made possible by the generosity of the following financial contributors: the Natural Sciences and Engineering Research Council of Canada, the University of Alberta and A.C.A. Biodiversity Fund, the Province of Alberta Graduate Fellowship and the University of Calgary, Alberta Environmental Protection, Canadian Pacific Corporation, Banff National Park Wildlife and Highways Divisions, Dr Margaret P. Hess, Mr Edward L. Alexander, the Alberta Sport-Recreation-Parks and Wildlife Foundation, Employment Canada, the Western Forest Carnivore Society and the Paquet Fund for Wildlife Biology.

References

Alexander, S.M. (2001) A Spatial Analysis of Road Fragmentation and Linkage Zones for Multi-Species in the Canadian Rocky Mountains: A Winter Ecology Study, PhD, Department of Geography, University of Calgary, Alberta, 350pp.

Alexander, S.M. and Waters, N.M. (1999) Decision support methods for assessing placement and efficacy of road crossing structures for wildlife, in Evink, G.L. *et al.* (eds) *Proceedings of the Third International Conference on Wildlife Ecology and Transportation*, Missoula, MT, ICOWET, September, pp. 237–46.

Alexander, S.M. and Waters, N.M. (2000a) Modelling wildlife movement and planning mitigation in highway transportation corridors, CD-ROM publication, Fourth International Ecological Modelling Conference, Banff, Alberta, September.

Alexander, S.M. and Waters, N.M. (2000b) The effects of highway transportation corridors on wildlife: a case study of Banff National Park, *Transportation Research Part C.*

Alexander, S.M, Paquet, P.C. and Waters, N.M. (2002) *Quantification of the Road Barrier Effect on a Wildlife Community in the Rocky Mountains of Canada*, in preparation.

Anderson, D.R., Burnham, K.P. and Thompson, W.L. (2000) Null hypothesis testing: problems, prevalence, and an alternative, *Journal of Wilderness Management*, **64**(4): 912–23.

Andren, H. (1994) Effects of habitat fragmentation on birds and mammals in landscapes with different proportions of suitable habitat: a review, *Oikos*, **71**: 355–66.

Banff-Bow Valley Study (1996) Banff-Bow Valley: At the Crossroads, Technical Report of the Banff-Bow Valley Task Force, prepared for the Honourable Shiela Copps, Minister of Canadian Heritage, Ottawa, ON.

Bartel, A. (2000) Analysis of landscape pattern: towards a 'top down' indicator for evaluation of landuse, *Ecological Modelling*, **130**: 87–94.

Bascompte, J. and Solé, R. (1996) Habitat fragmentation and extinction thresholds in spatially explicit models, *Journal of Animal Ecology*, **65**: 465–73.

Carroll, C., Paquet, P., Noss, R. and Strittholt, J. (2000) *Modeling Carnivore Habitat in the Rocky Mountain Region: A Literature Review and Suggested Strategy*, prepared for WWF Canada.

Conroy, M.J. and Noon, B.R. (1996) Mapping of species richness for conservation of biological diversity: conceptual and methodological issues, *Ecological Applications*, **6**: 763–73.

Dragicevic, S. and Marceau, D.J. (2000) A fuzzy set approach for modelling time in GIS, *International Journal of Geographical Information Science*, **14**(3): 224–45.

Eastman, J.R. (1999) *Idrisi 32. Guide to GIS and Image Processing, Volume 1,* Clark Labs, Clark University, MA, p. 170.

Ellison, A.M. (1996) An introduction to Bayesian inference for ecological research and environmental decision-making, *Ecological Applications*, **6**(4): 1036–46.

Evink, G.L., Garrett, P. and Zeigler, D. (eds) (1998a) *Proceedings of the Second International Conference on Wildlife Ecology and Transportation (ICOWET)*, FL-ER-73-99, Florida Dept. of Transportation, Tallahassee, Florida.

Evink, G.L., Garrett, P. and Zeigler, D. (eds) (1999) *Proceedings of the Third International Conference on Wildlife Ecology and Transportation (ICOWET)*. Florida Dept. of Transportation, Tallahassee, FL, Missoula, MT.

Evink, G.L., Garrett, P. Zeigler, D. and Berry, J. (eds) (1998b) *Proceedings of the International Conference on Wildlife Ecology and Transportation, FL-ER-69-98*, Florida Dept. of Transportation, Tallahassee, Florida, p. 263.

Fahrig, L. and Merriam, G. (1985) Habitat patch connectivity and population survival, *Ecology*, **66**(6): 1762–8.

Forman, R.T.T. and Alexander, L.E. (1998) Roads and their major ecological effects, *Annual Review of Ecology and Systematics*, **29**: 207–31.

Gordon, J. and Shortliffe, E.H. (1985) A method for handling evidential reasoning in a hierarchical hypothesis space, *Artificial Intelligence*, **26**: 323–57.

Harrison, S. and Bruna, E. (1999) Habitat fragmentation and large-scale conservation: what do we know for sure? *Ecography*, **22**: 225–32.

Holling, C.S. (1992) Cross-scale morphology, geometry and dynamics of ecosystems, *Ecological Monographs*, **62**: 447–502.

Jackson, S. (1999) Overview of transportation related wildlife problems, in Evink, G.L. *et al.* (eds) *Proceedings of the Third International Conference on Wildlife Ecology and Transportation (ICOWET)*, FL-ER-73-99, Florida Dept. of Transportation, Tallahassee, Florida, pp. 1–4.

Khattatov, B.V., Lamarque, J.-F., Lyjak, L.V., Menard, R., Levelt, P.F., Tie, X.X., Gille, J.C. and Brasseur, G.P. (2000) Assimilation of satellite observations of long-lived chemical species in global chemistry-transport models, *Journal of Geophysical Resources*, **105**: 29135.

Lenton, S.M., Fa, J.E. and Perez del Val, J. (2000) A simple non-parametric GIS model for predicting species distribution: endemic birds in Bioko Island, West Africa, *Biodiversity and Conservation*, **9**: 869–85.

Mace, R.D., Waller, J.S., Manley, T.L., Ake, K. and Wittinger, W.T. (1999) Landscape evaluation of grizzly bear habitat in western Montana, *Conservation Biology*, **13**: 367–77.

Maurer, M. (1999) Development of a community-based, landscape-level terrestrial mitigation decision support system for transportation planners, in Evink, G.L. *et al.* (eds) *Proceedings of the Third International Conference on Wildlife Ecology and Transportation (ICOWET)*, FL-ER-73-99, Florida Dept. of Transportation, Tallahassee, Florida, pp. 99–110.

Noss, R.F. and Csuti, B. (1997) Habitat fragmentation, in Meffe, G.K. and Carroll, R.C. (eds) *Principles of Conservation Biology.* Second edition, Sinauer Associates, Sunderland, MA, pp. 269–304.

Noss, R., Quigley, H., Hornhocker, M., Merrill, T. and Paquet, P. (1996) Conservation biology and carnivores: conservation in the Rocky Mountains, *Conservation Biology*, **10**(4): 949–63.

Paquet, P.C. (1993) *Summary Reference Document – Ecological Studies of Recolonizing Wolves in the Central Canadian Rocky Mountains*, prepared by John/Paul and Associates for Parks Canada, BNP Warden Service, Banff, AB, 215pp.

Pearce, J. and Ferrier, S. (2000) Evaluating the predictive performance of habitat models developed using logistic regression, *Ecological Modelling*, **133**: 225–45.

Pearl, J. (1987) Evidential reasoning under uncertainty, in Shrobe, H.E. (ed.) *Exploring Artificial Intelligence: Survey Talks from the National Conferences on Artificial Intelligence,* Morgan Kaufmann, Palo Alto, CA, pp. 381–490.

Peek, J.M. (1986) *A Review of Wildlife Management*, Prentice-Hall, Englewood Cliffs, NJ.

Richards, J.A. and Jia, X. (1999) *Remote Sensing Digital Image Analysis: An Introduction*, Springer, New York.

Singleton, P. and Lehmkuhl, J. (1999) Assessing wildlife habitat connectivity in the Interstate 90 Snoqualmie Pass Corridor, Washington, in Evink, G.L. *et al.* (eds) *Proceedings of the Third International Conference on Wildlife Ecology and Transportation (ICOWET)*, FL-ER-73-99, Florida Dept. of Transportation, Tallahassee, Florida, pp. 75–84.

Smith, D. (1999) Identification and prioritization of ecological interface zones on state highways in Florida, in Evink, G.L. *et al.* (eds) *Proceedings of the Third International Conference on Wildlife Ecology and Transportation (ICOWET),* FL-ER-73-99, Florida Dept of. Transportation, Tallahassee, Florida, pp. 209–30.

Spellerberg, I.F. (1998) Ecological effects of roads and traffic: a literature review, *Global Ecology and Biogeography Letters*, **7**: 317–33.

Srinivasan, A. and Richards, J.A. (1990) Knowledge-based techniques for multi-source classification, *International Journal of Remote Sensing*, **11**(3): 505–25.

Swart, J. and Laws, M.J. (1996) The effect of habitat patch connectivity on samango monkey (*Cercopithecus mitis*) metapopulation persistence, *Ecological Modelling*, **93**: 57–74.

Thomson, I.D., Davidson, I.J., O'Donnell, S. and Brazeau, F. (1988) Use of track transects to measure the relative occurrence of some boreal mammals in uncut forest and regeneration stands, *Canadian Journal of Zoology*, **67**: 1816–23.

Tischendorf, L. and Fahrig, L. (2000) On the usage and measurement of landscape connectivity, *Oikos*, **90**: 7–19.

Trombulak, S.C. and Frissell, C.A. (2000) Review of ecological effects of roads on terrestrial and aquatic communities, *Conservation Biology*, **14**(1): 18–30.

Urbanski, J.A. (1999) The use of fuzzy sets in the evaluation of the environment of coastal waters, *International Journal of Geographical Information Science*, **13**(7): 723–30.

Van Dyke, F.G., Brocke, R.H. and Shaw, H.G. (1986) Use of road track counts as indices of mountain lion presence, *Journal of Wilderness Management*, **50**(1): 102–9.

Verbyla, D. and Chang, K. (1994) Potential problems in using GIS for wildlife habitat research, *GIS World 1994 Conference Proceedings*, Vancouver, BC, February.

Wade, P. (2000) Bayesian methods in conservation biology, *Conservation Biology*, **14**(5): 1308–16.

Warntz, W. and Waters, N.M. (1975) Network representations of critical elements of pressure surfaces, *Geographical Review*, **65**: 476–9.

Weaver, J.L., Paquet, P.C. and Ruggiero, L.F. (1996) Resilience and conservation of large carnivores in the Rocky Mountains, *Conservation Biology*, **10**(4): 964–76.

Zadeh, L.A. (1965) Fuzzy sets, *Information and Control*, **8**: 338–53.

Part Four

NATIONAL SPATIAL PLANNING

14

Modelling Migration for Policy Analysis

Phil Rees, A. Stewart Fotheringham and Tony Champion

Abstract

This chapter reports on the development of an applied model of internal migration in the UK to support central government policy making. A two-stage modelling procedure was developed and parameters calibrated on a time series of historical data. Details of the data sets and the modelling are presented, together with illustration of the software package named MIGMOD that was constructed as a planning support system to assist government officials in predicting the impact of alternative scenarios.

14.1 Introduction

This chapter describes the results of a project carried out by a team (see Acknowledgements) based in the Universities of Newcastle upon Tyne and Leeds for the UK government during 2000 (Phase 1) and 2001 (Phase 2). Phase 1 was led by Tony Champion and Phase 2 by Stewart Fotheringham. The UK government department sponsoring the project (see Acknowledgements) changed its name and composition twice during the project and was known successively as the Department for Environment, Transport and the Regions (DETR), the Department of Transport, Local Government and the Regions (DTLR) and the Office of the Deputy Prime Minister (ODPM). For simplicity we will refer to the various organisations as 'the Department'. The chapter is published with the Department's permission but represents the views of the authors and not the Department.

Applied GIS and Spatial Analysis. Edited by J. Stillwell and G. Clarke
 ISBN: 0-470-84409-4

The chapter describes the aims of the project, the type of spatial model constructed and calibrated and explains the choices in model design that were made during the course of the project. Exposing the choices explored and the different model designs used in Phase 1 and Phase 2 will, we hope, provide valuable guidance to other researchers seeking to model the flows of migrants over time between a large number of origin and destination areas. The main output of the project is a user-friendly software system known as MIGMOD (Migration Modeller) which we also describe. The original aim of the project was that it might help inform policy decisions, and we have undertaken simple policy scenarios. These are illustrative and indicative of the model's performance, rather than a true reflection of the effect of policy changes. This account of the work complements the overview provided by Champion *et al.* (2002).

The chapter is divided into a further seven main sections. Section 14.2 places migration within the UK in the broader context of population change and explains why the Department needed a model of migration. Section 14.3 explains the aims that the model was designed to achieve. Section 14.4 outlines the 'state space' of the model: what spatial units were used, what migrant groups were studied and what time periods were used in which to measure migration. Most of the choices were 'data-driven'. Section 14.5 describes the nature of the dependent variable being modelled and the set of determinant variables used in the two stages of the model (out-migration from origins, distribution from origins to destinations). Section 14.6 describes the model of out-migration used, known as the Stage 1 model, which is used to identify the factors that contribute to varying rates of out-migration from origins across England and Wales. Section 14.7 describes the Stage 2 model, a model of destination choice by migrants, which is used to determine the factors that lead to varying levels of destination attractiveness. In Section 14.8 we describe how the two models are integrated within a user-friendly software system for migration modelling termed MIGMOD. We also discuss the revisions to model design made between Phase 1 and Phase 2 of the project.

14.2 Context: The Role of Internal Migration in Population Change

Population change is often a good indicator of regional economic prosperity. Regional policy is often concerned with the direction of population shift: decline is a trend to be avoided because it leads to destruction of social and physical capital. Population change is also one of the main ingredients in household change, although the two are not necessarily highly correlated as evidenced by the decrease in average household size in recent decades in the UK leading to a greater household growth than population increase. Population change is driven by components which can have very different geographical patterns.

Table 14.1 presents estimated regional components of population change for Government Office regions of the United Kingdom in 1999–2000. Only three regions – Scotland, the North East and West Midlands – experienced such population decline, while London and the southern regions showed the highest rates of increase. The

Table 14.1 *Components of population change for Government Office regions (England) and countries, UK, 1999–2000*

Country and Government Office Region (England)	Rates (per 1000 population)				
	Natural increase	Net external migration	Net internal migration	Other changes	Total change
Wales	−1.0	2.0	2.1	0.0	3.1
Scotland	−1.2	0.8	−0.8	0.2	−1.0
Northern Ireland	4.1	0.5	−0.5	−0.6	3.5
North East	−0.8	0.9	−1.7	0.1	−1.6
NW & Merseyside	0.1	2.8	−1.1	0.1	1.9
Yorkshire & the Humber	0.6	1.2	−0.1	0.5	2.2
East Midlands	0.5	−0.6	3.8	0.3	4.0
West Midlands	1.1	0.2	−1.4	0.1	0.0
East	1.4	1.7	4.0	0.3	7.5
London	5.9	15.6	−9.1	−0.1	12.3
South East	1.1	1.5	2.1	0.0	4.6
South West	−1.2	3.2	6.2	−0.3	8.0
United Kingdom	1.0	3.2	0.0	0.1	4.3

Source: Computed from the Office for National Statistics, *Population Trends 108*, Tables 1.7 and 8.1 and *Key Statistics and Vital Statistics, Local Authorities, England and Wales, 2000*, Table 9a Estimated mid-2000 resident population, persons all ages: components of change.

Notes:

1. External migration includes asylum seekers and visitor switchers.
2. The components are based on mid-year population estimates current at the time of writing. The population estimates for mid-years from 1981 through 2000 will be revised in the light of the publication of the 2001 Census results. Prior to the 2001 Census the best estimate for the mid-2001 population was that in the 2000-based national projection (Shaw, 2002) at 59.987 millions. This compares with a 2001 mid-year estimate published in October 2002 of 58.837 millions based on the 2001 Census 'One Number' of 57.789 millions.

contributions of the three components of population change, natural increase (births minus deaths), net external migration (inflows from the rest of the world minus outflows to the rest of the world) and net internal migration (inflows from the rest of the UK minus outflows to the rest of UK) vary across the regions. Natural decrease characterises Wales, Scotland, the North East and the South West. Natural increase is low elsewhere except in Northern Ireland and London. Net external migration makes a positive contribution to population change in all but one region but a massive contribution to the growth of London's population. From Table 14.1 we can also see that internal migration is a vital component that shifts populations between regions because there are both gaining and losing regions. For example, Scotland, the North East, North West and West Midlands lost between 4000 and 8000 people through internal migration, while Northern Ireland and Yorkshire and the Humber had small losses. London lost substantially through internal migration though this was more than compensated for by its gain from overseas. The other regions, particularly the East, South West and East Midlands gained through internal migration.

If more balanced population growth is a policy goal, then this means improving survival chances in high mortality regions, encouraging couples to have more children in certain regions or spreading the influx of immigration across regions more evenly or reducing migration outflows from certain regions and decreasing migration inflows into others. Of these options, the latter would seem less controversial and more acceptable as a solution to population imbalances. In order to stem population losses from depressed regions, and to stem the increasing population pressures in economically successful regions, it is necessary first to understand how such migration imbalances arise. One of the best ways to achieve that understanding is to build a model of migration behaviour and to investigate the determinants of both out-migration and destination choice. Such a model is reviewed in this chapter.

14.3 Aims of the Project

14.3.1 Introduction

The primary aim of the project was to provide the Department with an initial migration model aimed at informing policy. The model was designed to enable the investigation of the first-round quantitative impacts of alternative economic and policy scenarios on gross flows of population between regions. For the longer term, the intention was that the model would be capable of further enhancement, notably in order to gain a better understanding of the role of migration in influencing household change. It was clear from the outset that there were insufficient reliable data for regions, counties or local authorities on the migration of whole or part households or individuals migrating between households to construct a model that would directly connect to the Department's Household Projection model. The Department was well aware that migration flows produced feedback effects on areas (spirals of decline resulting in out-migration sometimes create opportunities for renewal or heavy in-migration resulting in denser settlement which lowers the attractiveness of destinations) but the team and the Department agreed early on that such modelling was over-ambitious until simpler cross-sectional or time series models had been calibrated and verified.

The 'problems' to be solved were of two kinds. The first was to find ways of reducing heavy migration losses from northern and midland cities in the UK and from other areas with poor economic track records (former coalfields, the Thames Gateway region, rural Wales). Heavy out-migration from these regions means shrinking demand for housing and other consumption goods, leading to a downward spiral of economic and social development, and loss of valuable social capital (Walker, 2002). However, we should also recognise that inter-area migration is also a means of shifting people to areas of economic opportunity. The second problem was to avoid the consequences of over-stimulation of successful regional economies. High in-migration might put pressure on house prices and make housing unaffordable to workers outside the dynamic sector, as is very much in evidence in and around London currently.

14.3.2 Detailed Objectives

The principal objectives were to develop an initial policy model which:

- was capable of handling gross migration flows between regions;
- could be used to examine the impact of alternative scenarios;
- could provide advice on urban–rural shifts;
- was compatible with household projections methodology;[1]
- was flexible, easy to use and transparent; and
- was capable of further enhancement, especially at a finer spatial scale.

It was agreed at the outset that, in developing this initial model, there was a need for pragmatism, notably in relation to data availability and feedback effects. In particular, this initial model would not attempt to incorporate the impacts of migration or other changes on the determinant variables. *Phase 1* of the modelling work established the framework, set up the database, developed a prototype working model and produced a standalone version for use by the Department. *Phase 2* reviewed model robustness, conducted a full range of diagnostic tests of determinants and investigated some counterintuitive results from the prototype model in detail. The overall model structure remained the same in both phases but different model forms were adopted in Phase 2 after extensive winnowing of the very large determinant dataset and after experimentation with multiplicative or additive equations linking dependent migration and independent determinant variables and with linear and quadratic and exponential functions of the determinant variables.

14.3.3 The Modelling Framework

The central features of the approach were:

- the modelling of out-migration from each area by reference to a set of determinant variables (Stage 1 of the model);
- the use of a model to distribute migrants between destinations (Stage 2 of the model);
- the development of an operational, user-friendly combination of Stages 1 and 2 that enables the model user quickly to set up and run a range of 'what if?' scenarios and to view the large volume of inputs and outputs (Stage 3 of the model); and
- the development of a selection of scenarios of determinant variables reflecting desired policy options (Stage 4 of the model).

[1] The ODPM household and ONS population subnational projections use local authority (LA) zones, whereas the migration model described here uses FHSAs. FHSAs are aggregations of LAs for most of the 1983–98 period, though from 1996 Health Authority reorganisation meant small inconsistencies in the aggregation. Changes in Health Service organisation since 1998 mean that the data series will not extend and future work will need to employ the New Patient Migration data series (ONS, 2002). These new data are available directly for the LAs used in subnational projections.

The modelling team discussed and agreed this framework with the Department very early on in the project. In fact, the rationale for this two-stage approach was reviewed in Champion *et al.* (1998). The main argument for modelling the out-migration flow separately from the destination choice is that different data sets were available for the two types of decisions. In order to model out-migration determinants, the team had available to it 15 years of annual out-migration data which allowed some time effects to enter the modelling framework (see below). In order to model the determinants of destination choices, an origin–destination matrix of migration flows was needed and this was only available for seven time periods which was not sufficient to allow time-related variables to enter the model. There is also evidence that migrants often conceive of migration as a two-stage process with worsening conditions at an origin eventually reaching a threshold level at which people decide to leave and then conditions at various locations being examined in order to decide on a destination. The model structure adopted conforms to the consistency principles for spatial interaction models first defined by Wilson (1967, 1972). The equation structures used in Phase 1 and Phase 2 versions of the migration model are discussed below in Section 14.4.

The calibration and testing of the model has embraced several dimensions of space, time, age and gender. Decisions needed to be taken on the zones to employ in the model, the time horizon over which the model would be calibrated and the disaggregation of migration into sub-categories for which different models were fitted. These decisions are discussed in the next section of the chapter.

14.3.4 Additional Investigations

In the development of the initial migration model, the project included three discrete modules of optional work. These were (i) improving the treatment of international migration, (ii) providing a population projection framework and (iii) examining the household dimension of migration. Details of this additional work can be found in the Phase 1 Final Report (Champion *et al.*, 2002).

14.4 The State Space

14.4.1 Spatial Units

For which areas should a migration model be developed? This is really two questions: what scale of spatial unit should be used and what collection of units should the model be calibrated for? Because the Department wished to know how migration affected household change, the ideal zone to have chosen would have been the local authority (LA). However, the UK did not have an annual time series of migration data at the LA scale until 1998 when migration data based on the patient registers came on stream. Data are available in the decennial censuses (1971, 1981, 1991 and soon 2001) at LA scale though the number and definition of LAs differs between

1971 and 1981 and between 1991 and 2001. The time period between the censuses is too long to allow any great understanding of the temporal influences on migration patterns.

An alternative is to use migration data available on an annual basis from the National Health Service Central Register (NHSCR). The zone system used in this series are the Family Health Service Authority (FHSA) areas as defined between 1983 and 1998 (see Figure 15.2 in Champion *et al.*, 2002) in England and equivalent authorities in Wales, Scotland and North Ireland. FHSAs were coincident with shire counties (England and Wales), metropolitan districts (England), groups of London boroughs, Area Health Boards (Scotland) and Health Boards (Northern Ireland). Ideally, the migration model would have been constructed using all of these areas which cover the whole of the United Kingdom. Unfortunately, there are gaps in the flow statistics between areas in Scotland for a number of years and between areas in Northern Ireland and in Great Britain. In Phase 1 of the model construction, therefore, it was decided to treat Scotland and Northern Ireland as two zones only. FHSAs in Wales were included as separate zones in the analysis. Results for Wales were not required directly by the Department because Wales was not its responsibility but the team argued strongly that migration links between places in England and places in Wales were strong and their exclusion would bias results for a good part of England. The flows between England, Scotland and Northern Ireland were much weaker and use of single zones for these countries would not seriously bias model results. The system to be modelled at the outset consisted of migration flows from 100 origins to 100 destinations disaggregated by age and sex.

In the event, one or two changes to this set of origins and destinations were necessary because of difficulties in estimating determinant variables for Scotland and Northern Ireland comparable to those for England and Wales. The statistics available for Scotland and Northern Ireland did not match sufficiently well to merit inclusion of these zones in Stage 1 of the model, so that only 98 zones were used. For Stage 2 there were fewer difficulties because the variables were cross-sectional in nature and only needed for 1996–97. However, major difficulties arose in the prediction of migration flows to Northern Ireland. With the set of determinant variables used in Stage 2 of the model, we consistently and massively overpredicted flows to Northern Ireland and therefore underpredicted flows to other zones. Missing from the model were variables that measured the relative attractiveness of zones based on perceptions of their political history and threat of external violence. In the absence of such measures, we confined the Stage 2 distribution model to 99 zones. In Phase 2 of the work these decisions were reviewed and the number of zones reduced to 98 in both stages of the model.

The effect of excluding some sub-national areas from the model calibration could not be tested with the data sets we were using but is an issue that could be addressed using 2001 Census data when they are available, as the 2001 Census Origin-Destination Statistics will provide comparable and comprehensive flow statistics between all LAs in the UK. There is also the question about what effect does immigration from, and emigration to, external zones have on migration between internal zones (whether those external zones fall within the UK or without). We did include estimated measures of external immigration and emigration (Stillwell *et al.*, 2001) in the internal migration model but their close correlation with variables measuring 'urbanness' meant their exclusion from the final determinant set to avoid problems of collinearity. The results of the model calibrations of immigration rates to destination zones suggest that previous immigrant stocks, the numbers of term-time students and population size of zone were the main attractors and that immigration and internal migration operated in different 'worlds'.

14.4.2 Time Series

For which years should the migration model be calibrated? Should we use as many years as the available information allow or strive to calibrate for particular periods? NHSCR flow data by origin, destination, age and sex are available for the 98-zone system from 1975–76 (mid-year to mid-year) until 1998–99 (at the time that the work was carried out). However, there are two major discontinuities in the series that make it necessary to confine analysis to the years 1983–84 to 1997–98. Prior to 1983–84, the data were published only as aggregations of the full origin–destination–age–sex (ODAS) array of flows, namely as three sub-arrays of migration flows by origin–destination (OD), by origin–age–sex (OAS) and by destination–age–sex (DAS). Between 1983–84 and 1997–98 the data were supplied as frequency counts classified by origin–destination–age–sex (ODAS). The age classification was by single years of age, whereas prior to then it had been by five-year age group. In the years 1996–98 local governments were reorganised successively in Scotland, Wales and parts of England. After 1 April 1998 the migration data are recorded for these new LAs (see Wilson and Rees, 1998, for details of the links between pre- and post-1998 LA geographies). So there were available 15 mid-year to mid-year counts of ODAS-classified migration flows. How could these data be used and with what kind of migration models?

The first idea was to model the out-migration flows or rates as a time series. All the advice of our econometric colleagues (Barmby, Tremayne – see acknowledgements) was against that approach: at least 40 years of data were necessary. A decision was therefore made to pool the data for the different years into one set of observations for model calibration purposes. However, because we decided after experimentation that it was best to have a one-year lag between the migration outcome and the time reference of the determinant variable, the time series of observations used dropped to 14 (1984–85 to 1997–98) for Stage 1 of the model (the origin prediction model). In Phase 1 of the model fitting we did not use a time trend as one of the explanatory variables but subsequent examination of the time series of out-migration rates for ages 16–19 and 20–24 revealed that a steady rise in out-migration took place from 1990–91 onwards, due primarily to the expansion of higher education in the 1990s. A linear time trend was therefore introduced to Stage 1 of the model in the Phase 2 work.

One argument that was put to us after several presentations of the model was that we should have explored the possibility that the determinants of migration would behave differently in times of economic boom and bust, and that different outcomes would result. Some analysts have interpreted the influence of the economic cycle on migration as follows: regions enter booms and busts at different times and the cyclic effects spread outwards from inception points. So, for example, the recession of 1981–3 saw downsizing of manufacturing industry and job losses first in the northern regions and then spread south increasing north to south transfers. The boom of 1988–1990 stimulated house price rises and outward migration in the South East which spread to the rest of the country tipping the north into positive migration gains for a year or so before the traditional pattern of north to south losses reasserted itself. These effects are ones we did not explore, given the need to develop a robust and operational model in a short time frame for the Department.

For Stage 2 of the model, time is less critical an issue. We calibrated the distribution model for a single year only but checks on alternative years indicated that the structure of migration distribution from origins to destinations is very stable, being dominated by the 'competing destinations' variables of distance (with a contribution from contiguity), destination accessibility and population size (Fotheringham, 1986, 1991). The year chosen for the final calibration was 1996–97 rather than 1997–98 because boundary changes were beginning to affect some migration flows in 1997–98.

14.4.3 Age Groups

The review of knowledge about the patterns of migration and the reasons for moving suggested that these would vary significantly with a person's position in the life course. It is not usually possible with aggregate data to place migrants into exact life course stages (e.g., retired from work or still working) but usually information is available on the age of migrants. Ages can then be grouped into approximate life-course stages. How many should there be and where should the boundaries be drawn between stages?

Since the NHSCR data were available for single years of age up to age 85+, it would theoretically have been possible to fit 86 different models, one per age group. However, previous empirical work (Stillwell *et al.*, 1992) had established that most of the variation between ages could be described using a small number of grouped ages. Initially, some eight age groups were proposed: 0–16, 17–19, 20–24, 25–29, 30–44, 45–59, 60–69 and 70+. These correspond to childhood/schooling ages, the ages at which adolescents leave home for higher education, the ages at which students leave higher education for their working and partnership careers, the ages when they look for career advancement, the family formation ages, the later working ages (quiescent in terms of migration), the ages of retirement when locations are reassessed and the older ages when events such as loss of spouse/partner or increasing infirmity precipitate migration. We examined the correlations between the flow matrices at ages around the bounds of these life-course stages and moved 16 year olds into the late adolescents group. We also amalgamated the last two ages as the flow matrix for the

70+ age group was too sparse for reliable calibration. The final age groups chosen for the model were therefore seven in number: 0–15, 16–19, 20–24, 25–29, 30–44, 45–59 and 60+.

14.4.4 Gender

There is a growing interest in the differences between men and women in their migration behaviour, though the differences are only expressed when they are not in marriage, partnership or in families. The Government Statistical Service now requires all analyses to include a gender dimension and this we did in the migration model, though the differences in results are generally not great.

14.4.5 Summary

Choosing the state space that we did meant that we would be calibrating the Stage 1 and Stage 2 parts of the migration model for seven age groups and two genders or 14 migrant groups in total. However, there were further choices to be made. Should the out-migration model be calibrated for each origin separately or for all origins together or for groups of origins?

The option of calibrating the Stage 1 model for each origin was ruled out first: there were too few observations (14) per origin to produce reliable parameter estimates. We calibrated models in Phase 1 with grouped origins using Government Office Regions as the grouping criterion and compared the results with a model calibrated on all origins collectively. As would be expected, the former produced more accurate predictions of out-migration rates but a lot of the improved predictive performance came from the ability to estimate separate intercept terms for each zone. How should the variation of these intercepts be interpreted? It appeared that much of the variation in the intercepts was masking variation in some of the relationships between out-migration rates and origin attributes which was thus misleading. It appeared that some origin attributes had little effect on out-migration rates but this was simply because the effect of differential levels of those origin attributes was subsumed within the varying intercept values. Hence we opted for an 'all zones together' calibration for the 14 migrant groups and somewhat reluctantly dropped the idea of separate models for different clusters of origins.

The situation for the Stage 2 destination choice model was different. Here we had, for each origin, 97 destination flows upon which to calibrate the distribution model. By selecting origin-specific distribution models we could allow the determinants to have different influences on the outcomes for each origin. For example, we found that house prices (in the Phase 2 model) had coefficients that were both negative and positive depending on origin (other things being equal), while distance has negative coefficients in all cases (no origin had migrants who preferred to move to distant places). For Stage 2, therefore, it was possible to calibrate 98 $\times$ 14 or 1372 separate models (for just one year).

14.5 Data

14.5.1 Database Assembly

The major element of the work in Phase I of the project was the assembly of the data needed for calibrating the Stage 1 and Stage 2 models. Three broad types of data were required: (i) data on migration flows, (ii) data on the distance between each pair of the 100 'zones' and (iii) data on the characteristics of the 100 'zones' and their populations. We were guided in selection of determinant variables by the review undertaken in Champion *et al.* (1998), paying particular attention to the inclusion of variables that were directly or indirectly connected to the Department's policies for urban, rural and regional development.

In practice, several obstacles were encountered. Few datasets are published for all four countries of the United Kingdom, so data either had to be obtained as a series of separate tasks or had to be estimated. No datasets present statistics for the 100 zones used, so either data had to be aggregated from smaller units or values imputed from larger areas containing the zones. Data sources did not always provide information for all years. Some variables were found to need denominator data from a separate source, thus doubling the task (e.g. for crime rate defined as total number of offences per household).

Despite these problems, data were obtained for 139 potential determinants of out-migration and 69 potential determinants of migration destination choice. Much of this achievement was due to the valuable assistance given by a considerable number of persons and agencies (see Acknowledgements). Even so, it proved necessary to resort to estimation in order to complete the data series for all 98 FHSAs of England and Wales and for all relevant years. The data series for Scotland and, even more so, those for Northern Ireland were less comprehensive. In Phase 2 of the project, the initial 139 determinant variables were reduced through careful qualitative review and examination of multicollinearity to 49 potential determinants for Stage 1 of the model and the 69 potential determinants of destination choice were reduced to 23. The variables in the two datasets are listed in Table 14.2, which also gives the source of the data collected on each variable. Both data sets represent a huge increase in size from those normally used in migration models.

14.5.2 Within-UK Migration

As outlined in Section 14.4, the migration data are derived from the NHSCR and made available by Office of National Statistics (ONS). For the purpose of this study, data files have been constructed for the 100 zones for the 14 age–sex groups for mid-year to mid-year periods as follows:

- the numbers of out-migrations and in-migrations for 15 years 1983–84 through to 1997–98 are converted to rates based on the respective population groups of each zone; and

Table 14.2 *List of variables used in the (Phase 2) Stage 1 model*

Variable name	S1	S2	Meaning	Source
AIR	✓		Principal component of no2nox, ozone (high values associated with areas of poor quality)	AEA Technology. Values taken dataset point nearest to FHSA population-weighted centroid
AIRPORTS		✓	Access to scheduled flights to non-domestic destinations in 1997	Input data from CAA.
ASDIVRT	✓		% divorced – age-specific	1991 Census
ASUNEM		✓	Age-specific unemployment rate	NOMIS database; ONS mid-year estimates
ASUNEM_L	✓		Age-specific unemployment rate lagged	As ASUNEM
ASUNEM_L_Y	✓		Y variable for age-specific unemployment rate lagged	ASUNEM & DIJNETW
CTAX		✓	Average council tax for households	DETR Local Government website
CLIMATE	✓	✓	Principal component of frosty days, sunny days, rainfall, July temperature (high values associated with warmer drier areas)	University of East Anglia. Values taken dataset point nearest to FHSA population-weighted centroid
COMMUT	✓		% long distance commuters	1991 Census
CONTIG		✓	Contiguity dummy, equals 1 for pairs of zones sharing a boundary, otherwise 0	Derived from map of zones
CRIME	✓	✓	Principal component of household insurance premiums, index of crime a serious problem, total rates of offences recorded by police (high values associated with high crime areas)	Insurance companies, police records
DESTACC		✓	Sum of alternative destination populations divided by network distance from origin	ONS population estimates and DIJNETW
DIJNETW		✓	Network distance between each pair of zones	Computed from district level map by Duke-Williams and Alvanides
EMPGRO		✓	Employment growth	1991 Census and Survey of Employment
EMPGRO_L	✓		Employment growth lagged	As EMPGRO
EMPR		✓	Employment rate	As EMPGRO
EMPR_L	✓		Employment rate lagged	As EMPGRO
EMPR_L_Y	✓		Y variable for employment growth lagged	EMPR & DIJNETW
GCSE_L	✓		% 16 year olds obtaining 5+ GCSEs at Grade C or above lagged	DfEE, Scottish Office, DENI
HHINC		✓	Household income	Estimated by Bramley & Smart 1996, Bramley 1996
HHINC_L	✓		Household income lagged	As HHINC
HHINC_L_Y	✓		Y variable for household income lagged	HHINC & DIJNETW

HPRICE		✓	House price	Nationwide Building Society & CURDS, University of Newcastle
HPRICE_L	✓		House price lagged	As HPRICE
HPRICE_L_Y	✓		Y variable for house price lagged	HPRICE & DIJNETW
LISTED	✓	✓	Listed buildings (divided by dwellings 1998)	National built heritage organisations
LISTED_Y	✓		Y variable for listed buildings	LISTED
LOWI_L	✓		Index of relative share of low earners lagged	Estimated from New Earnings Survey
LOWI_L_Y	✓		Y variable for index of relative share of low earners lagged	LOWI_L & DIJNETW
MGINTR_Z_L	✓		Building Society mortgage interest rate lagged	Building Societies average
NOCENT	✓		% no central heating	1991 Census
NOCENT_Y	✓		Y variable for % no central heating	NOCENT & DIJNETW
NONWH	✓		% non-white	1991 Census
NONWH_Y	✓		Y variable for % non-white	NONWH & DIJNETW
OCCMIG	✓		Occupational migration index (predicted number of migrants based on occupation divided by residents – age-specific	Based on analysis of 2% SAR, then applied to 1991 Census LBS.
PARDOM	✓	✓	Students at parental domicile divided by number of residents, average of start and end mid-year estimates	HESA, UCAS and 1991 Census (Bailey and Rees 2001)
PNBU	✓	✓	New housing development (units) on land in former urban use divided by total new housing development	DETR & Ordnance Survey Land Use Change Statistics (LUCS)
PNRL		✓	% net relets social sector	DETR HIP1 series
PNRL_L	✓		% net relets social sector lagged	DETR HIP1 series
POPN		✓	Number of residents, average of mid-year estimates	ONS Population estimates
PQPR		✓	% new build completions private sector	DETR HIP1 series
PQPR_L	✓		% new build completions private sector lagged	DETR HIP1 series
PQPR_Y_L	✓		Y variable for % new build completions private sector	PQPR & DIJNETW
PQSR		✓	% new build completions social sector	DETR HIP1 series
PQSR_L	✓	✓	% new build completions social sector lagged	DETR HIP1 series
PQSR_Y_L	✓		Y variable for % new build completions social sector lagged	PQSR & DIJNETW
PVAC		✓	% vacant all sectors	DETR HIP1 series
PVAC_L	✓		% vacant all sectors lagged	DETR HIP1 series
PVAC_Y_L	✓		Y variable for % vacant all sectors lagged	PVAC & DIJNETW
RGDPCH_Z_L	✓		Real GDP annual % change lagged	ONS Economic Trends, Wilcox Housing Finance Review
RGURB	✓	✓	Land area subject to urban development of all types divided by population	DETR & Ordnance Survey Land Use Change Statistics (LUCS), and ONS mid-year estimates
RGURB_Y	✓		Y variable for land area subject to urban development	RGURB & DIJNETW

Table 14.2 *Continued*

Variable name	S1	S2	Meaning	Source
RLAPSC	✓	✓	Index of private sector stock in poor condition	Index used in HNI system, based on EHCS and proxies. England only.
RLASC	✓	✓	Index of local authority stock in poor condition	Index used in GNI system by DETR, based on stock profile and EHCS. England only
RLASC_Y	✓		Y variable for index of Local Authority stock in poor condition	RLASC & DIJNETW
SCENIC	✓	✓	Access to National Parks, Areas of Outstanding Natural Beauty and Heritage Coasts	Provided by DETR. Weighted averages of ward values.
SINGELD	✓		Proportion of single elderly households	1991 Census
SINGLE	✓		% single	1991 Census
TER	✓		% housing terraced	1991 Census
TERMT		✓	Students at term-time address divided by number of residents, average of start and end mid-year estimates	HESA, UCAS and 1991 Census (Bailey and Rees 2001)
TERMT_L	✓		Students at term-time address divided by number of residents, average of start and end mid-year estimates, lagged	HESA, UCAS and 1991 Census (Bailey and Rees 2001)
TIME_Z	✓		Time trend (year counter)	1 = 1984, 2 = 1985 etc.
TPOPN_Y_L	✓		Y variable for total population lagged	ONS Population estimate, TPOPN & DIJNETW
UNEMP_Z_L	✓		National claimant unemployment rate % lagged	DfEE, ONS population estimates
UNIVPL_Z_L	✓		Total number of students in England and Wales lagged	HESA Statistics
URBAN_L	✓		Principal component of bright lights indicator, % flats, % shared dwellings, population weighted average ward density of population, immigration rate, persons per room, HWCH, urbanisation factor (high values indicate more urbanised areas)	See text and Champion *et al.* (2001)
VACDRL	✓	✓	Vacant and derelict land divided by Total dwellings 1998 (HIP1)	Based on new National Land Use Database. Covers England only.
VACDRL_Y	✓		Y variable for all vacant and derelict land	VACDRL & DIJNETW
VISITS		✓	Visitor numbers to tourist attractions in 1996 divided by 1991 Census population	British Tourism Authority
WIDOW	✓		% widowed	1991 Census

Notes:
1. The suffix _L indicates a lagged variable. The suffix _Y indicates a regional variable. The suffix _Z indicates a national variable.
2. S1 = Stage 1 of the model: out-migration from origins. S2 = Stage 2 of the model: destination choice.

- the numbers of out-migrations from each of the 100 zones moving to each of the other 99 zones for each of the seven years 1990–91 through to 1996–97.

This NHSCR migration dataset is derived from the re-registration of individuals with family doctors (GPs) in another FHSA area. It therefore has the advantage of providing a continuous record of migration within the UK. As such, it has for many years been the main source of within-UK migration data used by ONS for its official population estimates for England and Wales. This is not to say that the NHSCR data provide a full and perfect record of migration. There is a well-recognised undercount of young adult males in the NHSCR datasets caused by this group of migrants being relatively less likely to re-register with their local GP. This causes the measured total out-migration values for these cohorts from many FHSAs to be smaller than they should be. If not corrected, this will lead to biased model calibrations if the undercount is not uniform across all FHSAs (which it is not).

To appreciate the nature of the problem, consider the comparison in Table 14.3 of the ratios of males/females in the SMS (Special Migration Statistics compiled as part of the 1991 Census) and the 1990/91 NHSCR for four sample FHSAs. These figures support the suspicion that young males have a tendency to postpone re-registering with a GP until later years: the ratios of males/females are much lower in the NHSCR dataset than in the SMS for the age groups 16–19, 20–24 and 25–29. The undercount, however, is not spatially stable and varies considerably across the FHSAs. Consequently, a model of out-migration totals for each migrant group will have problems predicting values for males in the age groups 16–19, 20–24 and 25–29 given the problems in the recorded values. There might be an associated problem of males re-registering at a destination after several years and hence inflating the figures in later age groups. There is some evidence for this in the Table 14.3 for Newcastle and North Tyneside, for example. Can we use the information on sex ratios in the SMS (which are presumed to be unbiased) to produce better migration matrices from the NHSCR data?

To achieve better estimates of migration from the NHSCR data, we assume that there is little or no bias in the recording of female migration in the NHSCR and that the ratio of total out-migrations by males/females for each origin in the NHSCR

Table 14.3 *Male/female ratios in migration outflows by age in Census and NHSCR data for 1990–91 for four FHSAs*

FHSA	Dataset 1990–91	Ages 0–15	16–19	20–24	25–29	30–44	45–59	60+
Newcastle	SMS	.92	.98	.99	1.12	1.18	1.16	.58
	NHSCR	.97	.65	.75	1.03	1.3	1.31	.67
North Tyneside	SMS	.93	.90	.87	1.12	1.23	1.14	.64
	NHSCR	1.02	.70	.67	.91	1.36	1.08	.68
Barnet	SMS	1.08	.75	.78	1.02	1.16	1.11	.65
	NHSCR	1.09	.61	.65	.76	1.07	1.00	.71
Kensington & Chelsea	SMS	1.06	.87	.73	.86	1.17	1.38	.76
	NHSCR	1.10	.69	.50	.60	.93	1.21	.81

should be the same as that observed in the SMS. This means that for each age group and for each FHSA the following equality holds:

$$M(n)/F(n) = M(s)/F(s) \qquad (14.1)$$

where $M(n)$ and $F(n)$ are the numbers of male and female out-migrants, respectively, from the NHSCR and $M(s)$ and $F(s)$ are the respective values from the SMS. Consequently, we can construct a better estimate of $M(n)$, $M(n)^*$, by deriving separately for each origin the revised number of male out-migrants:

$$M(n)^* = [M(s)/F(s)]F(n) \qquad (14.2)$$

For example, in the case of Newcastle, the 1990/91 NHSCR records a total of 214 male out-migrants. Using the above adjustment, the new total would be 0.98 × 330 (the number of 16–19 female movers from Newcastle recorded in the NHSCR) giving a value of 323. In a similar way, the total number of males 20–24 leaving Newcastle would rise from 1416 to 1869. This process was repeated for each origin for all seven age groups of migrants and across all 15 periods of the NHSCR. The assumption is made that the ratios of males/females is relatively constant through time and is recorded relatively accurately via the SMS.

These adjusted matrices were then used in the Stage 1 modelling procedure described below. To examine the impact of the adjustment, we calibrated a version of the Stage 1 out-migration model with the two sets of out-migrations (unadjusted and adjusted). Although the results are broadly similar, there are some differences in the estimated parameters and in seven cases different conclusions would be reached about the effect of individual variables on out-migration. In five of these seven cases the parameters change in terms of being significant at the 95% level and in the other two the sign of the parameter changes.

14.5.3 Migration Determinants

The data on the determinants of migration, both out-migration and destination choice, fall into three broad groups: (i) a relatively small set of *national* economic indicators; (ii) a much larger set of purely cross-sectional *zonal* variables; and (iii) a set of variables that are measured both over space and over time. The selection of many of the variables was informed by the findings of the review *The Determinants of Migration Flows in England* (Champion *et al.*, 1998), which identified seven factors influencing migration rates and places' attractiveness to migrants. These were demographic, cultural and social, labour market, housing, environmental, public policy and impedance. Below we provide a brief justification for seeking variables to represent each of these categories in the models.

14.5.4 Types of Determinant Variables

The two most important distinguishing features of the data assembled to represent the determinants of migration are whether the variable is national or local, and whether the observation is in the form of a time series or a single point in time. For any cross-sectional variable we can also compute a *regional* version of the variable (see below) which measures the incidence of that variable in regions other than the one from which out-migration rates are measured. In this way, in order to model out-migration rates, we can capture the pull effects of conditions in other regions as well as the conditions from which out-migration takes place.

All the national-level variables are time-series, with observations for each of the 14 years being modelled in Stage 1. In Table 14.2 these are denoted with the suffix 'Z'. A number of the local-level variables are also time-series in nature. Where this is the case, we have used a one-year time lag for the equivalent variable in the Stage 1 modelling to avoid problems of circularity. Lagged variables are denoted with the suffix 'L'. For the remainder of the local variables, there is only one observation time point. The most common reason for this is that the data are taken from the Population Census and therefore relate to 1991. Fortunately, most of the variables affected are ones that change only rather slowly and the year 1991 is around the mid-point of the Stage 1 modelling period.

Regional variables (labelled with Y suffix) are calculated for the Stage 1 out-migration model. These variables are meant to capture the possible pull effects on out-migration caused by conditions elsewhere in the country. After experimentation with alternatives, the following formula was used for the regional variables:

$$Y_i = \left[\sum_{j\, j \neq i} (X_j / X_i) d_{ij}^{\beta} \Big/ \sum_{j\, j \neq i} d_{ij}^{\beta}\right] \tag{14.3}$$

where X_i is the value of X at location i and X_j represents the value of X at one of the other FHSAs. The formula produces a distance-weighted average ratio of X_j to X_i where nearby locations are weighted more heavily in the calculation than more distant ones. The value of β was taken as −2 which, from experience, gives a reasonably differentiated surface of Y values. Values of β less negative than this give a surface which is smoother; values of β more negative will give a spikier surface. Values of $Y_i > 1$ indicate that X_i is generally *smaller* than its neighbours. Values of $Y_i = 1$ indicate that X_i is generally very similar to its neighbours. Values of $Y_i < 1$ indicate that X_i is generally *larger* than its neighbours. One point to note with the formula for the regionalised variables is that the value of Y will tend to infinity as X_i tends to zero. In the few instances where this occurred, small values were substituted for zero. Also, the formula should not be used if the X values take both negative and positive signs.

14.5.5 National Indicators

A number of factors can be considered to affect the overall volume of migration taking place in the UK. These are time-series variables that are used only in the Stage 1 (migration generation) modelling. They are primarily related to the state of the economy: although we considered a wider set in Phase 1 of the work. The national variables that survived for input to the Stage 1 model in Phase 2 were GDP change (RGDPCH_Z), unemployment rate (UNEMP_Z_L), mortgage interest rate

(MGINTR_Z) and the number of university places (UNIVPL_Z_L), together with a time trend indicator (TIME_Z). These variables influence people's willingness and ability to move house or to set up home independently.

14.5.6 Demographic Variables

These variables are area-specific, with each zone being potentially able to take a different value. To varying extents, most of them are considered relevant to both model stages, because they are considered to influence both the extent to which people are likely to leave a zone and, for these migrants, the choice of one destination zone over another. The most important demographic factor influencing the propensity to move house – the age of a zone's population, with its linkage to stage in the life course – is already accounted for by out-migration being modelled for the 14 separate age–sex groups.

It was, however, considered important to have a direct measure of students in higher education, because of their high level of longer-distance mobility. Considerable effort was therefore put into generating two variables for students (Bailey and Rees, 2001). One variable, PARDOM, measures their presence at their parental/vacation address (from which most would be leaving for university and to which many might be expected to return). The other, TERMT, is for their term-time address (to serve initially as destination and then, at the end of their studies, as origin). Note that such special treatment does not need to be given to Armed Forces personnel, the other distinctive high-mobility subgroup, because their movements are not covered by the NHSCR.

The other specifically demographic variables included in the model relate to total population size (POPN), rate of immigration from outside the UK (incorporated in the URBAN composite variable), incidence of divorce (ASDIVRT), distribution of the population by marital status (SINGLE, WIDOW) and composition by household type (SINGELD).

14.5.7 Socioeconomic and Cultural Variables

The likelihood of people moving house, and therefore possibly leaving an area, is strongly related to their socioeconomic and cultural attributes. In particular, past research has highlighted the migration differentials between groups based on ethnicity, occupation/social class and income. One key variable presented under this heading, therefore, is the proportion of the population that is non-white in ethnic origin (NONWH, NONWH_Y), used in the Stage 1 model. Several measures were assembled to represent occupation/social class including the proportions in groupings by Social Class and by Socioeconomic Group, but after a data reduction exercise just one was used for the Stage 1 (migration generation) modelling. An age–gender specific index (OCCMIG) has been generated to represent the relative chances of out-migration from each zone based on the occupational structure of its population. Several measures of income were assembled and, after investigation for multicollinearity, measures of average household income (HHINC_L and HHINC_Y_L) and the concentration of those on low incomes (LOW_L, LOWI_Y_L) were input to the Stage 1 model calibration and HINC was used in the Stage 2 model. We also included an educational measure, GCSE_L, in the Stage 1 model inputs.

14.5.8 Labour Market Variables

Labour market factors are seen as potentially important both in prompting out-migration from an area and in influencing people's destination choices. They would normally be interpreted as measures of the overall economic environment of the area, but some can also stand as factors affecting the propensity of individuals to move away from an area. A few variables were assembled to represent directly the labour market situation, including levels of employment (EMPR_L and EMPR_Y_L at Stage 1 and EMPR at Stage 2) and changes in jobs (EMPGRO_L at Stage 1, EMPGRO at Stage 2) and unemployment (ASUNEM_L and ASUNEM_Y_L at Stage 1 and ASUNEM at Stage 2). Beyond this, a number of the variables included above (under the 'social and cultural' heading), particularly those related to occupation and income, can also serve as indicators of the economic performance of areas, as can measures of the tightness of the housing market.

14.5.9 Housing Variables

Housing factors form a critical element underlying migration patterns in the UK, but they have complex interactions with migration and need especially careful treatment. On the one hand, some housing measures such as high house prices and low vacancy rates can reflect the strong economic performance of an area and indeed of neighbouring areas within commuting distance. On the other hand, these factors can directly influence the opportunities for in-migration, with high house prices acting as a deterrent and high vacancy rates as an attraction. Even here, however, the effect may not be as simple, as high house prices may lead to more rapid in-migration in anticipation of future price rises and very high vacancy rates may serve to undermine confidence in an area and deter people from moving there. The size, composition and quality of the housing stock can also influence both the level and the type of migration. Housing tenure, besides being a reflection of the social composition of an area, is also known to affect migration patterns, most notably the well-documented problems that people moving between local authority areas have in accessing council housing. A large number of housing variables were therefore assembled for the model. House prices are measured by HPRICE_L and HPRICE_Y_L at Stage 1 and by HPRICE at Stage 2. Vacancy rates are captured by PNRL_L, PVAC_L, PVAC_Y_L at Stage 1 and by PNRL, PVAC at Stage 2. House building is represented by PQPR_L, PQPR_Y_L, PQSR_L, PQSR_Y_L at Stage 1 and PQPR and PQSR at Stage 2. Housing condition is measured by NOCENT, NOCENT_Y,

RLAPSC, RLASC and RLASC_Y at Stage 1 and RLAPSC, RLASC at Stage 2. Housing tenure is input through TER at Stage 1.

14.5.10 Environmental Variables

In the current era within advanced economies, environmental factors play a major role in people's residential moves, both in prompting exits from areas and in acting as 'pull' factors. The term is used here in its broadest sense, covering all the physical, economic, social and political aspects that affect both the everyday quality of life and the longer term trends in life chances. This heading can therefore be considered to include most of the factors mentioned under the other headings above, insofar as they bear upon the overall quality of an area and of the neighbourhoods that it comprises.

Among those not specifically considered above are variables relating to derelict and vacant land (VACDRL and VACDRL_Y at Stage 1 and VACDRL at Stage 2), variables relating to the pattern of development, such as the proportion of new housing on brownfield land (PNBU at Stages 1 and 2), variables relating to population density, settlement size and level of urbanisation (combined into a composite variable URBAN), variables relating to crime (CRIME at Stages 1 and 2), variables relating to climate (CLIMATE at Stages 1 and 2) and air quality (AIR at Stage 1). Also included under the environment theme are variables relating to physical attractiveness, such as accessibility to scenic areas (SCENIC at Stages 1 and 2), and listed buildings (LISTED at Stages 1 and 2 and LISTED_Y at Stage 1), number of visitors (VISITS at Stage 2), the extent to which an area acts as a dormitory for commuters to relatively distant jobs (COMMUT at Stage 1) and accessibility to international air passenger connections (AIRPORTS at Stage 2). A 'bright lights' indicator measuring access to theatres and concert halls was incorporated into the URBAN composite.

14.5.11 Public Policy Variables

Public policy variables relevant to migration behaviour include not only direct interventions such as migration incentives and immigration policy but also indirect influences through the uneven effects of government grants, local taxes, defence spending, higher education expansion and the amount and location of land approved for house building. Two variables relating to expenditure on local government services, namely overall expenditure per capita and expenditure relative to the Standard Spending Assessment, were explored in Phase 1 but dropped because of multicollinearity. Level of council tax (CTAX) was included in the Stage 2 model.

In general, it was believed more satisfactory to estimate the role of public policy, past or anticipated, by reference to variables representing the aspects that public policy seeks to alter. For instance, the migration impact of a regional development initiative can be assessed by reference to, for instance, the number of extra jobs, while the impact of a policy that alters the availability of land for housebuilding can be studied via changes to the number of housing completions. This is the approach adopted for assessing the consequences for migration of alternative policy scenarios.

14.5.12 Impedance Variables

The term 'impedance' is used here to refer broadly to the friction of distance, which is the single most important factor affecting movers' choice of destination. For a whole set of reasons, people changing address tend to move over short distances; indeed, three-fifths of movers recorded by the 1991 Population Census moved less than 5 kilometres. All other things being equal, places that are further apart therefore tend to have less to do with each other than places that are close together. Secondly, areas that share a boundary tend to have more migration between them, because this migration will include a proportion of short-distance moves from one side of the boundary to the other. Thirdly, areas situated in high population-density regions are likely to be less attractive to migrants, everything else being equal, because of increased spatial competition between destinations.

These three spatial processes are allowed for in the Stage 2 (migration distribution) component of the model, using respectively:

- a network-weighted distance variable, DIJNETW, based on distances between the population 'centroids' of pairs of zones, calculated on the basis of shortest surface route rather than straight-line distance, so as to take account of the effect of estuaries;
- a contiguity variable, CONTIG, taking the value of 1 for zones that share boundaries with each other, 0 for other pairs of zones;
- a destination accessibility variable, DESTACC, which measures the degree of spatial competition faced by a destination from nearby destinations.

The distance variable was used indirectly in the Stage 1 model as it is used in the computation of all regional variables.

14.6 Stage 1: A Model to Predict Total Out-migration from Each Origin

The model consists of two stages which recognise that migration results from two factors: the desire to leave an origin and the choice of destination. The model therefore consists, firstly, of a model to predict migration outflows from each area and, secondly, of a model to allocate the flows of migrants from each origin area to each of the alternative destinations. This section outlines the Stage 1 model. The volume of results from this exercise is substantial and beyond the limitations of this chapter. They are left for detailed discussion elsewhere. However, the end product of the project was the combination of both the Stage 1 and the Stage 2 model results into a user-friendly software package. We describe some sample results from this software to show how the overall package operates. We also describe the changes in both the

Stage 1 and Stage 2 modelling procedures as we moved from the prototype in Phase 1 to the final product in Phase 2.

14.6.1 The Phase 1 Out-migration Model

The volume of out-migration from an origin zone i is predicted as

$$O_{it}^{m} = omr_{it}^{m} P_{it}^{m} \tag{14.4}$$

where O_{it}^{m} is the total out-migration of migrant group m from zone i in time interval t, omr_{it}^{m} is the out-migration rate from origin i in time unit t for migrant group m (one of the 14 age–sex groups) and P_{it}^{m} is the population of migrant group m at risk of migrating from origin i during time interval t (usually estimated as the mid-interval population or the average of start of interval and end of interval populations). The out-migration rate is in turn modelled, in general form as

$$omr_{it}^{m} = f\left(X_{it/t-1}^{m}, Y_{it/t-1}^{m}, Z_{t-1}\right) \tag{14.5}$$

where omr_{it}^{m} is the out-migration rate for migrant group m from zone i in time interval t, $X_{it/t-1}^{m}$ is a vector of origin attributes in either year t or $t-1$ (lagged by one year); $Y_{it/t-1}^{m}$ is a vector of distance-weighted attributes describing the situation in other areas in either year t or $t-1$ and Z_{t-1} is a vector of attributes describing the national economic situation as it affects the overall volume of migration in year $t-1$ (lagged by one year).

In total, 14 models were calibrated, one for each migrant group defined on the basis of seven age groups and the two sexes. The models were calibrated using NHSCR data on total outflows from each of the 98 FHSA areas in England and Wales. Once calibrated, the Stage 1 model allows the total volume of out-migration from each area to be predicted, given conditions of that area, of surrounding areas and of the national economy.

What specific form should equation (14.5) take? Should the form be multiplicative or additive? Should we use logged functions of the variables (the multiplicative option) or leave them unlogged? Should non-linear forms of the variables such as quadratic forms be considered? In Phase 1, we adopted a multiplicative model with the specific form:

$$omr_{it}^{m} = \exp(K^{m}) A^{m} V_{1}^{a1m} \ldots V_{53}^{a53m} \exp(a_{54}^{m} V_{54}) \ldots \exp(a_{58}^{m} V_{58}) \tag{14.6}$$

where the term $\exp(K^{m})$ is the intercept in the log–log regression, specific to each age and sex group. The age and sex specific adjustment factor A^{m} is necessary because of a statistical bias in the intercept estimate in a log–log regression. The adjustment factor is obtained by constraining the total predicted out-migration of a specific migrant group for all 98 origins to equal its observed value for the 1996/97 time period. The model contained 58 explanatory variables. The variables V_{1i} to V_{53i} were explanatory variables which became log functions in the log–log regression, while the variables V_{54i} to V_{58i} could not be logged due to negative values and were represented in the exponential form. The regression parameters, represented by power coefficients, a_v^m, are specific to each variable v, and to each migrant group m.

The Stage 1 model developed in Phase 1 has a number of drawbacks, which were addressed in Phase 2 of the project. The first was the strong possibility of multicollinearity between determinant variables. This was addressed by both qualitative and quantitative means resulting in a much-reduced set of potential explanatory variables and a small set of composite indices constructed using principal components analysis. The second difficulty was the rather awkward combination of power and exponential relationships. This was addressed by switching to a straightforward linear model. The third problem was the assumption that a monotonic relationship existed between a dependent variable and determinant variables. This was addressed by introducing quadratic forms of each variable entered into the models. The fourth drawback of the Phase 1 model was a tendency for out-migration from London FHSAs to be underpredicted. This was handled through the introduction of a London dummy variable. The London effect we think results from the different nature of internal migration in London, with migration flows being within a labour/housing market area rather than between such market areas. The fifth was that the model form made no allowance for possible time trends in migration (not accounted for by the set of time varying determinants used). A time trend was introduced to address this problem. The sixth difficulty was the arbitrary way variables were selected for inclusion in the model. In Phase 1, we had aimed at a common model across all migrant groups and selected the variables based on a majority of migrant group models reporting significance for particular variables. In Phase 2, stepwise inclusion based on significance of parameters was used for each age–sex group to determine a set of 14 age–sex-specific final models.

14.6.2 Investigations of the Phase 1 Out-migration Model Problems

Modelling Time Trends

To gain an understanding of the level of variation in out-migration volumes both across space and over time, we graphed out-migration volumes over time for each of the seven age groups (males and females combined) for all 98 FHSAs. Figure 14.1 illustrates trends in out-migration by the age groups used in the model for one FHSA, Newcastle. The main findings from the 98 graphs are as follows.

- There is a great deal of stability over time in the out-migration rates for the age groups 0–15; 30–44; 45–59; and 60+. This is true across the country and for many FHSAs the relationships are essentially flat. The main variations in the rates for these age groups are cross-sectional.
- Migration rates tend to be highest for the age groups 20–24, 25–29 and 16–19 and lowest for the age groups 60+ and 45–59.

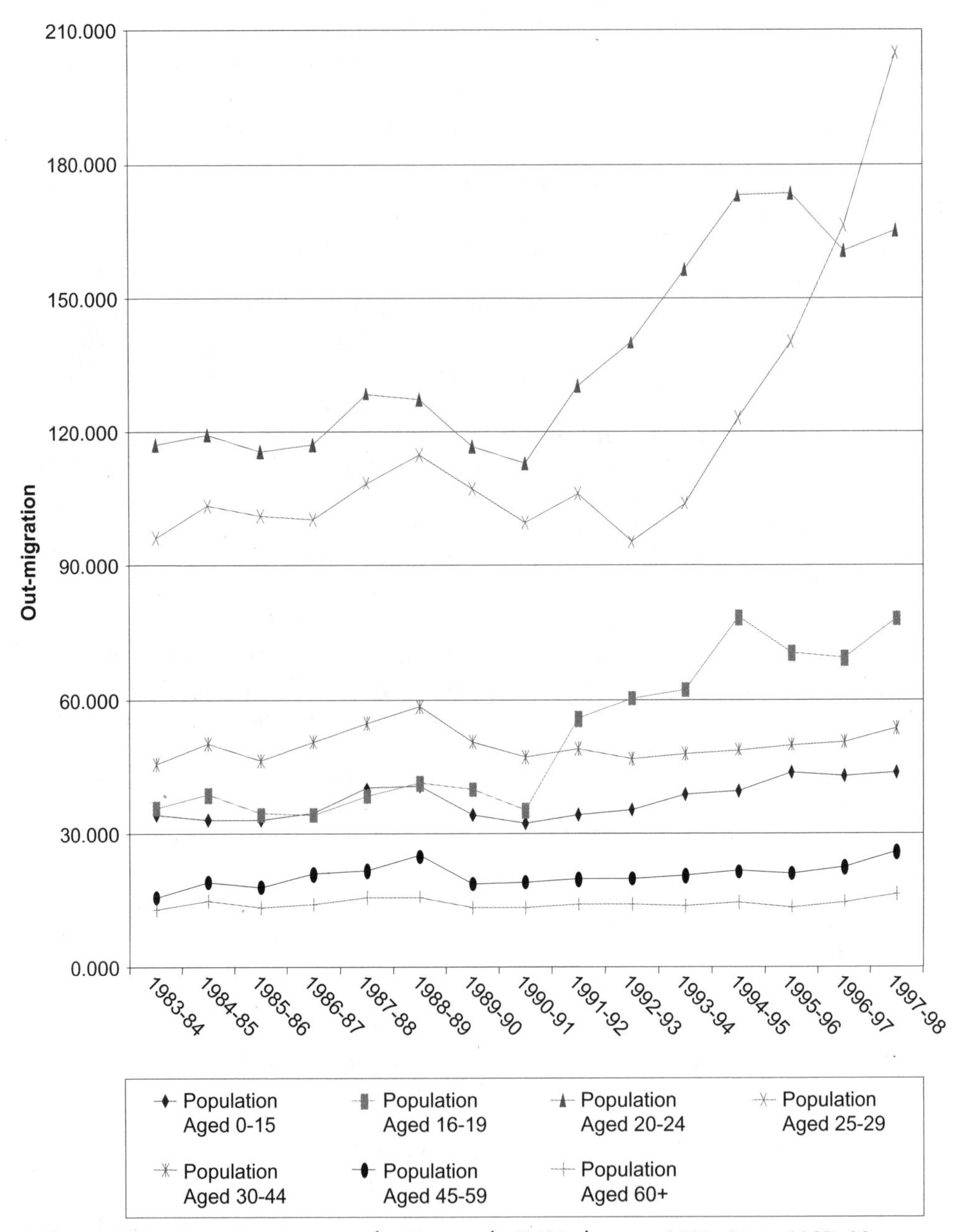

Figure 14.1 *Out-migration rates for Newcastle FHSA, by age, 1983–84 to 1997–98*

- There has been a rapid and strongly linear rise for almost all FHSAs in out-migrations by the 16–19 age group since 1990. Before this date, the level of out-migration for 16–19 year olds was relatively constant. The rise is due to the rapid expansion of higher education in the 1990s. We therefore added a variable to the Stage 1 model, UNIVPL_Z, the total number of university places available in a year.
- For FHSAs with relatively large student populations, such as Newcastle, concomitant with the rise in out-migration by the 16–19 age group through the 1990s, has been a rise in out-migration by the 20–24 and the 25–29 age groups. This is highly likely to be due to graduated students leaving university.
- There is little evidence to support any particular time trend otherwise in out-migration rates over time.

As well as plotting out-migration rates over time, all the other time-varying variables were also plotted over time to get a better 'feel' for the data. Many trends appeared to be highly erratic over time which, to anticipate the results of the Stage 1 model calibrations, probably explains the lack of explanatory power in some of these variables. The Stage 1 model contains some time series variables that might need special treatment in the model because of trends in some of the observations and some of the determinants. In Phase 1, no special treatment was given to these variables.

There are different strategies, of varying degrees of complexity, for handling time series variables in regression. To a large extent, the choice of a particular strategy depends on asymptotic properties of the time series variables, both dependent and independent, that are difficult to assess, especially in the case here with only 15 different time periods. Basically, the issue is whether any trends that are observed are stochastic or deterministic. Different strategies need to be followed depending on the answer to this question: stochastic trends are better handled using a differencing approach; deterministic trends are better handled using a de-trending approach. Although there is a vast literature on this subject, there is no real consensus on how to proceed in our situation where we have only 15 time periods and no real way of assessing the nature of any time-dependency. This severely limits what we can do. It should also be noted that although we are using 15 years of data to estimate certain relationships in the out-migration model, the majority of determinants are cross-sectional. However, despite these caveats, we conducted the following experiment.

- We added a linear trend to the out-migration model. This is the simplest and perhaps most effective way to account for time trends in both the dependent and independent variables. A non-linear trend, such as a quadratic, could be added if the data suggest such a trend exists although there is no real evidence for such a function.
- We transformed all time-varying variables into first differences by subtracting the value in time $t - 1$ from the value in time t. One obvious result of this procedure is that we were left with only 14 time periods. Variables that are constant over time were not first differenced.

A 'plausible' Stage 1 model was constructed and the experiment run with males 30–44 migrants (the largest volume of migrants). The following calibrations of this model were then undertaken:

- a 'baseline' model with no time trends taken into account;
- a model with a linear time trend added; and
- a model using first differences.

The results demonstrated the superiority of using a linear time trend. The use of first differences gave a very poor model fit and most of the cross-sectional variables were statistically insignificant. There was a modest improvement in goodness-of-fit in using a linear time trend over using no time trend. As anticipated, the two models yield broadly similar results for the parameter estimates of the cross-sectional variables but there are some variations in parameter estimates for the time-varying variables, which are more accurately estimated in the model including the time trend. Several other forms of the Stage 1 model were calibrated with and without the time trend and all the results supported the conclusion that adding a linear time trend provided slightly better results. Consequently, the Stage 1 models results used in the MIGMOD software were obtained from models containing a linear time trend.

OLS or WLS?

In Phase I, the Stage 1 models were calibrated with WLS with extra weight given to data pertaining to health areas with large populations. The justification for this was that the dependent variable in the regressions is migration rate which might be less reliable for origins with small populations. However, the justification may not be particularly strong in this case as the FHSAs all have fairly large populations. Consequently, we examined the efficacy of WLS compared with OLS on a sample of models. The results of applying the two calibration techniques were virtually identical so that there is little rationale for using WLS over OLS. Consequently, we employed OLS for the Stage 1 model calibrations, which avoids the need for a somewhat speculative definition of the weighting variable.

Unlogged or Log–Log Format?

In Phase I, the Stage 1 models were calibrated in log–log format. This had the disadvantage of producing a biased estimate of the intercept which had to be corrected after the calibration had taken place by the addition of an adjustment factor to the model format. This was done originally because the log–log format appeared to give slightly better predictions of out-migration. To examine whether this advantage still held, given the change in the Stage 1 models, we performed the following experiment.

Two plausible Stage 1 models were identified that contained the variables listed in Table 14.2. These models differed only in that one contained a linear time trend and the other did not. Both models were then calibrated by unlogged regression and then

Table 14.4 Goodness of fit (R^2) for model tests

Model	Unlogged	Log–Log
Without time trend	0.924	0.907
With time trend	0.928	0.918

again in a log–log format. The adjusted R^2 values from the four calibrations are shown in Table 14.4. These results suggest that an unlogged regression is superior to a log–log format. This also has the added advantage of producing an unbiased estimate of the intercept so that an adjustment factor is not needed in Stage 3 (the front-end) to balance the predicted values. Results from the calibrated models in the form of scatterplots of the standardised residual against the standardised predicted value confirmed the validity of the unlogged form of the model. The plots exhibit no evidence of heteroscedasticity and no evidence of any non-linearity.

Lagged or Unlagged Time-Varying Variables?

Some of the variables in the Stage 1 dataset vary over time. It is an issue as to whether these variables are entered in the model with their current year values (unlagged) or the previous year's values (lagged). The answer relates to whether migrants respond more to current conditions or to conditions in the previous year (the justification for the latter being a time lag between the stimulus to move and the move itself). It is difficult, if not impossible, to argue convincingly for either approach and so the question then becomes an empirical one – which form of the variables generates the more accurate model?

To answer this question a plausible Stage 1 model was calibrated for Males 30–44 using the variables described in Table 14.2. In one version of the model the time-varying variables were lagged; in another they were unlagged. The lagged model produced an adjusted R^2 value of 0.952, while the unlagged model R^2 value was 0.935. It was therefore decided to use lagged versions of the time-varying variables.

14.6.3 The Phase 2 Out-migration Model

From the resolution of the set of issues discussed above, the final form of Stage 1 of the model was defined. The Stage 1 model relates the adjusted out-migration rate for migrant group m in FHSA i at time t to a series of independent variables which describe the economic, social and geographic conditions of i. The independent variables are of three types: cross-sectional variables (some of which also vary over time); regional variables; and national time series variables. The linear model of out-migration rates has the basic form:

$$omr_{it}^{m} = \kappa^{m} + \sum\nolimits_{p} \alpha_{p}^{m} X_{pit}^{m} + \sum\nolimits_{q} \beta_{q}^{m} Y_{qit}^{m} + \sum\nolimits_{r} \gamma_{r}^{m} Z_{r}^{m} + \varepsilon_{it}^{m} \tag{14.7}$$

To this were added quadratic terms, a linear time trend and a London dummy:

$$omr_{it}^{m} = \kappa^{m} + \sum_{p} \alpha_{p}^{m} X_{pit/t-1}^{m} + \sum_{q} \beta_{q}^{m} Y_{qit/t-1}^{m} + \sum_{r} \gamma_{r}^{m} Z_{rt-1}^{m} + \sum_{p} \delta_{p}^{m} (X_{pit/t-1}^{m})^{2} + \sum_{q} \eta_{q}^{m} (Y_{qit/t-1}^{m})^{2} + \sum_{r} \theta_{r}^{m} (Z_{rt-1}^{m})^{2} + \psi^{m} T_{t} + \zeta^{m} LD_{i} + \varepsilon_{it}^{m} \quad (14.8)$$

where omr_{it}^{m} is the predicted out-migration rate for migrant group m in FHSA i in time interval t, X represents a cross-sectional variable (some of which can also be time-varying); Y represents a regional variable; and Z represents a national time-varying variable, T is a time counter and LD stands for London dummy (= 1 if the FHSA is in London, = 0 otherwise). The regressions coefficients are denoted by Greek letters and ε_{it}^{m} is the error term for each zone, time and migrant group combination. The model is calibrated separately for each of 14 migrant groups (males and females in each of seven age groups).

14.7 Stage 2: A Model of Destination Choice

14.7.1 The Model

The Stage 2 model is a migration destination model that distributes the total number of out-migrants from FHSA i to each of the 97 destination FHSAs based on the characteristics of each destination FHSA and the separation between the origin and each destination. It is a spatial choice model of the destinations chosen by migrants from an origin. The model to be calibrated has the general form:

$$M_{ij}^{m} = O_{i}^{m} \prod_{p} X_{pj}^{\alpha pim} d_{ij}^{\beta im} \Big/ \sum_{j} \prod_{p} X_{pj}^{\alpha pim} d_{ij}^{\beta im} \quad (14.9)$$

where O_{i}^{m} is the volume of out-migration of type m from origin FHSA i (known when the model is calibrated, predicted by the Stage 1 model when predictions are being made at Stage 3); X_{pj} represents an attribute of FHSA j that affects the choice of j by migrants from i; and d_{ij} is the distance between i and j. The X variables are raised to powers, α_{pim}, specific to each variable p, origin i and migrant group m, while the distance variable is raised to the power β_{im} specific to each origin i and migrant group m. The parameters of this model indicate the sensitivity of migration flows to particular destination characteristics: in essence they indicate what features of a destination make it attractive to migrants and which features make it unattractive. For example, a relatively large score on an attribute with a positive parameter estimate would make a destination attractive to migrants, *ceteris paribus*, while a relatively large score on an attribute with a negative parameter estimate would make a destination unattractive to migrants, other things being equal. This model has a long history in the analysis of spatial interaction patterns (see *inter alia*, Fotheringham and O'Kelly, 1989, and Fotheringham *et al.*, 2000, Chapter 9).

The model is calibrated separately for each of the 98 origins and each of the 14 migrant groups. For any origin, $O_i^m/\Sigma_j \Pi_p X_{pj}^{\alpha pim} d_{ij}^{\beta im}$ will be a constant (k_i^m) so that the origin-specific model is then simply,

$$M_{ij}^m = k_i^m \prod_p X_{pj}^{\alpha pim} d_{ij}^{\beta im} \quad (14.10)$$

Note that there will be 98 × 14 = 1372 sets of model parameters to be calibrated in this highly specific spatial interaction model.

14.7.2 Poisson Regression or OLS?

In Phase 1, we calibrated the Stage 2 model by Ordinary Least Squares regression by taking logs of both sides of the equation to make it linear-in-parameters. Although convenient, this has two problems:

- The dependent variable M_{ij}^m is a count of migrants and therefore is likely to be Poisson distributed rather than normally distributed. This means that one of the key assumption of OLS is probably not met. This is particularly a problem where the migration matrices contain lots of small flows as is the case here.
- A practical problem arises in taking logs of both sides of the model when M_{ij}^m is zero (as are many of the recorded flows). Taking the log of 0 yields negative infinity. The solution to this used in Phase 1 was to add a small increment (0.5) to all flows before taking logs. Although tests showed this does not appear to bias the results greatly, nevertheless it is a somewhat arbitrary procedure.

An alternative, and perhaps superior method of calibration, is to use Poisson regression which assumes the conditional mean of the variable we are trying to model has a Poisson distribution and avoids the need for making some approximation to zero flows. The use of Poisson regression for calibrating migration distribution models has been demonstrated in a number of papers (*inter alia*, Flowerdew and Aitkin, 1982; Flowerdew, 1991).

In order to compare the performances of Poisson and OLS regression for migration destination choice models, we conducted a series of experiments comparing Poisson and OLS calibration results on the same datasets. A relatively simple, but quite plausible, Stage 2 model was postulated containing the following explanatory variables (the model was constructed from the original set of variables from Phase 1): Population, Destination accessibility, Distance, Contiguity, Urban, Beach, Rain, Council Tax and Household Income. This model was calibrated separately for each of the 98 origins using migration data for males 30–44. The performance of the calibrated models in replicating migration flows was assessed using the modified psi statistic, ψ (Knudsen and Fotheringham, 1986), which demonstrated the superiority of Poisson regression in terms of goodness-of-fit to the observed data. Consequently, all the Stage 2 models were calibrated with Poisson regression.

14.7.3 Variables for Stage 2

In Phase 1, the Stage 2 dataset contained 69 variables. This set was greatly reduced following a qualitative assessment of each of the variables in the dataset and an examination of multicollinearity amongst the independent variables. This left a final set of 27 variables in the Stage 2 model which are listed in Table 14.2.

14.7.4 Migration Data

As with Phase 1, the migration data used in all calibrations of the Stage 2 model were for the 1996/97 time period. Previous results in Phase 1 suggested that the calibration results were either stable over time or that the 96/97 period provided a good average of the years since 1990 (see Champion *et al.*, 2001).

14.7.5 Modelling Procedure

The procedure for variable selection was different from that employed in Phase 1. In Phase 1 initially, a separate model was calibrated for all origins for each migrant group using all 69 explanatory variables. Then, to determine the final model form for each migrant group, variables were selected from the 69 if they were significant for >9 origins and the model containing only those variables was re-calibrated. Instead of this largely *ad hoc* procedure, in Phase 2 we calibrated the model containing the 27 variables in the Stage 2 dataset for each of the 98 origins and for each migrant group. This avoids the need for any *ad hoc* selection procedure and allows us to make comparisons of the parameter estimates across the 14 migrant groups.

One final note should be mentioned on how the Stage 2 modelling procedure differed from that in Phase 1. It was agreed with the Department that we should drop the modelling of migration flows from Scotland so that we had only 98 origins for which migration flows are estimated endogenously in Stage 3 of the procedure.

14.8 Stage 3: The MIGMOD Software

The out-migration and destination choice models described above (Stages 1 and 2) are combined within a user-friendly software package termed MIGMOD built around the GeoTools mapping software developed at the University of Leeds. MIGMOD contains all the data described above as well as the estimated parameters from the calibrations of the Stage 1 and 2 models. As shown below, the user has access to these in both tabular and, where feasible, mapped form at the click of a button. We now describe some of the features of the MIGMOD software.

The parameters from the calibrated Stage 1 model in equation (14.8) and the calibrated Stage 2 model in equation (14.10) were entered into MIGMOD along with data on observed migrations and the determinants of both out-migration and destination choice. Within MIGMOD the level of out-migration from each origin is then computed by first estimating the out-migration rate from each origin and for each migrant group within that origin and then multiplying these rates by the relevant population size. This produces estimates of out-migration volumes for each migrant group within each origin. These out-migration numbers are then allocated across the destinations on the basis of the attributes of these destinations according to the Stage 2 model. The user can then alter conditions in any origin or set of origins and/or in any destination or set of destinations and examine the impact such changes have on the resulting migration patterns.

The starting point of MIGMOD is the initial menu as shown in Figure 14.2. This gives the user four options:

- view any of the coefficients from either the Stage 1 or the Stage 2 models;
- examine various aspects of the observed and predicted migration matrices;
- develop policy-related scenarios in which some combination of the independent variables in the Stage 1 and 2 models is changed and a new set of predicted migration flows is obtained which can then be compared with the original set of flows; and
- view any of the time-varying variables in a series of graphs over time.

For instance, selecting 'View Coefficients' produces a set of options on which coefficients you want to view. The user has a choice of a table of Stage 1 parameter estimates, as shown in Figure 14.3, or a matrix of origin-specific Stage 2 parameter estimates for any one of the 14 migrant groups as shown in Figure 14.4 for the

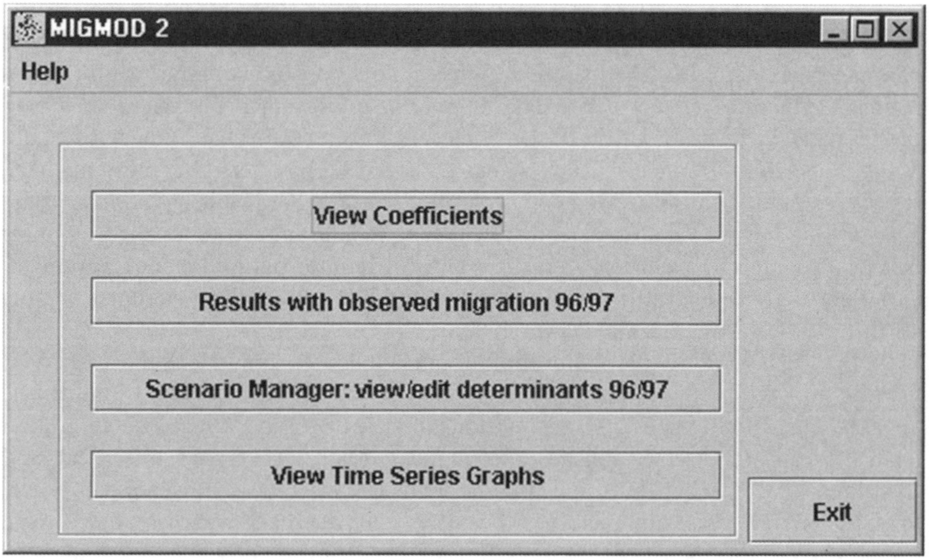

Figure 14.2 *MIGMOD main menu*

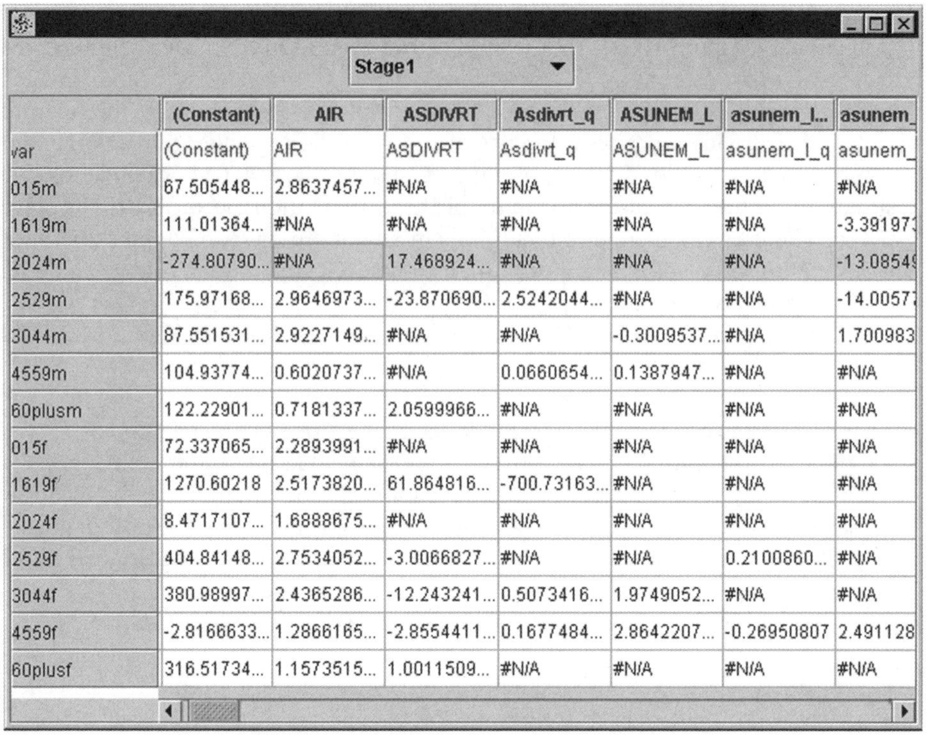

Stage1

	(Constant)	AIR	ASDIVRT	Asdivrt_q	ASUNEM_L	asunem_l...	asunem_
var	(Constant)	AIR	ASDIVRT	Asdivrt_q	ASUNEM_L	asunem_l_q	asunem_
015m	67.505448...	2.8637457...	#N/A	#N/A	#N/A	#N/A	#N/A
1619m	111.01364...	#N/A	#N/A	#N/A	#N/A	#N/A	-3.39197
2024m	-274.80790...	#N/A	17.468924...	#N/A	#N/A	#N/A	-13.0854
2529m	175.97168...	2.9646973...	-23.870690...	2.5242044...	#N/A	#N/A	-14.0057
3044m	87.551531...	2.9227149...	#N/A	#N/A	-0.3009537...	#N/A	1.700983
4559m	104.93774...	0.6020737...	#N/A	0.0660654...	0.1387947...	#N/A	#N/A
60plusm	122.22901...	0.7181337...	2.0599966...	#N/A	#N/A	#N/A	#N/A
015f	72.337065...	2.2893991...	#N/A	#N/A	#N/A	#N/A	#N/A
1619f	1270.60218	2.5173820...	61.864816...	-700.73163...	#N/A	#N/A	#N/A
2024f	8.4717107...	1.6888675...	#N/A	#N/A	#N/A	#N/A	#N/A
2529f	404.84148...	2.7534052...	-3.0066827...	#N/A	#N/A	0.2100860...	#N/A
3044f	380.98997...	2.4365286...	-12.243241...	0.5073416...	1.9749052...	#N/A	#N/A
4559f	-2.8166633...	1.2866165...	-2.8554411...	0.1677484...	2.8642207...	-0.26950807	2.491128
60plusf	316.51734...	1.1573515...	1.0011509...	#N/A	#N/A	#N/A	#N/A

Figure 14.3 *Stage 1 parameter estimates viewed through MIGMOD*

migrant group 'Males 30–44'. A very useful feature of MIGMOD is that if any column of data or parameter estimates relates to a set of regions, as the Stage 2 parameter estimates do, then simply double clicking on the column heading automatically generates a map of the data or parameter estimates. For example, a map of the spatial distribution of the house price parameter in the Stage 2 model for Males 30–44 is shown in Figure 14.5. This shows the interesting pattern that migrants from relatively wealthier origins primarily in the south of England view high house prices at a destination as an attractive feature whereas migrants from relatively poorer origins view high house prices as a deterrent. More information on such parameter variations is given in a future publication and can be found in the government report on this project (Fotheringham *et al.*, 2002).

Rather than viewing the calibration results, the user can select the option of viewing the original migration data, in which case the window shown in Figure 14.6 appears. This allows the user several options to view different facets of the migration data. Basically the user establishes two migration data sets termed 'Observed' and 'Baseline'. The 'Baseline' button is used to define one migration set which will henceforth be referred to as 'The Baseline'. The user is given a set of options ranging from the observed data, the predicted migration matrix from the baseline model and the predicted migration matrix from other runs of the model using different scenarios of

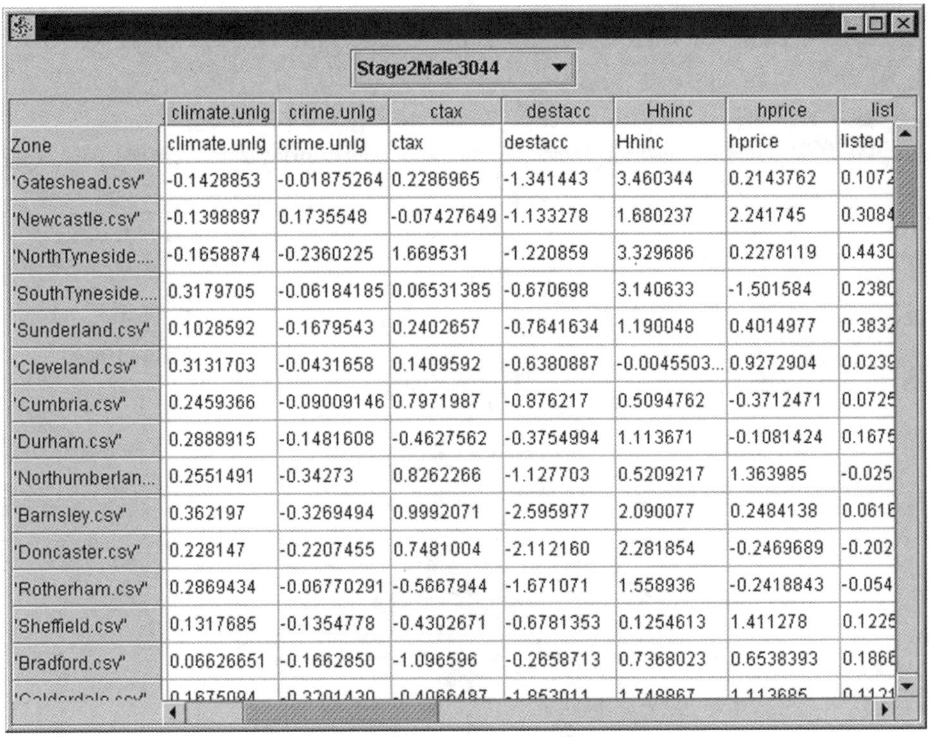

Stage2Male3044

	climate.unlg	crime.unlg	ctax	destacc	Hhinc	hprice	list
Zone	climate.unlg	crime.unlg	ctax	destacc	Hhinc	hprice	listed
'Gateshead.csv'	-0.1428853	-0.01875264	0.2286965	-1.341443	3.460344	0.2143762	0.1072
'Newcastle.csv'	-0.1398897	0.1735548	-0.07427649	-1.133278	1.680237	2.241745	0.3084
'NorthTyneside....	-0.1658874	-0.2360225	1.669531	-1.220859	3.329686	0.2278119	0.4430
'SouthTyneside....	0.3179705	-0.06184185	0.06531385	-0.670698	3.140633	-1.501584	0.2380
'Sunderland.csv'	0.1028592	-0.1679543	0.2402657	-0.7641634	1.190048	0.4014977	0.3832
'Cleveland.csv'	0.3131703	-0.0431658	0.1409592	-0.6380887	-0.0045503...	0.9272904	0.0239
'Cumbria.csv'	0.2459366	-0.09009146	0.7971987	-0.876217	0.5094762	-0.3712471	0.0725
'Durham.csv'	0.2888915	-0.1481608	-0.4627562	-0.3754994	1.113671	-0.1081424	0.1675
'Northumberlan...	0.2551491	-0.34273	0.8262266	-1.127703	0.5209217	1.363985	-0.025
'Barnsley.csv'	0.362197	-0.3269494	0.9992071	-2.595977	2.090077	0.2484138	0.0616
'Doncaster.csv'	0.228147	-0.2207455	0.7481004	-2.112160	2.281854	-0.2469689	-0.202
'Rotherham.csv'	0.2869434	-0.06770291	-0.5667944	-1.671071	1.558936	-0.2418843	-0.054
'Sheffield.csv'	0.1317685	-0.1354778	-0.4302671	-0.6781353	0.1254613	1.411278	0.1225
'Bradford.csv'	0.06626651	-0.1662850	-1.096596	-0.2658713	0.7368023	0.6538393	0.1866

Figure 14.4 *Stage 2 parameter estimates viewed through MIGMOD*

explanatory variables. The 'Observed' button is used in a similar manner to define a different set of migration data to that defined as 'The Baseline' and this is termed 'The Observed'. The 'Operation' bar then allows the following matrices to be displayed:

- View the 'Observed' Matrix;
- View the 'Baseline' matrix;
- View the 'Observed' – 'Baseline' matrix;
- View the absolute differences between 'Observed' and 'Baseline'; and
- View the percentage differences between the 'Observed' and 'Baseline'.

To this point all the operations are based on the cells of the full matrix. The button labelled 'Matrix' gives the user the option of viewing outflow totals, inflow totals and net flows for any of the migration matrices. The button labelled 'All Ages' allows the user to define which of the seven age groups should be displayed and one option here is to use 'All Persons'. The final option entitled 'Persons' allows the user to define whether the migration data displayed are for males or females or both. The use of MIGMOD for developing scenarios for policy evaluation is described elsewhere (Champion *et al.*, 2002; Fotheringham *et al.*, 2002).

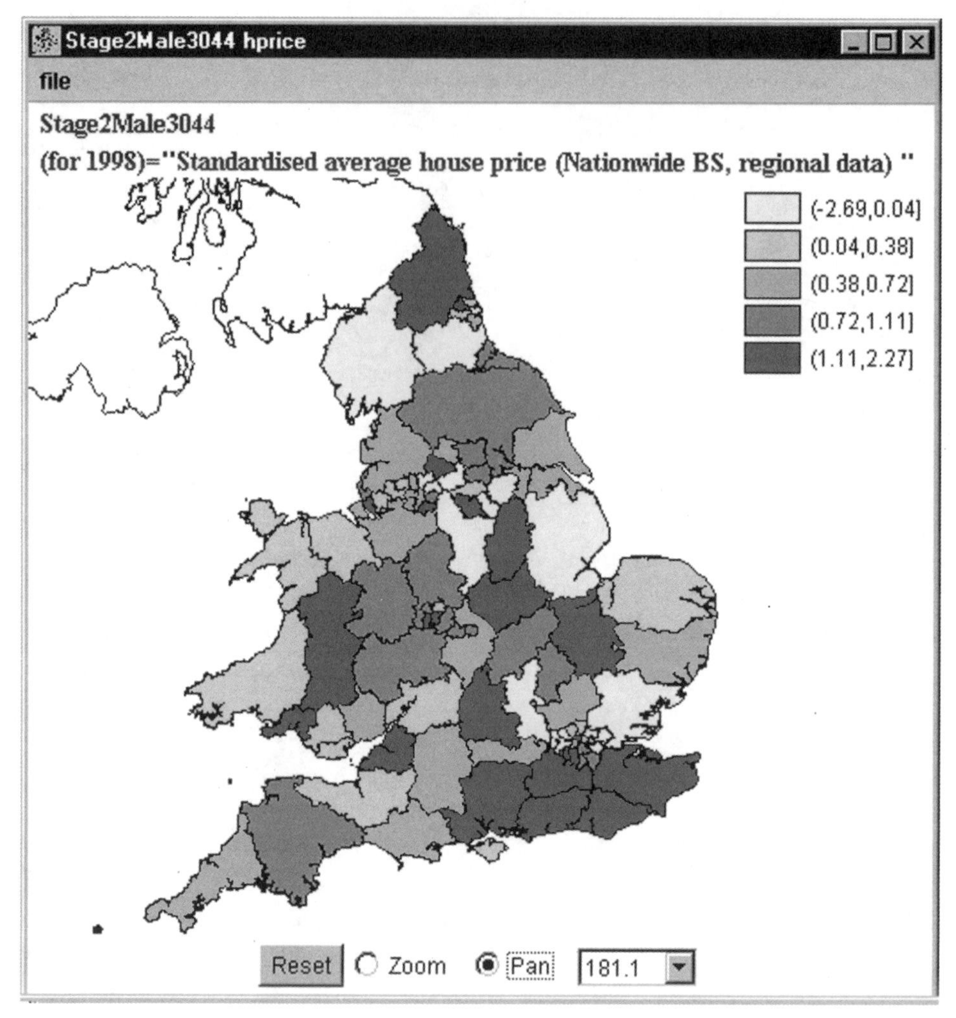

Figure 14.5 The spatial distribution of the house price parameter in the Stage 2 model

14.9 Summary

This chapter has described the structure of a very large spatial interaction model, constructed to explain the flows of migrants between 98 zones in England and Wales. The model was constructed in two phases and many changes were made in the second phase to overcome difficulties present in the first phase model. The two model stages, prediction of out-migration flows from origins and their distribution to destinations, were combined in Stage 3 into an operational model for the sponsoring Department so that scenarios related to policy could be implemented. Full details are provided in the Final Reports to the Department (Champion *et al.*, 2001; Fotheringham *et al.*, 2002) and in Champion *et al.* (2002). The authors believe that the model of

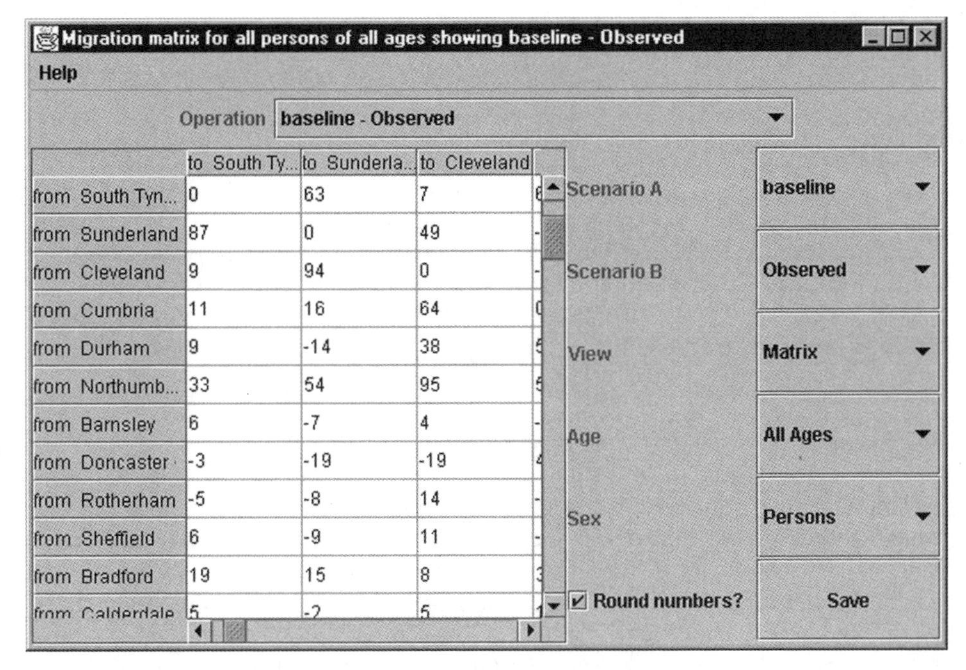

Figure 14.6 *The migration data viewer in MIGMOD*

migration constructed here to be one of the most comprehensive and meticulously researched such endeavours, which we hope will find utility in the formulation of English regional development and housing development policy in the future.

Acknowledgements

The Team

The migration model was constructed by a research team from four organisations: the University of Newcastle upon Tyne, University of Leeds, Heriot-Watt University/Edinburgh College of Art and Greater London Authority (formerly London Research Centre). The authors were supported in their work by other members of the team in data assembly, modelling and scenario development: *Newcastle* – Tony Champion (who led Phase 1), Seraphim Alvanides, Tim Barmby, Chris Brunsdon, Martin Charlton, Mike Coombes, Simon Raybould, Andy Tremayne; *Leeds* – Adrian Bailey, Oliver Duke-Williams, Heather Eyre (now with Education Leeds), James Macgill, John Stillwell; *Edinburgh* – Glen Bramley (who was responsible for scenario development), Hilary Anderson, Tania Ford; *Greater London Authority* – John Hollis and *University of Southampton* – Tom Wilson (now at the University of Queensland, Brisbane). The authors would also like to thank DETR's steering group led by Andrew Parfitt in 2000 and DLTR's steering group led by Stephen Rosevear in 2001, who managed the contract so efficiently, and members Dorothy Anderson and Jenny Wood (ONS) who were helpful in identifying and supplying data.

The Sponsor

This chapter describes work carried out for the Department of the Environment, Transport and the Regions (DETR), restructured and renamed Department of Transport, Local Government and the Regions (DTLR) in June 2001 under contract RADS 5/9/22 (2000) and a variation (2001). The full report on Phase 1 and Phase 2 has been published by the Office of the Deputy Prime Minister (ODPM, 2002).

The Operational Model (MIGMOD, version 2.0)

An interface to the model was developed at the University of Leeds, primarily by James Macgill of the Centre for Computational Geography with help from Heather Eyre and Phil Rees and delivered to the Department on CD as a self-install and standalone software suite with accompanying observed migration datasets, the determinant variables, the model parameters, a set of predicted migrations and a system for developing alternative scenarios. MIGMOD is the intellectual property of the Department (currently ODPM). Readers who would like a copy of the operational model should seek permission from the ODPM.

Data Contributors

The research team is indebted to the many individuals and organisations that provided advice and help in the assembly of data. We are especially grateful to those at the Department who played a facilitating role in putting us in touch with data holders: Andrew Parfitt, Dominic Bryant, Russell Harris, Stephen Rosevear. In addition to those listed below against individual variables, Kerrick Macafee (the Department) and Glenn Everett (DTI) were also very helpful. We would also like to acknowledge the help and advice provided by members of the national statistical offices, especially in relation to data on international migration: *ONS*: Roma Chappell, Lucy Vickers, Alastair Davies, Denise Anderson, Brett Leeming; *GRO-Scotland*: Francis Hay; *NISRA*: Robert Beatty, Maire Rodgers, Maureen Ferguson. The following persons helped with particular variables. They are listed by variable with affiliation and name: agric, ONS (AES/ACE), Notice Ref 195R, via NOMIS (DETR), Carole Sutton, Simon Connell; air, AEA Technology, Emma Linehan; asunem, Employment Service/LFS via NOMIS, Andrew Craven (NOMIS); bank, ONS (AES/ACE), Notice Ref 195R, via NOMIS (DETR), Carole Sutton, Simon Connell; cars, DETR, Andrew Ledger; climat, Hadley Centre (UEA), David Viner; constr, ONS (AES/ACE), Notice Ref 195R, via NOMIS (DETR), Carole Sutton, Simon Connell; divrate, ONS, John Haskey; dlgexp, DETR, Welsh Office, Scottish Executive, Somerset CC, CIPFA, Stephen Greenhill, Ade Akinfolajimi, Justin Vetta, Stuart

Neil, Ingrid o'Malley, John Rae; earn, ONS (NES), Derek Bird, Claire Nichol; ecact, Employment Service/LFS via NOMIS, Andrew Craven (NOMIS); empgro, ONS (AES/ACE), Notice Ref 195R, via NOMIS (DETR), Carole Sutton, Simon Connell; empr, ONS (AES/ACE), Notice Ref 195R, via NOMIS (DETR), Carole Sutton, Simon Connell; expssa, DETR, Welsh Office, Scottish Executive, Somerset CC, CIPFA, Stephen Greenhill, Ade Akinfolajimi, Justin Vetta, Stuart Neil, Ingrid O'Malley, John Rae; frosty, Hadley Centre (UEA), David Viner; gcse, DfEE, Mark Franks; Catherine Blackham; hhinc, ONS (Regional Accounts), Dev Virdee, David Lacey; ijsa, DSS (ASD), Katie Dodd, Kath Williams, Jacky Javan; ilchip, DETR (1991 ILC), Gillian Smith, Sam Mason; immigr, ONS, Roma Chappell; jlytemp, Hadley Centre (UEA), David Viner; jvacs, Employment Service/LFS via NOMIS, Andrew Craven (NOMIS); listed, CADW, DETR, DoE(NI), RCME, Welsh Assembly Office, Christopher Brown, Lisa Moralee, Gerry Cash, Donnie Mackay, Adam Al-Nuaimi, lowi, ONS (NES), Derek Bird, Claire Nichol; manuf, ONS (AES/ACE), Notice Ref 195R, via NOMIS (DETR), Carole Sutton, Simon Connell; no2nox, AEA Technology, Emma Linehan ; ozone, AEA Technology, Emma Linehan; pnbu, DETR (LUCS), Bob Garland, Barinder Hundle; pnrl, DETR (LA HIP1 returns), Welsh Ofice, Trevor Steeples, Shadan Yusuf, Henry Small, Fiona Leadbitter;pqpr, DETR, Oscar Yau; pqsr, DETR, Oscar Yau; psr, DETR (LA HIP1 returns), Welsh Ofice, Trevor Steeples, Shadan Yusuf, Henry Small, Fiona Leadbitter; punfps, DETR (LA HIP1 returns), Welsh Office, Trevor Steeples, Shadan Yusuf, Henry Small, Fiona Leadbitter; pvac, DETR (LA HIP1 returns), Welsh Ofice, Trevor Steeples, Shadan Yusuf, Henry Small, Fiona Leadbitter; rain, Hadley Centre (UEA), David Viner; regcmp, SBS, Ian Dale; rgdppc, ONS (Regional Accounts), Dev Virdee, David Lacey; rgurb, DETR (LUCS), Bob Garland, Barinder Hundle; rlapsc, DETR (GNI/HNI Indicators), Tina Golton; rlasc, DETR (GNI/HNI Indicators), Tina Golton; rpdi, ONS (Regional Accounts), Dev Virdee, David Lacey; rtoff, Home Office (Crime Statistics), David Covey; scenic, DETR, Stephen Hall; sehcr, DETR (SEH), Jeremy Groves; stmort, ONS, Tim Devis, Justine Fitzpatrick; sunny, Hadley Centre (UEA), David Viner; unempr, Employment Service/LFS via NOMIS, Andrew Craven (NOMIS); vacdrl, DETR (NLUD), Bob Garland; vaclb, DETR (NLUD), Bob Garland; vacul, DETR (NLUD), Bob Garland; voting, Plymouth University, Brian Cheal.

References

Bailey, A. and Rees, P. (2001) Students in the migration model – the estimation process. Technical Appendix 6 in Champion *et al.* (2001) *Development of a Migration Model: Final Report to the Department of the Environment, Transport and the Regions under Contract RADS 5/9/22*, Department of Geography, University of Newcastle, Newcastle upon Tyne.

Champion, T., Bramley, G., Fotheringham, A.S., Macgill, J. and Rees, P. (2002) A migration modelling system to support government decision-making, in Stillwell, J. and Geertman, S. (eds) *Planning Support Systems in Practice*, Springer Verlag, Berlin, pp. 257–78.

Champion T., Coombes M., Fotheringham A.S., Eyre H., Macgill J., Rees P., Stillwell J., Wilson T., Bramley G. and Hollis J. with S. Alvanides, M. Charlton, T. Ford, S. Raybould, A. Bailey, O. Duke-Williams and H. Anderson (2001) *Development of a Migration Model: Final Report to the Department of the Environment, Transport and the Regions under*

Contract RADS 5/9/22. March 2001. Department of Geography, University of Newcastle, Newcastle upon Tyne. (The Phase 1 Final Report. Published in Office of the Deputy Prime Minister (2002)).

Champion, T., Fotheringham, A.S., Rees, P., Boyle, P. and Stillwell, J. (1998) *The Determinants of Migration Flows in England: A Review of Existing Data and Evidence*, Department of Geography, University of Newcastle, for the Department of Environment, Transport and the Regions. Accessible via www.geog.leeds.ac.uk.

Flowerdew, R. (1991) Poisson regression modelling of migration, in Stillwell, J. and Congdon, P. (eds) *Migration Models: Macro and Micro Approaches,* Belhaven Press, London, pp. 92–112.

Flowerdew, R. and Aitkin, M. (1982) A method of fitting the gravity model based on the Poisson distribution, *Journal of Regional Science,* **22**: 191–202.

Fotheringham, A.S. (1986) Modelling hierarchical destination choice, *Environment and Planning A,* **18**: 401–18.

Fotheringham, A.S. (1991) Migration and spatial structure: the development of the competing destinations model, in Stillwell, J. and Congdon, P. (eds) *Migration Models: Macro and Micro Approaches,* Belhaven, London, pp. 57–72.

Fotheringham, A.S., Brunsdon, C. and Charlton, M. (2000) *Quantitative Geography: Perspectives on Spatial Data Analysis,* Sage, London.

Fotheringham, A.S., Barmby, T., Brunsdon, C., Champion, T., Charlton, M., Kalogirou, S., Tremayne, A., Rees, P., Eyre, H., Macgill, J., Stillwell, J., Bramley, G. and Hollis, J. (2002) *Development of a Migration Model for England and Wales: Analytical and Practical Enhancements. A Report on the Modelling Procedures and Results Connected with the Research Contract Variation to RADS 5/9/22.* March 2002. Department of Geography, University of Newcastle, Newcastle upon Tyne (The Phase 2 Final Report).

Fotheringham, A.S. and O'Kelly, M.E. (1989) *Spatial Interaction Models: Formulations and Applications*, Kluwer Academic, Dordrecht.

Knudsen, D.C. and Fotheringham, A.S. (1986) Matrix comparison, goodness-of-fit, and spatial interaction modeling, *International Regional Science Review,* **10**: 127–47.

Office of the Deputy Prime Minister (2002) *Development of a Migration Model.* The University of Newcastle upon Tyne, The University of Leeds and The Greater London Authority/London Research Centre. Office of the Deputy Prime Minister, London. ISBN 1 85112 583 3.

ONS (2002) Report: internal migration estimates for local and health authorities in England and Wales, 2001, *Population Trends,* **109**: 87–99.

Shaw, C. (2002) 2000-based national population projections for the United Kingdom and its constituent countries, *Population Trends*, **107**: 5–13.

Stillwell, J., Rees, P., Eyre, H. and Macgill, J. (2001) Improving the treatment of international migration. Technical Appendix 7 in Champion *et al.* (2001) *Development of a Migration Model: Final Report to the Department of the Environment, Transport and the Regions under Contract RADS 5/9/22*, Department of Geography, University of Newcastle, Newcastle upon Tyne.

Stillwell, J, Rees, P. and Boden, P. (eds) (1992) *Migration Processes and Patterns. Volume 2, Population Redistribution in the United Kingdom,* Belhaven, London, p. 307.

Walker, D. (2002) Get off yer bike, *The Guardian*, 30 July, p. 13. This draws material from Geographic Mobility, a presentation by John Muellbauer to the Cabinet Office (accessible via www.piu.gov.uk).

Wilson, A.G. (1967) A statistical theory of spatial distribution models, *Transportation Research,* **1**: 253–69.

Wilson, A.G. (1972) *Papers in Urban and Regional Analysis*, Pion, London.

Wilson, T. and Rees, P.H. (1999) Linking 1991 population statistics to the 1998 local government geography of Great Britain, *Population Trends*, **97**: 37–45.

15

Modelling Regional Economic Growth by Means of Carrying Capacity

Leo van Wissen

Abstract

In this chapter a new model of spatial economic growth is proposed. It is based on the ecological concept of *carrying capacity*, defined for a geographical system of regions. The use of ecological concepts refers to the dynamic evolution of populations over time. Here, we view the development of economic markets as an ecological process of growth towards the carrying capacity level. The carrying capacity is given an economic interpretation, and is equal to the market demand for a product. In addition, a spatial dimension is introduced that gives, for each location in a spatial system, the disequilibrium between carrying capacity (demand) and supply. The chapter shows a number of characteristics of the model and demonstrates how the model may be used empirically. Some empirical results are given from an application of the model in the field of industrial demography.

15.1 Introduction

Regional economic growth can be modelled in many ways. The question why certain regions grow faster than others has attracted many geographers and economists in the past. Amstrong and Taylor (2000) give a concise overview of a number of these models and methods, such as the regional input–output model, the neo-classical regional growth model, the regional multiplier model, or the economic base method. Most existing models share one common characteristic: they are essentially

Applied GIS and Spatial Analysis. Edited by J. Stillwell and G. Clarke
 ISBN: 0-470-84409-4

macro-economic models, applied at the meso-scale of the region. As a result, they share many of the characteristics of macro-economic models: in particular they are either implicitly or explicitly based on the notion of equilibrium. This equilibrium view of the world has been challenged in recent years from different viewpoints. After all, it seems more logical that regional models of economic growth should focus on change instead of equibrium and structure. Change in these models is a necessary but short episode between two states of equilibrium, or can only be deduced from the difference between two equilibrium states.

One of the most important criticisms against this essentially neo-classical view of the world was raised by Nelson and Winter (1983), who advocated a new evolutionary paradigm in economics. Their models were based on an evolutionary concept of economic change. It emphasises change instead of equilibrium, process, instead of structure. Important economic processes are selection, adaptation, inertia and competition. Within sociology, a somewhat similar field has developed, which focuses on processes of change in populations of organisations. This field is called organisational ecology, or simply demography of firms or industrial demography (Hannan and Freeman, 1989; Carroll and Hannan, 2000), in which populations of firms evolve through selective processes of entry and exit.

Demography of firms is also becoming more widespread in economic geography (van Dijk and Pellenbarg, 2002; van Wissen, 2002). In principle it may offer an alternative view on regional economic growth, based on regional variations in birth, death, migration and firm growth processes. In the past it has been used largely as a descriptive tool for regional employment change (see e.g., Gudgin, 1978), but more recently analytical models of regional economic growth have been developed as well (van Wissen, 2000).

In analytical models of firm population evolution in a regional economic setting, there are a number of problems that have to be solved. These problems derive from (1) the translation of demographic processes to the level of firm populations; (2) the introduction of economic mechanisms of demand and supply in demographic processes; and (3) introducing the spatial dimension in firm demographic models (van Wissen, 2002). The first problem involves issues such as definitions, model structure, the modelling of firm birth, death and other demographic components, and the impact of age, period and cohort, or other primary variables on demographic events. Many of these issues have been treated in van Wissen (2000). This chapter deals with a possible solution to the second and third problems.

The second problem deals with the notion of the market. Even if equilibrium is not taken as the starting point, there is still the impact of the 'invisible hand of the market' on firm founding and failure, and growth. One of the most important differences between human and firm demography is the existence of demand and supply interactions or, in ecological terms, a feedback mechanism between the number of firms in a population and population growth. In short, the larger the population of firms and the larger the supply of products in the market, the fiercer the competition and the lower the population growth. Population ecologists call this feedback between population size and growth the density dependence model (Carroll and Hannan, 2000). The ecological notion of *carrying capacity* is a fruitful starting point for introducing an invisible market hand into these models. *Carrying capacity* is an ecological concept. In an ecological framework a population (generally of animals) will grow if its current size is smaller than the carrying capacity for that population. If the actual size is larger than the carrying capacity, the population will decrease in size. We will build upon this ecological concept and extend it to apply to a multi-sector economy.

The third problem involves introducing the spatial dimension into the model. In most regional economic models, as well as in most population ecology models, the notion of space is not very well developed. In the simplest case, regions are small-scale variants of nations, with only imports and exports as signs of life outside of the region. In other cases, such as in the multi-regional input–output table, a limited number of regions interact, but the influence of distance remains implicit in the model. In recent years, Krugman and colleagues developed a series of economic models that take the spatial dimension seriously (Krugman, 1995; for an introduction see Neary, 2001). Although these models have many attractive theoretical properties, it is difficult to apply them other than in a highly stylised world, where only two regions exist, or which has a linear or circular form. The ecological model to be developed also contains an explicit spatial dimension. In population ecology, populations interact with each other. These interactions occur between different populations as well as between different regions. The spatial dimension essentially implies that nearby populations have higher propensities to interact than others.

In short, this chapter proposes a model of regional economic change, based on the ecological concept of *carrying capacity*. It also contains elements of regional input–output models, thus establishing a link with more traditional regional economic growth models. Thirdly, the model involves inter-sectoral as well as spatial interactions. Ironically this resulting model has a much larger applicability than the field of firm demography. Although its structure is based on an ecological concept, the resulting model can be used in a population ecology or evolutionary framework, but it may also be used as a regional economic model of change without reference to these fields.

This chapter is organised as follows. First, in Section 15.2, we introduce the basic concept of *carrying capacity*. In the following sections, we extend this concept to a multi-sector economy (Section 15.3) and include a spatial dimension (Section 15.4). In Section 15.5 a general matrix notation is given. A number of model characteristics are investigated in Section 15.6 using a simulation framework. The concept of spatial carrying capacity is not only relevant from a theoretical point of view. On the contrary, it can be applied in an empirical setting as well. Section 15.7 contains empirical evidence of the usefulness of this approach. Section 15.8 concludes.

15.2 The Concept of Carrying Capacity

Carrying capacity denotes the maximum number of users that can be sustained by a given set of resources in a specified territorial unit. In organisational ecology it is defined as the maximum number of organisations of a specified type that the market, or society, can sustain (Hannan and Freeman, 1989). The term can be given a clear

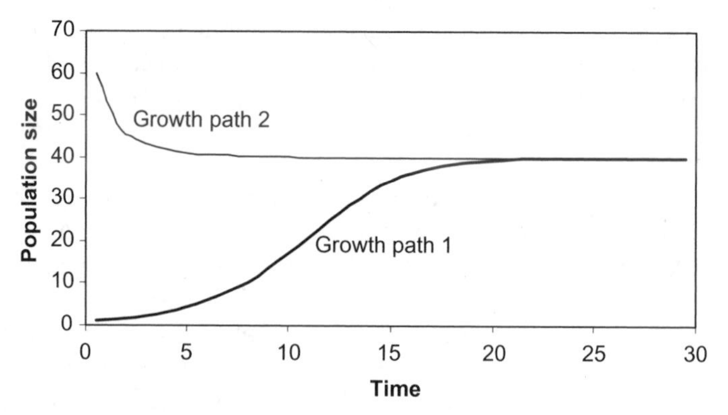

Figure 15.1 *Population growth following a logistic growth curve towards the level of the carrying capacity, starting from two different initial values*

mathematical interpretation, most notably in the example of the logistic growth curve describing the temporal growth path of the number of members of a population in a closed environment; for instance, the number of rabbits on an island. If natural growth of a population in discrete time between time t and $(t + 1)$: $X(t + 1) - X(t)$, denoted as $\dot{X}(t)$ is proportional to the size of that population then we have geometric growth:

$$\dot{X} = g(X(t)) \tag{15.1}$$

where g is the growth rate of the population.[1] If growth is positive this leads to an ever-increasing size of the population. It is clear that this growth curve is not realistic for most, if not all, species. A population cannot grow with a fixed growth rate g indefinitely. Given its dependence upon external resources, such as food or living space, a population is limited to a certain maximum, the level of which is conditional on the level of the resources and the technological level of the population to exploit these resources. Let the maximum size of the population be K, then the geometric growth model may be modified to the logistic growth model by making the growth rate not only dependent on the intrinsic growth parameter g but also dependent on the difference between the actual size of the population, X, and the maximum size, or carrying capacity, K:

$$\dot{X}(t) = gv(K - X(t))X(t) \tag{15.2}$$

The additional parameter v is the adjustment speed of the process. Figure 15.1 depicts two possible growth logistic paths resulting from this equation for fixed g. If $X(t)$ is very small compared with K then $K - X(t)$ is large and growth is high. With increasing size of the population over time, the gap between the maximum and actual size

[1] Formally the growth rate should be multiplied by the average size of the population between t and $t + 1$: $\frac{1}{2}(X(t) + X(t + 1))$, but for simplicity we use the initial population.

of the population narrows and consequently the growth rate $g(K - X(t))$ becomes smaller. If $X(t)$ is equal to K then growth is equal to zero. If, for some reason, the actual size of the population is larger than the carrying capacity, $K - X(t)$ is negative and the growth rate is negative as well.

Hannan and Freeman (1989) applied this logistic growth model (15.2) to longitudinal data on growth in the number of units of organisations. In their approach they included a number of explanatory variables in their model but, at the same time, they circumvented the problem of modelling the carrying capacity K explicitly. Instead, they estimated a second-degree polynomial to longitudinal growth data in order to find an empirical estimate of the maximum level of the population. This can easily be shown to be possible if we rewrite equation (15.2) as follows:

$$\dot{X}(t) = gvKX(t) - gvX(t)^2 = aX(t) - bX(t)^2 \tag{15.3}$$

The value of K may be found as $K = a/b$. Its precision depends on the precision of the coefficients a and b and the estimate is therefore in general not very reliable. But even if the value of the carrying capacity could be estimated precisely, there are a number of objections to model equation (15.3). First, the level of the carrying capacity is not static, but conditional on the size of resources and the level of technology. These change over time and therefore the size of the carrying capacity cannot be treated as static either. In other words: K should be $K(t)$. Secondly, there is no explicit content in this interpretation of the carrying capacity. It is simply a coefficient to be estimated from observed growth curves. Instead, an explicit model of carrying capacity is necessary that captures the essential underlying factors. More specifically, we require a model of carrying capacity based on economic notions.

15.3 An Economic Interpretetion of Carrying Capacity

The starting point is the population growth function equation (15.2). Firms operate in a multi-sector economy where they interact with each other. Interactions occur between industries in the form of intermediate supply and demand relationships, other interactions involving information, transfer of knowledge, and so on. Interactions also occur between industries and consumers. The latter type of relationship is self-evident but the inter-industry linkages are important as well. Each firm, as a member of a certain industry, has a potential network of linkages with other industries. Some linkages are beneficial for the firm: for instance, when these linkages are deliveries from one firm to the other. Some linkages are of the competitive type: the same firms compete in one market. The existence of other industries in the neighbourhood with whom it may interact is therefore a significant growth stimulating or inhibiting factor for this firm. At the same time, the growth of this particular firm is in itself again a potential growth-enhancing factor for other industries with inter-industry linkages with this firm, or a growth-limiting factor for similar firms who are in competition with this particular firm in the same market. In short, the size and composition of the population of firms in the neighbourhood is a determining factor for the growth potential for an individual firm. The effect of the population of firms in the neighbourhood upon the individual firm is realised through interaction linkages between firms. The concept of carrying capacity, as defined here, captures this growth potential of a local population of firms due to the size and composition of firm populations in the neighbourhood.

The concept of carrying capacity is here defined formally as the level of demand exerted by intermediate and final demand categories in the market. This notion is close to the economic interpretation of the concept as given by Carree and Thurik (1999). They propose a model of entry and exit of firms in markets as adjustments to a disequilibrium situation. The equilibrium situation is called the carrying capacity (Carree and Thurik, 1999, pp. 986–7). It is the benchmark number of firms which drives the direction of the change in the actual number of firms. Carree and Thurik relate the carrying capacity in a local market to the level of consumer demand. Here we use a specification, which is close to an input–output model, involving both intermediate and final demand. First, we present the situation involving only one region.

The actual level of supply will adjust to market demand. If supply is less than demand, there is an opportunity for increased production: birth rates go up, death rates go down, and incumbent firms will grow. If supply exceeds market demand, then the level of production will decrease: firm birth levels will drop, death rates will increase and incumbent firms will diminish in size. The relation between change and the level of production, $\dot{X}(t)$ and $K(t)$ is then given by equation (15.2) above. However, in a multi-sector economy the carrying capacity is specific for each sector s ($s = 1, \ldots, S$). Thus, the level of $K_s(t)$ is determined by both intermediate and final demand in the region:

$$K_s(t) = \sum_{r=1}^{S} a_{sr} X_r(t) + c_s B(t) \qquad (15.4)$$

where a_{sr} is the technical coefficient in the input–output table of sector r on sector s, c_s is the per capita final consumption of goods produced by sector s, and $B(t)$ is the size of the population at time t. Equation (15.4) is an elementary input–output equation without investments and exports, that specifies the interrelations among producers and between producers and consumers in the national economy. Note that here final consumption is explicitly related to the size of the population B using the consumption coefficients c. Equation (15.4) does not include exports and investments. They might be added as additional terms. Note that imports are implicit in the input–output specification. Sectors that rely partially on imports will have technical coefficients that do not add up to 1: the demand for this sector will only partially be met by internal supply.

The parameters of the demand equations of the input–output model may be derived from the national input–output table and the national accounts. So, equation (15.4) gives the level of the carrying capacity as a function of intermediate and final demand. Since the input–output model is an equilibrium model, $K(t)$ and $X(t)$ are equal to each other, and small changes in $B(t)$ will lead to a new equilibrium value for the $X_s(t)$. Here, we allow both $B(t)$ and the $X_r(t)$ to change, and specify a logistic adjustment function of $X_s(t)$ to its equilibrium value (carrying capacity) $K_s(t)$:

$$\dot{X}_s(t) = g_s v_s (K_s(t) - X_s(t)) X_s(t) \tag{15.5}$$

Note that the growth rate g and the speed of adjustment v are sector-specific.

15.4 Introducing Spatial Interactions into Carrying Capacity

There is another source of disequilibrium in the market. In a multi-regional economy, local variations may exist in supply and demand. The input–output model is generally used to determine levels of production given an increase in one of the final demand categories. For economic demography with a spatial dimension, we need an instrument to determine demand–supply ratios at the local level. Of course we cannot assume that each region or local area is an island without interactions with other zones. Producers will deliver their goods to retailers and other producers in other regions; consumers will make shopping trips to retailers in other zones as well; in other words, there exist complicated spatial patterns of transport flows of goods between producers and other producers (inter-industry linkages), and between producers and consumers. In order to deal with these spatial demand–supply interactions, we introduce the notion of the *spatial demand field*. A spatial demand field defined at location j ($j = 1, \ldots, I$) is a probability density function that gives, for every location i ($i = 1, \ldots, I$) in the neighbourhood of j, the probability that the good m, that is required at location j as either an intermediate input by a producer or as final input by a consumer, is produced at location i. This conditional probability $p_m(i|j)$ is not uniform over all production locations i, but decreases with distance between i and j. In other words, the form of the demand field is in general not flat but a dome with its top in the centre located at the demand origin j and decreasing with increasing distance from the centre. In addition, since the good is produced somewhere, it holds that $\Sigma_i p_m(i|j) = 1$. A function that is often used for modelling probability functions is the logit function. Used as a spatial interaction function with distance as its only argument, it has the following form:

$$P_m(i|j) = \frac{\exp{-\beta_m d_{ij}}}{\sum_k \exp{-\beta_m d_{kj}}} \tag{15.6}$$

where d_{ij} is the distance between locations i and j, and β_m is a distance deterrence parameter which is specific to the good m being demanded. If instead of the linear distance the log of the distance is used, the exponent in both nominator and denominator is replaced by $d_{ij}^{-\beta_m}$. A high value of β_m indicates a strong distance decay effect and a steep slope of the spatial demand field, implying that with increasing distance it soon becomes highly unlikely that the required good m is produced at i. A small value of β indicates a small distance decay effect and thus a relatively flat spatial demand field: even at larger distances between i and j it is still likely that goods demanded by producers and consumers at j are produced in i. A value of $\beta = 0$ implies a perfectly flat spatial demand field and the absence of any distance effect: at every location i it is equally likely that the good m demanded in j is produced in i. It should be clear that β is specific for every economic good m, since the form of the demand field varies with the character of the good. The spatial demand field of grocery products is very small (high β), whereas the demand for specialised services is often exerted at the national or even international market (β is very small or zero). Note that this system, just like the input–output model, is demand-driven. Firms and households require inputs or consumer goods m, and look around for suppliers to meet their demand. In which locations they look around for producers in alternative locations i is specified by the demand field $p_m(i|j)$. For practical reasons we do not work with individual goods m, but with goods produced by the aggregate sector s.[2]

How does this relate to the model of carrying capacity introduced here? First we note that the concept of carrying capacity refers to production capacity. In the spatial framework just introduced we make a distinction between locations i, where production takes place, and locations j, where goods are demanded or consumed. Of course in the recursive structure of the input–output framework producers are at the same time demanding intermediate goods as well, so each location may belong to both categories. In order to distinguish between production and consumption characteristics of locations, we introduce the total level of demand exerted by consumers and firms in location j:

$$D_s(j,t) = \sum_r^S a_{sr} X_r(j,t) + c_s B(j,t) \tag{15.7}$$

Note that this is quite similar in form to equation (15.4). There is one important difference, however. In the spatial framework of equation (15.7) we make a distinction between the level of demand and the level of production in each region, because, unlike the closed system of equation (15.4), they are not equal. The link between demand $D_s(j,t)$ and production capacity $K_s(i,t)$ is a spatial interaction relation, as given by the spatial demand field (15.6). The carrying capacity $K_s(t)$ introduced in equation (15.4) is the aggregate market demand, summed over all intermediate and final consumers. Thus, in a spatial setting, if we sum over all locations j, where these intermediate and final demanders are located, we arrive at the total market demand for goods produced by sector s in location i. The required production capacity of sector s at location i for intermediate and final demand in j at time t is given by:

$$K_s(i,j,t) = p_s(i|j) D_s(j,t) \tag{15.8}$$

and the total required production capacity, or carrying capacity at location i is the sum over all locations j:

$$K_s(i,t) = \sum_j^I p_s(i|j) D_s(j,t) \tag{15.9}$$

[2]Formally: s is the index of the class of goods Ω_s to which the single good m belongs: $m \in \Omega_s$.

By comparing the carrying capacity $K_s(i,t)$ with the actual level of production $X_s(i,t)$ in region i at time t the likely direction of local economic development of sector s located in i may be derived. The direction and size of growth can be modelled in terms of a logistic growth path as follows:

$$\dot{X}_s(i,t) = g_s v_s (K_s(i,t) - X_s(i,t)) X_s(i,t) \quad (15.10)$$

This equation states that growth in production in sector s in location i in the unit time interval t, $\dot{X}_s(i,t)$, is determined by the intrinsic growth rate g_s of sector s, multiplied by the overall speed of adjustment of the process, v_s, and modified by the effect of the gap between carrying capacity $K_s(i,t)$ and actual level of production $X_s(i,t)$.

The inter-sectoral and spatial interaction linkages developed here reflect a specific form of spatial agglomeration effect. A certain mix of production in a region may increase the level of the carrying capacity of the region and therefore the growth rate of the regional economy. Other mixes will only increase the actual level of production, while keeping the carrying capacity level constant, which results in lower growth rates in the region. The model therefore produces local variations in growth rates.

In summary, the advantages of this definition of spatial carrying capacity for economic demography are the following:

- the spatial input–output representation is consistent with the national input–output table;
- if the national input–output table represents an equilibrium situation, then the spatial input–output model gives for each location i the local disequilibrium in demand and supply for each sector. Since the system is consistent with the national input–output table, excess demand in some locations will be mediated by excess supply in other locations;
- inter-sectoral linkages as well as spatial interaction linkages between sectors are included in the model;
- the model presents a link between spatial economic disequilibrium on the one hand and macro-economic indicators of the national economy on the other;
- the information concerning $K(i,t)$ and $X(i,t)$ is of interest, since it describes the spatial distribution of demand and supply, as well as local market disequilibrium for each economic sector;
- in addition K and X are instrumental in driving the local spatial economic development for each sector, as hypothesised in equation (15.10); and
- the combination of inter-sectoral and spatial interaction linkages reflects the mechanisms of spatial agglomeration effects. As a result of this, local variations in economic growth rates will occur as an endogenous process.

We will test some of these ideas in Sections 15.6 and 15.7 below. First, however, the dynamic concept introduced in equation (15.10) will be discussed a little further in the next section, since it can be reformulated in other terms, giving it a more familiar structure.

15.5 The Model as a Dynamic Difference Equation for its Subpopulations

Equation (15.10) is a specific example of a system of logistic growth equations for interacting populations (Wilson, 1981, pp. 54–5). The system may be formulated in matrix terms as a set of difference equations. For convenience we make no explicit distinction between intermediate and final demand. In matrix terms, the system has the following form:

$$\dot{\mathbf{X}}(t) = \mathbf{G}[\mathbf{K}(t) - \mathbf{X}(t)]\mathbf{X}(t) \tag{15.11}$$

where $\dot{\mathbf{X}}(t)$ is an $S \times I$ growth matrix with elements $\dot{X}_s(i,t)$ and $\mathbf{G}$ is an $S \times S$ diagonal matrix of sector-specific relative growth rates and adjustment speeds $g'(s,s) = g_s v_s$. The matrix of carrying capacity $\mathbf{K}(t)$ with dimensions $S \times I$ is calculated as:

$$\mathbf{K}(t) = [\mathbf{AX}(t) + \mathbf{cb}(t)]\mathbf{M} \tag{15.12}$$

Here, $\mathbf{A}$ is an $S \times S$ matrix of technical coefficients a_{sr} from the input–output table, $\mathbf{c}$ is an $S \times 1$ vector of consumption coefficients, and $\mathbf{b}(t)$ is a $1 \times I$ population distribution vector. The elements of the $I \times I$ matrix of spatial discounting factors $\mathbf{M}$ are calculated as:

$$m(i,j) = p(j|i) \tag{15.13}$$

where $p(j|i)$ is the probability that a good demanded in j is produced in i; the functional form is given in equation (15.6). The interpretation of equations (15.11)–(15.13) is similar to (15.6)–(15.10). Note, however, that in this matrix formulation the distance deterrence function is generic for all economic sectors. In more realistic formulations this is not tenable. Equation (15.11) is the logistic growth equation for $S \times I$ submarkets. Equation (15.12) is the model for carrying capacity. $\mathbf{AX}(t) + \mathbf{cb}(t)$ is the demand exerted by firms and consumers, located in j, for goods produced by sector s. The matrix $\mathbf{M}$ defines the spatial demand field around j. The row elements of $\mathbf{M}$ sum to 1.

Difference equation systems of the form of (15.11) have been studied extensively in the sphere of organisational ecology (see, for example, Tuma and Hannan, 1984). Although each equation is linear, the resulting behaviour of the system is intrinsically non-linear and not always stable. A complete treatment of all properties of this system will not be given here. General accounts of coupled difference equations involving spatial interactions are provided in Wilson (1981). A number of features of the model should be discussed here, since they are related to the nature of sectoral and spatial interactions and the consequences for overall system stability. From equation (15.11) it is clear that equilibrium occurs whenever the carrying capacity is equal to supply: $[\mathbf{AX}(t) + \mathbf{cb}(t)]\mathbf{M} = \mathbf{X}(t)$. Unfortunately analytical solutions cannot be found for this system of equations, except for trivial cases. In the next section some results of a series of computer simulations with varying characteristics of intersectoral and spatial interactions will be given and discussed. The simulation results are illustrative for the structural characteristics of the model.

Table 15.1 Input parameters of simulation

Parameter		Value
ν	=	0.02
G	=	diag [0.1, 0.2, 0.4]
A	=	$\begin{bmatrix} 0.30 & 0.07 & 0.12 \\ 0.45 & 0.40 & 0.15 \\ 0.15 & 0.13 & 0.31 \end{bmatrix}$
c	=	[0.20, 0.40, 0.60]
β	=	−1.00
M	=	See equation (15.6) for specification. Distances between zones based on 4 × 4 square pattern, spaced 1 unit apart horizontally or vertically, and √2 units diagonally 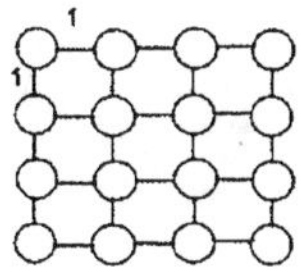

15.6 Computer Simulations of the Model

For the computer simulations we used a three-sector economy plus population for 16 regions, configured in a four by four square regional pattern. Interregional distances were calculated using Euclidean distance. The input parameters for the system are given in Table 15.1. We will focus on the following key parameters: v is the speed of the system, the sectoral interactions a_{sr}, and the spatial interactions, determined from the logit function (15.6), where the β_s coefficient determines the demand field.

Figure 15.2 gives the initial distribution of the population and each of the three sectors over the 16 zones. Figure 15.3 presents the equilibrium configuration after 300 iterations. If the evolution of the system is smooth, the equilibrium distribution does not reflect the initial distribution of the sectors. Using the speed of adjustment as given in Table 15.1, the evolution of the system is indeed smooth, and the initial highly skewed spatial distribution of the sectors is replaced by a more balanced distribution that reflects the combined effects of different parameters. A number of conclusions may be drawn from the equilibrium distribution. First, the distance effect favours economic development in more centrally located zones over peripheral zones. Second, the equilibrium configuration of each sector reflects, to a certain extent, the population distribution or final demand. The population impact is stronger for sector 3 than for the other two, since this sector is more dependent on final demand. The correlation between the population and sector 3 distribution is 0.78, against values around 0.69 for the other two sectors. The final distributions of each of the sectors are very similar, although the level is different, with correlations around 0.98. Formally, the spatial distribution of each of the sectors is a weighed average of the distribution of the other sectors, with weight matrix **A**.

As can be observed from Figures 15.2 and 15.3, the chosen input parameters have resulted in a model that has evolved to a steady state. This steady state is equal to the level of the carrying capacity for each (i,s) combination. However, if the speed of adjustment is increased (or, equivalently, the intrinsic growth rate of sectors), the

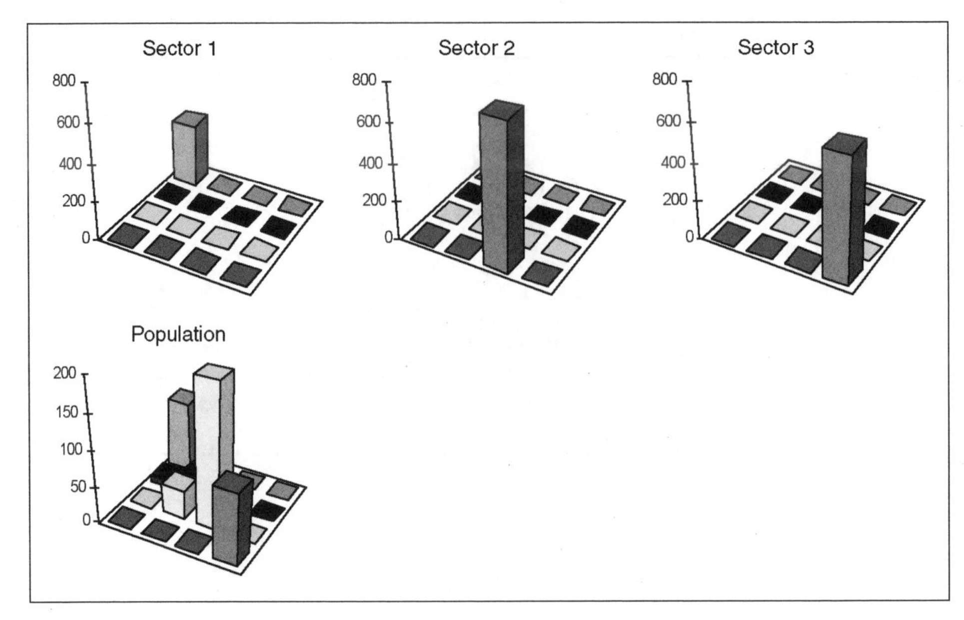

Figure 15.2 Initial distribution of sectors and population

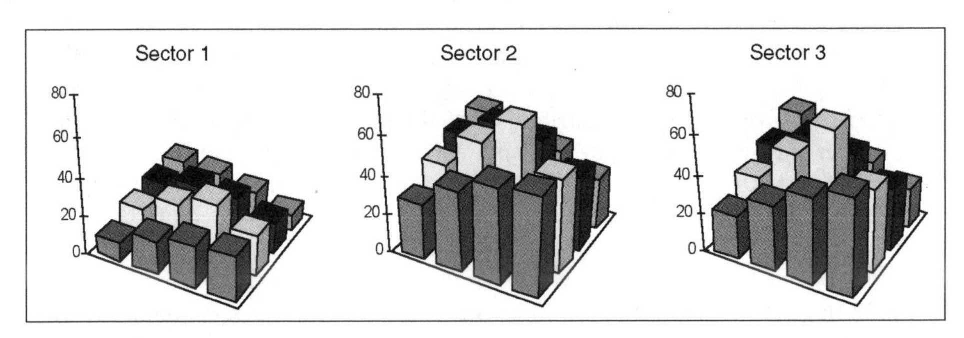

Figure 15.3 Equilibrium configurations of sectors

resulting growth of each sector is no longer guaranteed. Figure 15.4 shows an interesting example for one region. In the slow mode with a system speed of 0.02 the region develops from an initial value of 0.5 to the equilibrium value of 60. Increasing the speed to 0.10 results initially in smooth but fast growth up to size 50. At this point, growth decreases substantially and small oscillations set in that become larger after reaching the slow mode equilibrium of 60. After this point, growth oscillates around the value of 60. With speeds higher than 0.10, oscillations set in from the start and are much larger. Oscillations may also occur in a regular pattern; for instance with a certain periodicity. These oscillations are the result of an overreaction of the subsector to a large carrying capacity. This overreaction leads to a growth

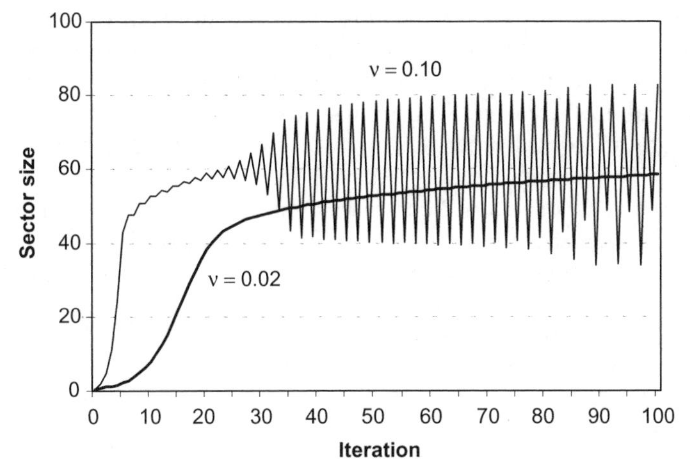

Figure 15.4 *Development towards a steady state and oscillatory behaviour using different speeds of adjustment*

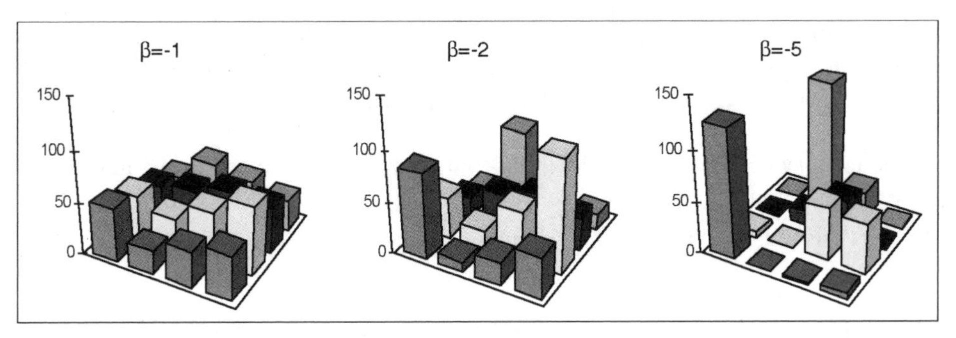

Figure 15.5 *The effect of distance friction on the equilibrium configuration of sector 3*

of the subsector above the carrying capacity. In the next period this has to be compensated by a large negative growth, and so on. Thus, the growth speed of a sector may have serious impacts upon the stability of the regional economic system as a whole.

Increasing the effect of distance will result in the emergence of a few locations where most of the activities are concentrated (Figure 15.5). Interestingly, the regions where concentration takes place are not the regions where initially activities were concentrated. Moreover, with increasing distance friction, the system no longer converges towards the steady state, but starts to oscillate. The configuration shown for the case of $\beta = -5$ in Figure 15.5 is not an equilibrium state. Conversely, interactions without distance friction will result in a completely flat surface and smooth behaviour towards the steady state.

From Table 15.1, which shows the technical coefficients used in the simulations and per capita final consumption, it is clear that sector 3 is most directly related to final demand, whereas sector 1 is the most independent sector of the three. Sector 2 is dependent on sector 1 for deliveries but also dependent on population. The effect of a change in technology will change the spatial distribution. For instance, a change of the technical coefficient $a(2,1)$ from 0.45 to 0.10, implying that for the production of one unit of good 1 much less input of good 2 is needed than before, has the result that the equilibrium value of sector 1 increases on average by 30%, whereas the equilibrium value of sector 2 decreases on average by 20 per cent. The effect on sector 3 is modest: a slight increase in equilibrium value of 7%.

These examples show that a spatial equilibrium of a multi-sectoral economy depends crucially upon the key interaction parameters of the economic system: speed of adjustment and intrinsic growth speed, sectoral interactions, and spatial interactions. The speed of adjustment of economic activities and the spatial deterrence effect have potentially highly disturbing effects, including non-equilibrium situations. In the next section we will relate some of these findings to empirical observations of individual firm dynamics.

15.7 Empirical Implementation of the Model

The concept of carrying capacity as developed in the previous sections may be implemented in empirical modelling. The required amount of information is only limited. The parameters necessary for calculation of the spatial carrying capacity are:

- the technical coefficients of the input–output table a_{sr} collected in the matrix $\mathbf{A}$;
- per capita consumption coefficients c_s, collected in the row vector $\mathbf{c}$; and
- sector-specific distance parameters β_s.

In order to calculate the level of the carrying capacity $\mathbf{K}(t)$ we also need:

- the matrix $\mathbf{X}(t)$ of production of activity s at location i at time t;
- the population distribution over regions at time t, $\mathbf{b}(t)$; and
- inter-zonal distances $d(i,j)$.

Technical coefficients and per capita consumption coefficients are readily available in most countries, even at the regional scale. The distance function parameter β_s may be estimated using some form of optimisation. In an empirical application in the Netherlands, which will briefly be discussed below, a simple line search algorithm was used for the distance parameter for each sector, using the Pearson correlation coefficient between sectoral growth $\dot{X}_s$ and $V_s = K_s - X_s$ as the criterion variable. The results are presented in Table 15.2.

All values of β are positive, which implies that interaction between sectors decreases with increasing distance between them (the function uses $-\beta$ as the coefficient value). The smallest coefficients are found for the recreational sector, transport and banking. The value of 0 would imply a completely flat demand field. This is plausible for transport, but less so for banking and certainly not correct for the recreational sector. The largest value, and a very small demand field, is found for

Table 15.2 Estimated distance deterrence parameters β_s and resulting correlations between sectoral growth $\dot{X}_s$ and $V_s = K_s - X_s$

Economic sector	Coefficient	Correlation *r*
1 Agriculture	0.25	0.26
2 Energy, mining	0.30	0.09
3 Industry	0.70	0.45
4 Construction	2.00	0.14
5 Wholesale	2.00	0.75
6 Retail	5.00	0.18
7 Recreation, hotels, restaurants	0.00	0.02
8 Transport communications	0.00	0.81
9 Banking, insurance, business services	0.00	0.32
10 Other services	5.00	−0.07
11 Public sector	0.15	0.07

retailing and other (domestic and health) services, which is again quite plausible. Industry has an intermediate value of 0.75. The resulting correlations between sectoral growth and the difference between carrying capacity and the actual level of production are in general positive and significant, but not very high. These results are in line with the results obtained by Feser and Sweeney (2000). Although using a different approach, they found empirical evidence that economic clustering, as indicated by the technical coefficients of the input–output table, is associated with spatial clustering as well.

The model of carrying capacity was applied at the regional level in the Netherlands (see van Wissen, 2000 for details). Equation (15.11) is a deterministic difference equation defined at the meso level of a regional multi-sector economy. The spatial level in the empirical application was the municipal level ($I = 685$), and 11 economic sectors were distinguished. However, the model was given a micro interpretation. The growth rate of sector s at location i at time t may be viewed as an underlying latent variable that determines the growth potential of individual firms of sector s in location i. Firms in this application are equal to business establishments. The data were taken from a register of business establishments in the Netherlands, called LISA. LISA is a longitudinal administrative register that includes all addresses in the Netherlands where paid work is being carried out. Although the agricultural sector and self-employment are not completely represented, and the province of North Holland is not covered, this register is used as the most important source for spatial information about business establishments in the Netherlands (see Kemper *et al.*, 1996). Here we used the data from the years 1990 and 1991. Individual firms are the agents of economic growth through new firm formation, firm closures, relocations and firm growth. In this application economic growth was operationalised in terms of birth and death probabilities of firms, as well as firm growth for incumbent firms. Within this demographic framework, birth and death probability functions, as well as firm growth functions, were estimated in which the variable $V_s(i,t) = K_s(i,t) - X_s(i,t)$ was used as an explanatory variable. The coefficient should be positive for firm birth and firm growth and negative for firm death. The results of this application, as presented by van Wissen (2000), were moderately favourable for the model. For firm birth and growth the expectations were indeed confirmed, with positive and significant coefficients. However, for firm death the coefficient was negative, as it should be, but insignificant.

15.8 Conclusions

In this chapter we introduced the notion of spatial carrying capacity, based on a model of sectoral and spatial interactions between submarkets of the economy. The resulting model can be viewed as a dynamic spatial model of growth in a multi-sector economy. The characteristics of this model were illustrated in a simulation context using a deterministic difference equation. The speed of adjustment of the process, the spatial deterrence function and the inter-sectoral interactions turn out to be crucial variables in shaping the spatial equilibrium of the economic system. Moreover, the existence of equilibrium is not guaranteed. The simulations showed that a very dynamic economy, with high speed of adjustment, may result in oscillatory behaviour of the level of economic production. A very strong distance effect has the same oscillating effect. Moreover, with high distance friction, economic activities tend to concentrate in just a few regions, leaving other regions largely empty. The simulations also showed that a change in technology, resulting in a reduction of the technical coefficients and thus in reduced inter-sectoral interactions, changed the equilibrium distribution without entering instability.

The concept of carrying capacity can be applied empirically. The required amount of information is limited, and usually readily available. Sector-specific distance parameters have to be estimated. In an empirical application in the Netherlands the estimated distance parameters were to a large extent, although not completely, according to theoretical expectations. Moreover, there are first signs that the concept of carrying capacity as developed here may be useful in explaining and predicting regional sectoral growth.

In conclusion, the concept of carrying capacity as defined in this study deserves more attention, both theoretically and empirically. From a theoretical point of view the model has many interesting properties and through the relation with the input–output table the link with macro-economic variables is established easily. The model is one way of making the concept of agglomeration economies operational, by explicitly taking into account inter-industry linkages as well as their spatial representation. The model may be used at the meso level of sectoral and spatial submarkets. The usefulness of the approach at this level is not studied in depth yet, but from the theoretical, simulation and empirical results reported in this chapter, further explorations in that direction look promising.

REFERENCES

Amstrong, H. and Taylor, J. (2000) *Regional Economics and Policy*, Third Edition. Blackwell Publishers, Oxford.

Carree, M.A. and Thurik, A.R. (1999) The carrying capacity and entry and exit flows in retailing, *International Journal of Industrial Organization*, **17**: 985–1007.

Carroll, G.R. and Hannan, M.T. (2000) *The Demography of Corporations and Enterprises*, Princeton University Press, Princeton, NJ.

Feser, E.J. and Sweeney, S.H. (2000) A test for the coincident economic and spatial clustering of business enterprises, *Journal of Geographical Systems*, **2**(4): 349–74.

Gudgin, G. (1978) *Industrial Location Processes and Regional Employment Growth*, Saxon House, Westmead.

Hannan, M.T. and Freeman, J. (1989) *Organizational Ecology*, Harvard University Press, Cambridge, MA.

Kemper, J., Kloek, W. and van Wissen, L . (1996) Databronnen voor economische demografie, *Planning Methodiek en Toepassing*, **48**: 3–11.

Krugman, P. (1995) *Development, Geography and Economic Theory*, MIT Press, Cambridge, MA.

Neary, J.P. (2001) Of hype and hyperbolas. Introducing the new economic geography, *Journal of Economic Literature*, **39**(3): 36–61.

Nelson, R. and Winter, S. (1983) *An Evolutionary Theory of Economic Change*, Cambridge University Press, Cambridge, MA.

Tuma, N.B. and Hannan, M.T. (1984) *Social Dynamics. Models and Methods*, Academic Press, London.

van Dijk, J. and Pellenbarg, P. (eds) (2002) *Demography of Firms Spatial Dynamics of Firm Behaviour*, Nederlandse Geografische Studies no. 262, KNAG Utrecht/Groningen.

van Wissen, L.J.G. (2000) A micro-simulation model of firms: applications of concepts of the demography of the firm, *Papers in Regional Science*, **79**: 111–34.

van Wissen, L.J.G. (2002) Demography of the firm: a useful metaphor?, *European Journal of Population*, **3**: 263–79.

Wilson, A.G. (1981) *Catastrophe Theory and Bifurcation, Applications to Urban and Regional Systems*, Croom Helm, London.

16

Planning a Network of Sites for the Delivery of a New Public Service in England and Wales

Mike Coombes and Simon Raybould

Abstract

The Family Law Act (1996) aimed to introduce across England and Wales a new service of Information Meetings (IMs) for people considering divorce. Attendance would be compulsory, which meant that the service would have to be provided in places that left few people objecting to attending one. The research reported here identified a suitable pattern of sites at which IMs could be held. Several methods of analysis were needed: a regression model estimates the spatial distribution of potential IM attenders; survey data analysis measures people's willingness to travel to attend IMs; multi-criteria analysis establishes the list of sites which meet the necessary criteria; and location-allocation modelling identifies which of these sites would minimise the travelling which would be required of attenders. The chapter summarises this mapping and modelling research programme, and highlights the key elements of sensitivity in the results. One important part of the process is to identify which are the issues to be resolved by the researchers and which are the decisions that remain with the client government department.

16.1 Introduction

When the Family Law Act was passed in 1996 some of its most innovative aspects were put 'on hold' until a programme of research studies had provided the basis for deciding how to implement these innovations (Walker, 2001). In particular, Part 2 of the Act made provisions for Information Meetings (IMs) that would be compulsory

Applied GIS and Spatial Analysis. Edited by J. Stillwell and G. Clarke
 ISBN: 0-470-84409-4

for married people intending to initiate divorce proceedings. The compulsory nature of the IMs implies that they would have to be delivered in such a way that few people would be likely to object to attending one. The research programme involved evaluating a range of alternative *pilot* IMs – for example, varying the length, style and content of the meeting – in different regions. This chapter summarises one distinct research strand, identifying a set of potential IM sites whose locations would make it unlikely that many people would object to attending IMs specifically because they could not readily reach any of the IM sites.

The research described below aimed to provide the information needed by the Lord Chancellor's Department (termed 'the Department' here) in order to decide the number – and location – of sites at which IMs would need to be provided. It should be noted at the outset that the term 'site' here denotes an approximate location (e.g., the Victoria area of Westminster): the research did *not* aim to choose specific premises as venues for IMs. In practice, the Department's policy decision would centre on judgements as to the likely impact on married people if, for example, the government felt it was no more obliged to provide all areas with a local IM site than it is to provide a university in every locality. By providing IMs at relatively few sites the implementation costs should be minimised, but at the risk of increasing the number of objections by people who found the sites hard to reach. The research challenge is thus to identify how many sites would be needed – and where these should be – so that few people would have to travel far to reach one.

There are many existing public services which are provided through networks of local sites: examples range from the various kinds of court provided by the Department to other local public services provided by central government (e.g., Job Centres). These networks have typically evolved over many years, gradually adjusting to alterations in the services provided and also to changes in the pattern of demand such as the development of new towns. Each of these networks evolved differently to meet its own service's needs, so none of them provides an ideal blueprint for the selection of IM sites. Although the networks of sites used by other public services evolved over long periods, sophisticated methods of analysis have increasingly become available to help maximise the sites' accessibility to their users. For example, Thomas *et al.* (1991) illustrated the value of a location-allocation model in helping to inform decisions on which of the North West region's courts to close while minimising any increase in the distances people have to travel to reach their nearest court. The basic elements of a location-allocation modelling strategy are:

- mapping the spatial distribution of demand for the service;
- measuring people's reluctance to travel to access the service;
- identifying candidate sites for providing the service; and
- selecting the optimal sites from these candidates.

It is the final step which uses the location-allocation model itself and here it becomes most clear how the results' optimality depends on certain assumptions, including those which are embedded in the three previous steps.

This chapter reports location-allocation modelling to tackle the highly unusual proposal to 'roll out' a full national network of sites rapidly for a new public service:

this is clearly more a case of a 'big bang' than the process of evolution which is characteristic of the public sector. The rapid development of a new network is rather more familiar in the private sector than in public services, although even the vigorously growing commercial services such as multiplex cinemas have in fact been 'rolled out' over several years. It is notable that location-allocation modelling has been extensively adopted for site selection in the private sector (Ghosh and Harche, 1993). In the private sector, the key requirement is usually that the sites are close enough to large concentrations of people so as to maximise the likely profitability of the network. By contrast, the public service emphasises more the minimising of differences in service provision across the country. The general framework of location-allocation modelling provides for choices to be made between such different objectives, so that each analysis can be 'customised' to meet the specific objectives of that particular application.

The analyses in this chapter build up the information that is needed for a location-allocation model relevant to identifying sites for IMs. By way of clarification, it should be stated at the outset that all the analyses – the three preparatory stages and then the location-allocation modelling itself – rely upon a geographic information system (GIS) which can provide a high level of resolution through recognising up to 10 000 separate areas across England and Wales (these areas are the local government wards existing at the time of the 1991 Census). Fotheringham *et al.* (1995) showed that the size and shape of the zones used in a location-allocation model can influence the results of the analysis, but here the level of detail provided by nearly 10 000 zones means that there is very little loss of accuracy in a national-level model when the analysis treats each ward's population as if it were all located at the ward's population-weighted centroid. Straight-line distances between the respective centroids are similarly assumed to provide robust estimates of the distances between wards. All models simplify some of the complexity and peculiarities of the real world, the important task is to portray accurately enough the major factors that will shape the evaluation of the alternative scenarios to be explored (Densham and Goodchild, 1990).

The first preparatory analysis is outlined in Section 16.2 and is termed *demand distribution* because it attempts to estimate, for each local area of the country, the number of people who may attend IMs. It is necessary to make broad estimates of the 'population at risk' of attending IMs because the Department has not suggested restricting attendance to an identifiable category of people, such as those people who have petitioned for divorce. A key problem to be faced is that there is insufficient information available on the current whereabouts of relevant population groups (e.g., couples who have been married between five and 10 years). The information that *is* available is modelled to provide plausible estimates at the highly localised scale; this level of detail is necessary for the location-allocation modelling to identify a set of sites that will be as near to as many people as possible. The second preparatory step – described in Section 16.3 – analyses data from the pilot IMs on the extent to which people are deterred from attending IMs if they have to travel further. The results of this *distance deterrence* analysis provide an essential input to the location-allocation model. If people are very strongly deterred, then the model is designed to focus on reducing the minority of very long journeys in more remote areas; if distance deterrence is not so severe, then the emphasis will be more on making smaller reductions to the journeys of very many more people through the provision of additional sites in heavily populated areas. Section 16.4 of the chapter describes the third preparatory set of analyses that provide a set of suitable *candidate sites* for the location-allocation model. For example, the mountainous centre of the Lake District is unlikely to be a suitable site to serve the population of Cumbria even though it is equidistant from all the county's main population centres (which lie in an arc around the periphery of the Lake District from Kendal through Barrow and Workington to Carlisle). The type of site favoured by people who attended IMs during the pilot phase of implementation provides the basis for identifying the criteria that sites need to satisfy. These candidate sites will then be input to the location-allocation model so that only suitable sites could be included in an eventual network. Section 16.5 describes the location-allocation modelling itself, then Section 16.6 provides a brief evaluation of the results. The chapter ends with some conclusions about this particular research, as an example of modelling in a dynamic policy context.

16.2 Estimating the Demand Distribution

The first step in the modelling research is to produce a set of local area predictions of the numbers of people likely to attend IMs in any year. An annual national total was estimated by the Department separately, and the local area predictions are used to 'share out' this total national demand between the local areas. For the modelling here to choose the best IM sites, this national total value is not actually relevant: the other inputs to the model ensure that central London would be the 'first choice' site regardless of whether the national annual demand was a hundred or a million people.

The best starting point for a model to estimate the likely local distribution of IM attenders is the number of divorces registered at each of the county courts across the country, so the modelling used the 180 county courts for which the number of divorces in the mid-1990s was known (between 165 000 and 170 000 a year in England). This dataset provided only a count of divorces at each court: there is no available dataset with postcoded addresses of divorce petitioners, so assumptions are needed about where people who divorce at any specific court actually live. The first step of this process was to define court catchment areas in terms of 1991 wards, for which areas it would then be possible to compile Population Census data as the basis for the model's independent variables. The catchment area definitions sought to reflect the guidance that the Department publishes, in which around 25 000 'places' are listed and allocated to individual courts. Substantial efforts were needed to allocate each ward to a court area, because there is no linkage between the 25 000 'places' listed in the directory and either wards or any other set of standard statistical or administrative areas. A further necessary step was to group together some county courts, because divorce petitions are not heard in every court. Information on other patterns of movement, notably commuting, was used for this grouping: for example, the Caerphilly county court area was grouped with Cardiff because this city is the predominant destination of out-commuters from Caerphilly and the adjacent

Table 16.1 *Sample of divorce petitioners at North East courts*

	Female (%)	Male (%)	All (%)
Petitioned in 'local' court	77.6	72.5	76.2
Petitioned in 'adjacent' court	17.3	12.3	15.9
Petitioned elsewhere in region	2.3	11.6	4.9
Petitioner from outside region	2.8	3.6	3.1
N	*353*	*138*	*491*

villages. The output from this step is an estimated catchment area of each divorce court.

Although there is no national dataset about where the people who divorce at each court in fact live, anecdotal evidence suggested that people do *not* always use their local court. The addresses of a sample of people who petitioned at courts in some of the North East pilot areas was postcoded as part of the research. Table 16.1 presents the evidence from this dataset as a test of the appropriateness of the catchment areas that have been estimated here. Over three-quarters of petitions were from people who did indeed live within the estimated catchment areas. The proportion was slightly less for male petitioners, but they are notably in the minority in practice (McCarthy and Simpson, 1991, established that men were more likely to move away from the marital home during the divorce process). Of those who petitioned at North East courts but did *not* live within that court's estimated catchment area, most lived just across that estimated catchment area boundary in one of the adjacent areas. A substantial degree of 'boundary hopping' is in fact inevitable with courts so closely spaced as are, for example, Newcastle and Gateshead (bearing in mind that both North and South Shields also have divorce courts). In many parts of the Tyneside conurbation neighbouring people have petitioned at different courts, so adjustments to the estimated catchment area boundaries could barely, if at all, reduce 'boundary hopping' in such areas.

This interpretation is borne out by the results in the other areas where a court sample was collected. Those areas that cover a similar urban/rural mix to the North had a very similar proportion of court sample members petitioning at their 'local' court. More heavily urbanised areas such as Greater Manchester and, most especially, London had noticeably higher proportions of petitioners 'boundary hopping' due to the very close spacing of courts there. All the same, the only practical assumption here is that the numbers of people moving around the country to petition for divorce roughly cancel each other out, so that the number registered at each court should be a fairly close approximation to the number of petitioners who live in that court area and – for the purposes of this research – also should provide a close approximation to the number of potential IM attenders.

Using these 180 court catchment areas, a regression model was developed which used the courts' number of divorces as the dependent variable, and their catchment areas' population size and characteristics as the independent variables. This model could then be used to 'shift scales' down to the ward by predicting the number of divorces likely to arise from each ward's population size and characteristics. These ward-level predicted values will then be turned into 'shares' of the national total of predicted values, which in turn will provide the basis for locally 'sharing out' the total number of people expected to attend IMs nationally.

The dependent variable is *the number of divorces per thousand married men*: the variables were generally defined in terms of male characteristics because some of the potentially important factors (such as unemployment) are not so well measured for women. There is not a huge literature on the factors that may be important in shaping the rate of divorce in this country, although Clarke and Berrington (1999) recently provided a valuable summary. Research on the subject has also been hindered by the limited official statistics available, because divorce registration procedures date back many years and do not, for example, record either party's date of birth and this hampers some longitudinal analyses of divorce (cf. Kiernan and Mueller, 1999).

The age of married people – and the duration of their marriage – are identified in the literature as major influences on the likelihood of divorce. The modelling experimented with a number of detailed age groups variables, with the final model eventually selecting a variable relating to the 16–44 age group. Ethnic minority populations were represented by four variables which measured the proportion of men aged 20–54 who were (a) Afro-Caribbean or (b) Pakistani/Bangladeshi or (c) Indian/Chinese or (d) Irish (see Berrington, 1996 and Beishon *et al.*, 1996, for the background research). As well as a variable on the proportion of men in low-skilled occupations, two alternative measures of male unemployment were made available to the model as options to represent the issue of poverty and/or insecure employment, which is known to be linked to divorce (Böheim and Ermisch, 2001). Three other employment-related variables were explored, all representing factors that the literature suggests to be positively associated with increased rates of divorce (e.g., Gibson, 1996):

- the proportion of employed men who have security-related occupations;
- the proportion of married women who either are self-employed or have full-time jobs; and
- the proportion of all households that included two earners but no children.

A variable on the proportion of younger men in the area who are married was among the variables to represent demographic factors that could represent contextual influences on the local level of divorce. The two final variables assessed were both more familiar features of area-based modelling. A housing tenure variable measured the proportion of households that are owner-occupying (this has been argued to be negatively associated with the probability of divorce). The other variable tested was a measure of the level of urbanisation in each area (Coombes and Raybould, 2001).

One important feature of area-based modelling such as this is that two variables which have similar distributions across the areas analysed can 'cancel each other out' in their effect on the dependent variable. As a result, modelling the difference between areas' populations can yield rather different results to modelling the behaviour of the individuals who make up those populations. A particularly clear example of this 'cancelling out' involves the ethnicity variables. It is known that Afro-Caribbeans have higher divorce rates than white people, whilst the Pakistani and Bangladeshi groups have low rates. At the scale of analysis undertaken here, the same areas – in practice, the more rural areas – tend to have populations that include very few people from either of these broad minority ethnic groups. As a result of this similarity of the geographical patterns of the Afro-Caribbean and the Pakistani/Bangladeshi variables, they tend to 'cancel each other out' in the court area analyses and so neither emerges as significant in the model.

After considerable experimentation on technical aspects of the modelling, and also on the precise detail of the independent variables, the robust model to emerge included significant variables, all taking the expected signs, on the age and class/occupation of local populations *plus* a factor lowering the rate according to the proportion of the younger men in the area who are currently married. The latter variable suggests a contextual effect that produces greater resistance to divorce in areas where the proportion of young men who are married is higher than the national average. To check the sensitivity of the modelling, the analyses were run on each of two separate years' data, as well as the two in combination. The results were encouragingly consistent in the selection of variables as significant, and in the calibration of those variables. There is also no clear pattern to the residuals from the model which would call for additional variables to be included. Yet in practice the model's predictions do not widely diverge from an alternative set of results based purely on the number of married people living in each area. Although it is rather disappointing that the efforts devoted to developing this model have not provided a dramatic improvement in the accuracy of predictions vis-à-vis those which could have been obtained by assuming a simple per capita divorce rate, it might also be seen as reassuring that the estimates here do not substantially depart from 'bedrock' demographic data. As a result, this model has been used here to predict the ward-level distribution of the numbers expected to attend IMs nationally: this then provides the demand measure for the location-allocation model.

16.3 Estimating the Level of Distance Deterrence

The next step is to model the degree to which people are reluctant to travel to reach an IM site. This reluctance can be estimated here by considering the behaviour of people in the pilot areas. For each person registering their interest in attending an IM during the pilot phase, their *nearest* IM site was identified by calculating straight-line distances between their home postcode and the locations of all IM sites which provided at least one IM after the date on which the person registered.

Table 16.2 examines both these distances to nearest sites, and the actual distances travelled to pilot IM sites. The first row presents the distribution of distances, for the

Table 16.2 Distance from meeting sites

	Distance band (km)					
	0–1	1–2	2–5	5–10	10–20	over 20
% people who attended IMs:						
distance to IM site attended	2.7	7.8	29.0	26.8	20.9	12.8
distance to nearest IM site	4.5	10.0	35.4	30.7	16.0	3.4
% non-attenders:						
distance to nearest IM site	4.7	10.9	36.1	29.7	15.9	2.8

people who *did* attend IMs, between the IM site which they attended and their home address. The next row shows the distances between the same people's home locations and their nearest meeting site. The difference between the values in the two rows is rather modest, suggesting that the vast majority of people attended their nearest meeting site (although nearly 10% did in fact travel over 20 km when there was a nearer meeting site). The third row shows the distance to the nearest site for those *non-attenders* whose home location is known. Comparing the second and third rows suggests that the distance to the nearest meeting is not necessarily the major influence on people's likelihood of attending: if non-attendance had been very substantially fuelled by remoteness of meeting sites, then a noticeably higher proportion of non-attenders would have been found to live in areas which are a longer distance away from the nearest IM site.

Table 16.2 can also be seen to provide an overview of the kinds of distances which people are prepared to travel to attend IMs. More than one in eight of the people attending pilot IMs travelled more than 20 km from their homes, and over a third attended sites at least 10 km away. People who had attended pilot IMs were interviewed, and given other opportunities to feed-back their views, but very few references were made to the remoteness of IMs posing a substantial difficulty. As a result, it might be hesitantly concluded here that for *most* people a journey of up to 20 km is unlikely to pose a major problem.

There was, perhaps surprisingly, little difference between the median distance travelled by women and men (6.3 and 6.6 km respectively). The median distance travelled by car users (7.3 km) was around 50% longer than that for the users of each of the other methods of travel except train, hinting at the higher levels of mobility which are commonplace among more affluent people. This issue is explored here by using the Townsend Index of deprivation (Townsend *et al.*, 1988), which measures average levels of deprivation (positive scores), or affluence (negative scores), for each ward in the country. People who registered an interest in attending one of the pilot IMs provided their home postcode for the research so the Index value of their home area can be identified. Table 16.3 draws upon these calculations for all people attending pilot IMs, grouping them according to their home area Townsend Index values. The first row shows a notable pattern, in which people living in the most deprived neighbourhoods are found much more often to live near to IM sites. This result will be

Table 16.3 *Home neighbourhood and likelihood of attending IMs*

	Townsend Index values					
	Up to −4	−4 to −2	−2 to 0	0 to +2	+2 to +4	Over +4
Distance from nearest pilot IM (km)	14.1	10.4	11.1	12.0	10.4	4.2
Probability that attended (% of all who registered)	0.85 (12.4)	0.83 (28.7)	0.82 (23.5)	0.81 (15.3)	0.77 (10.0)	0.69 (10.0)

mainly due to many deprived areas being in inner cities, which are thereby near to the pilot IM sites whose locations were mostly in town or city centres.

The relationship between the level of deprivation of people's home neighbourhood and the likelihood of them attending IMs can be examined directly with the evidence about people who had made themselves known to pilot projects. Table 16.3 shows that the majority of people known to the research are from non-deprived areas: the three left-hand columns, which cover people from non-deprived areas, include around 65% of all people in the pilot research dataset. The second row then reveals, for each type of area, the proportion of the people who expressed an interest in attending who did indeed attend a meeting. There is a consistent, if unspectacular, tendency for the more affluent groups to be more likely to attend a meeting whilst those from the most deprived areas are the least likely to have attended.

The extent to which people's reluctance to travel to attend IMs increased with their distance from the nearest site provided the basis for the distance deterrence model here. The first step was to categorise people who expressed an interest in attending a pilot IM according to the distance they lived from an IM (viz. the first category includes all who lived under 1 km from their nearest site, the second category includes those at least 1 km from a site but less than 2 km away, and so forth). For each category, an appropriate 'typical' distance was assumed (e.g., for the first category this was 0.3 km, and for the second 1.3 km). Not everyone who registered an interest did eventually attend an IM and, in practice, people who lived further from an IM site proved to be less likely to attend. The distance deterrence model uses the proportion of people, in each distance category, who did *not* attend an IM as the dependent variable in a non-linear regression against 'typical' distances (using SPSSx). The beta value from the regression is then taken as the distance deterrence coefficient in the location-allocation modelling.

Ever since the early gravity models, the default distance deterrence value has been 2 and, in practice, empirical evidence has quite often found that the model which best fits the data on people's resistance to travelling has the distance squared. Rather higher distance deterrence is found for very 'everyday' activities, whereas the 'one off' nature of IM attendance suggests people may be more willing to travel further. On this basis, a distance deterrence exponent value nearer to 1 might be expected here. In fact, the modelling found the exponent's value to be slightly below 1 when the whole research sample population was analysed.

The distance deterrence analyses were also run on sub-groups of the data: the research sample had been partitioned according to Townsend Index scores, so as to identify separately those who live in deprived/average/affluent areas. These sub-groups' exponent values show the variation which would have been expected, with the deprived group's value found to be 1.2 whilst the affluent group's was 0.9 (and the average group's is 1.0). Given that the policy concern is that the IM sites should be accessible to *all* groups, it seems appropriate to use the more deprived group's deterrence value in the location-allocation model. In other words, the IM sites were then selected on criteria which, in effect, assumed that everyone attending IMs has the higher resistance to travelling which was found among those living in more deprived areas. This approach also protects the analysis somewhat from the probability that many of the people included in the research data do not appear to be so deterred from travelling longer distance simply because they are 'volunteers' who come from more affluent backgrounds. Even so, the relatively low level of the deterrence values which have been found here were broadly to have been expected, given that IMs are concerned with such major issues in people's lives.

A footnote to these analyses is that the distances calculated here have all, in effect, assumed that people have travelled from home to IMs. Table 16.2 had shown that a notable minority of people did *not* attend the IM site nearest to their home, which suggests that the travelling to IMs may often be undertaken as part of a multi-purpose journey. In particular, people may quite often travel from work rather from home (e.g., in their lunch-break). As a result, people's decisions about 'how far is too far' may quite often be based on calculations which are rather different from the distance-from-home measures which have been estimated here.

16.4 Identifying Candidate Sites

One of the key concerns in identifying suitable IM sites is the more restricted level of mobility of some members of the population. A site in the countryside equidistant between three nearby towns might seem to be a geographically 'fair' solution, but would this be appropriate? In particular, would such a site be suitable for people without access to a car? In the absence of comprehensive public transport service information, the search for suitable IM sites *could* be restricted to places within a certain distance of a major road, but the assumption that all such roads have a reasonable level of public transport services would not be a particularly safe one. By contrast, it *is* likely that the 'high street' of a town and city centre, or a major suburban centre, will feature quite high levels of public transport services from its catchment area. This approach to site selection was reinforced by the findings from the pilots, in which people stressed their preference for IM sites located in busy areas to which people go for a variety of purposes.

Considerable data analysis was required to implement this strategy because there is no 'off the shelf' list of appropriately defined town centres. Following substantial experimentation, the criteria for a candidate site were set as the requirement for the location to have at least two of the following characteristics:

- it has a Benefits Agency office;
- it has a market (open or covered); and
- it is a substantial focus for commuting flows.

These characteristics were chosen partly because they were likely to be associated with places frequented by less affluent people and so be more likely to identify centres which have better public transport provision.

The list of Benefit Offices used here does not include 'neighbourhood' Benefits Agency offices, which would have provided a much longer list but would not have been suitable for indicating 'high street' locations. Market towns were identified as locations holding a retail market – on days other than Sundays or Bank Holidays – more often that once a month. No source provides an up-to-date comprehensive list, so the method adopted here was to include any town recorded as holding a market in one or more of four different sources (ranging from a general-purpose gazetteer of towns to the developing Internet site of the National Market Traders' Federation). This inclusive approach minimised the risk of 'missing' a town – which thereby might not be identified as a candidate site – purely due to incomplete data.

The identification of commuting foci was automated. The first step was to analyse the data on flows between wards in the 1991 Census commuting dataset. Of over two million such flows, the program distinguished those of at least 10 (note: these are in fact flows which involve over 100 people, due to the sample frame of this dataset). The next step was to identify wards whose centroids are so close together they can be considered to be parts of the same location. Wards whose centroids are less than 1.5 km apart are in effect, linked together. (This device side-steps the tendency in some towns for a major employer to be located a short distance away – and thus possibly in a different ward – from the High Street where the Benefit Office and/or market is likely to be located.) Wards whose centroids were no more than this distance apart were then grouped together for the purpose of counting the number of large flows to them. The criterion adopted was that a (grouping of) ward(s) had to have at least three large commuting flows to be considered a substantial commuting focus here. Figure 16.1 illustrates the results of the analysis in the North East region. It can be seen that this multi-criteria approach tends to find the 'high street' centres of most small to medium size towns. The locations qualifying with just one of the three criteria are mostly suburban parts of the larger towns (e.g., Newcastle). Over 2000 potential sites nationally possess at least one of these three characteristics, of which 647 had at least two and so make up the list of candidate sites. Together these 647 provide a very comprehensive coverage of the country, with only a few rather remote upland areas more than 25 km from all of them.

16.5 Location-Allocation Modelling

All the research detailed so far in the chapter has been necessary preparation for the location-allocation modelling. As the core of the challenge which was faced, it needs to be described here in a little detail. The location-allocation modelling used the

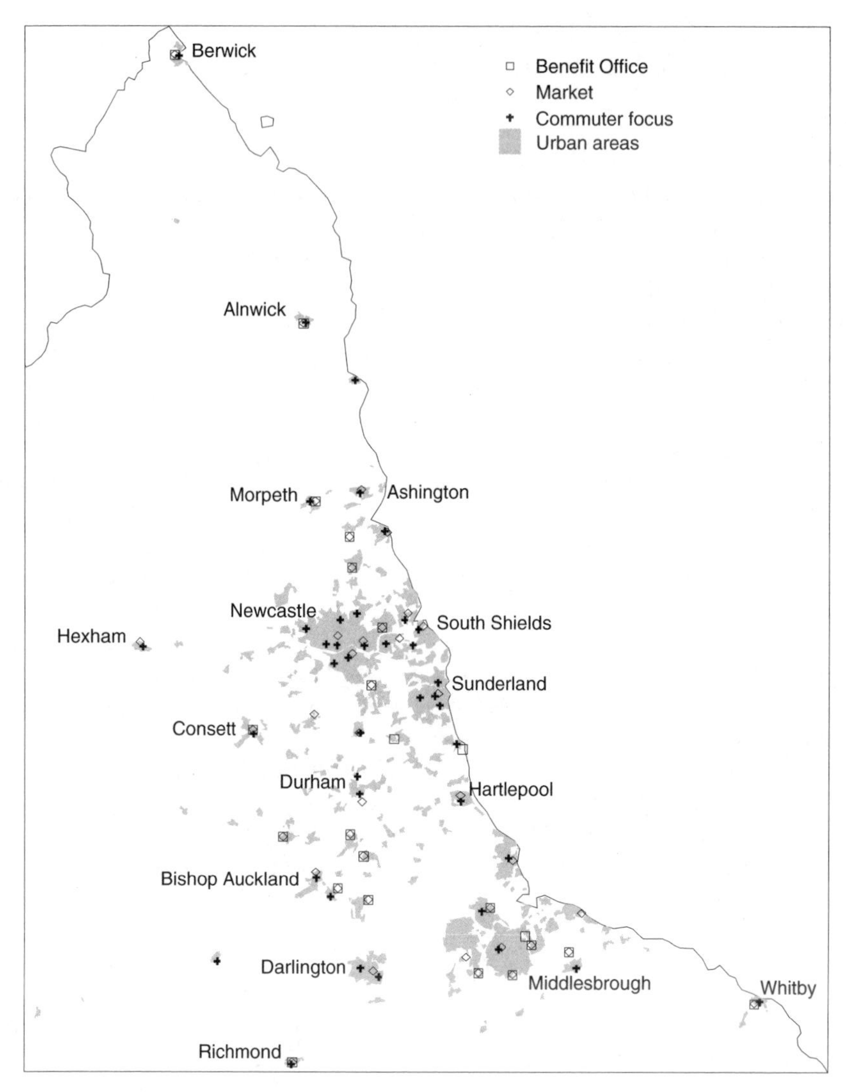

Figure 16.1 *Candidate sites in North East England*

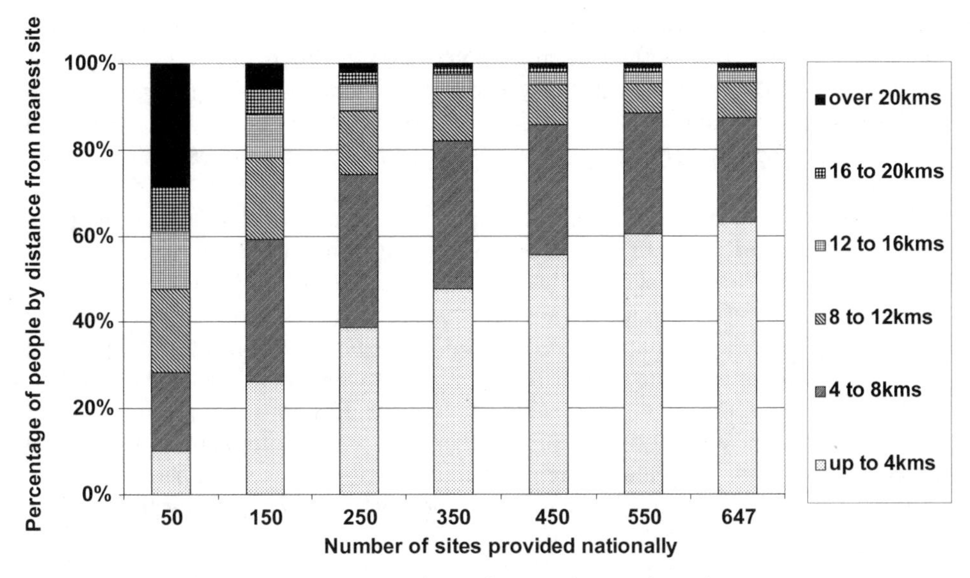

Figure 16.2 *Distance to nearest site, depending on the number of sites*

ARCPLOT module within the ARC/info package. The 647 candidate sites were input, along with the demand distribution model's estimate of demand from each ward, plus the distance deterrence coefficient. The analyses used *mindistpower* as the *locatecriteria*: this means that the analyses sought to choose sites which minimise the total distance travelled, but with these distances weighted by the distance deterrence coefficient. The optimisation procedure needed to be as fast as possible, given the very large number of wards and sites involved, so the default heuristic GRIA was applied. The allocation results were exported to Excel for post-processing and presentation. An additional point of information is that the Department did not set a minimum workload – a number of people to use each site over a year – although this could be expected to reduce costs (Densham and Rushton, 1996). For the record, it appears that this analysis (with nearly 10 000 separate demand locations and nearly 650 candidate sites) is one of the largest ever applications of location-allocation modelling.

The strategy for the modelling was to carry out many different analyses, each reflecting a different scenario from a wide range of options. As has been stressed from the outset, the key policy decision here is 'how many sites should IMs be provided at?' Figure 16.2 summarises the results from the analyses, in terms of the distances people would have to travel, depending on the Department's choice over the number of IM sites. It can be seen that increasing the number of sites much above 250 makes surprisingly little difference to the proportion of people attending who have to travel rather long distances. As an extreme extension to these analyses, it was found that over 2500 sites would have been needed for less than 5% of people to have to travel over 8 km (5 miles). The decision was made by the Department to explore in detail the 250 site option, which provides 95% of potential attenders with a site no more than 16 km (10 miles) from their home. One notable advantage of this decision for the modelling was that it then became unnecessary to build in site capacity constraints (cf. Murray and Gerrard, 1997), because this number of sites meant that even the busiest one was estimated to have fewer people attending it over the course of a year than had been separately estimated to be a maximum practical workload for one site.

There were no data available, on local patterns of journeys made for comparable reasons, on which to base the predictions of where people would go to attend IMs (cf. Hodgson, 1990). The basic modelling was rooted in the assumption that people would attend IMs after making a journey from home. As noted earlier, a substantial minority of the research sample of people appeared to have travelled to the IMs from work, while others combined the trip with a visit to shops or other facilities. This confirmed the wisdom of only allowing 'high street' locations as plausible sites; it also meant that the distribution of jobs, rather than resident population, provided a better 'starting point' for calculating the distances travelled by some people.

In the absence of a reliable estimate of the proportion of people travelling to IMs 'as if from work' (rather than from home), five scenarios were run with different break-downs of the total attending (viz. 100% 'as if from work' rather than from home; 75% 'as if from work' and 25% from home and so forth). It was notable that it was in the conurbations where the results were most sensitive to the choice of scenarios, with additional city centre locations more likely to be selected when the proportion assumed to be travelling from home was lower. Even so, the overall level of sensitivity to this parameter setting was quite low: town and city centres are usually the optimal locations for IMs *whether or not* people are travelling mostly from the suburbs, so the locations of the selected sites tend not to change greatly between the different scenarios. It was also found that slightly fewer people have to travel longer distances for options in which a proportion of the people are travelling 'as if from work' due to the fact that, on this basis, more of them are assumed to start their journeys in or near town and city centres.

Figure 16.3 shows the location of sites, chosen by any of the five scenarios – with differing proportions assumed to travel 'as if from work' – in the North East region. Sites shown by a diamond in a square were selected by all the five scenarios; those shown as just a square were selected by three or four, while sites represented by diamonds were selected by only one or two scenarios. The remaining candidate sites, shown simply by crosses, were not selected by any of these five analyses. As already indicated, the areas with the most instability tend to be in and around the larger cities. Even there, the effect was often simply to shift the chosen site a mile or two, helped by the fact that quite a large number of the 647 candidate sites are located just a short distance away from each other in the major cities. Looking at the national picture, it was unsurprising that the London area showed the highest level of instability, although again very many of the changes there simply shift the selected site a relatively short distance. Given these location-allocation modelling runs – each finding the optimal 250 sites based on one of five different scenarios in terms of the proportion of people travelling 'as if from work' – the final step was to identify the sites selected by at least three out of the five runs. Figure 16.4 shows the 249 sites selected in this way, and draws attention to the rather few parts of the country, most

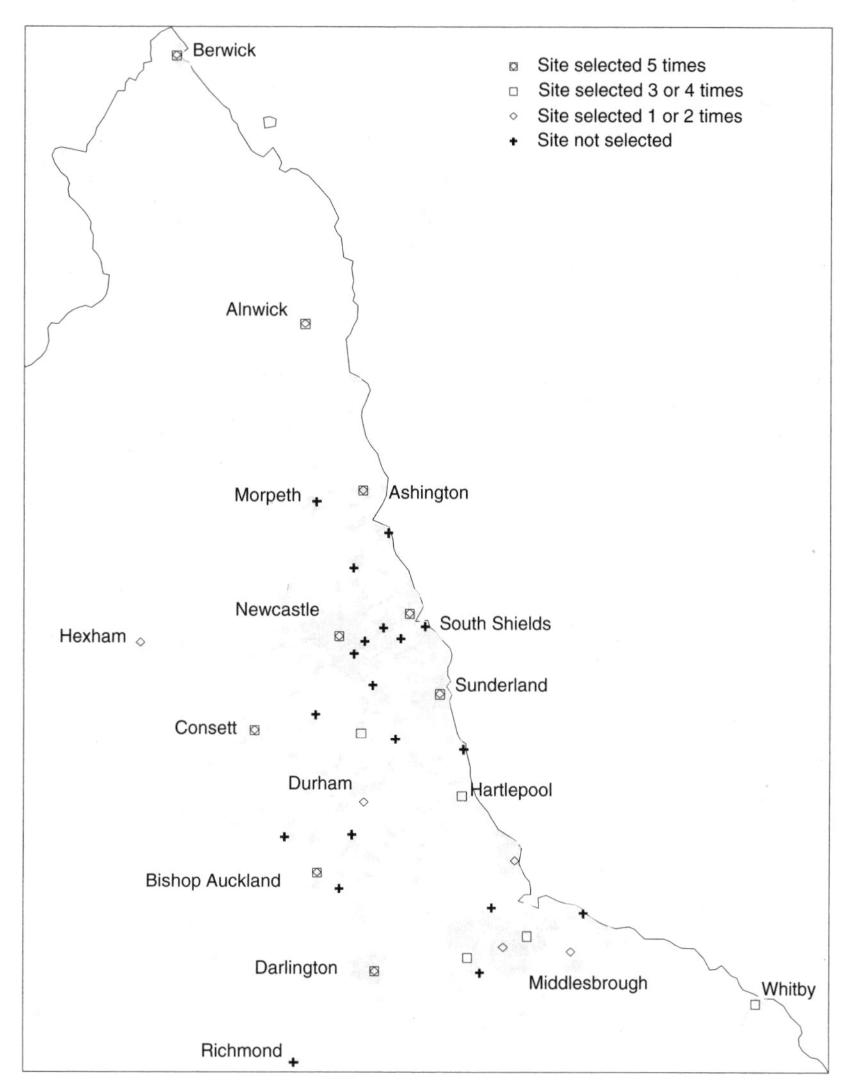

Figure 16.3 *Site selection in North East England*

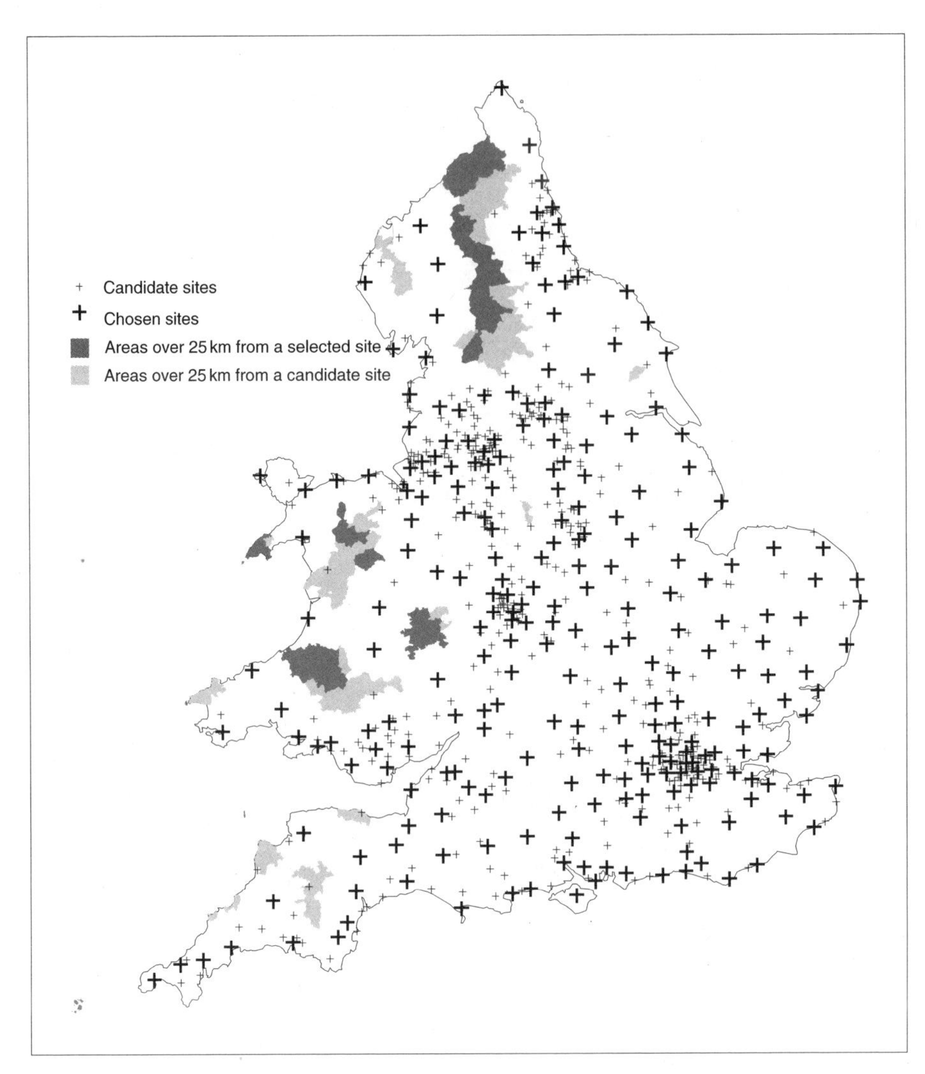

Figure 16.4 Candidate and selected sites

of which are in fact only thinly populated, where people would face journeys of over 25 km to their nearest IM site.

16.6 Evaluation of Selected Sites

One way of assessing the results from the modelling is to compare the 249 sites' accessibility with that of the network of Divorce Courts and, in addition, the Benefit

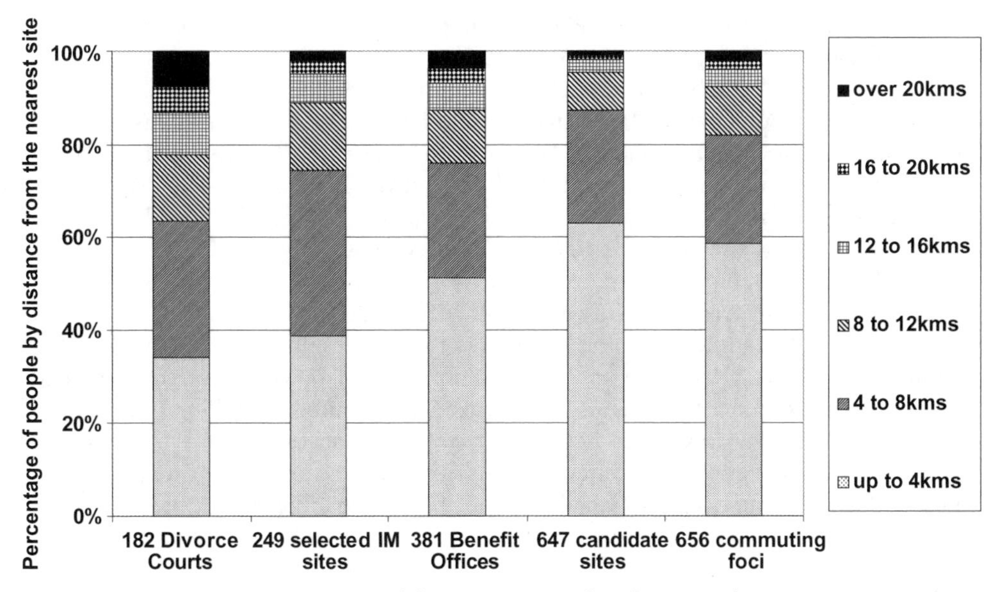

Figure 16.5 *Distance to nearest of the 249 sites, and to the sites of comparator networks*

Offices and commuting foci which were part of the analysis to identify candidate IM sites. Figure 16.5 shows that only the two sets of well over 600 sites substantially surpass the 249 selected IM sites in the proportion of people who would have less than 4 km to travel. In fact the location-allocation model has identified a network of sites which does seem to have minimised the number of people having to travel very long distances. These comparisons give a substantial degree of reassurance that the 249 sites identified here make up *at least* as 'fit for purpose' a network as do these long-established sets of sites whose levels of accessibility seems to attract little public criticism.

16.7 Conclusions

Evaluation suggests that the site identification method which has been developed proved able to identify an appropriate network of sites. This is not necessarily to argue that the set of 249 sites is definitely the best possible network of sites for IM provision: the aim was the more modest one of answering questions which are popularly phrased 'what if . . . ?'. More formally put, the results from the analyses are based on a specific set of hypotheses, assumptions and policy preferences. The modelling has been explicitly designed to allow different scenarios to be explored and, as a result, to assess the sensitivity of the results to changing from one option to another. Figure 16.2 provided the clearest illustration of this sensitivity testing, but many other permutations have been explored to build up the central case results which have been described here. The analytical challenge, which has been met, was to test scenarios with numbers of sites numbered in the hundreds – and nearly 10 000 'demand points' – when location-allocation models are more often applied to a single city or urban region (e.g., Yeh and Man, 1996).

For reasons unrelated to the modelling, the government has subsequently decided not to proceed with a national implementation of IMs. The new proposal is for a set of Family Advice and Information Networks: new research has begun to evaluate this alternative way of providing information to divorcing couples. The decision to change policy – and so not to implement the provision of IMs as outlined here – was based on concerns over just what couples need at times of possible family breakdown. The decision did not turn on any of the geographical issues discussed in this chapter. In its own terms, this research successfully met its objectives. Perhaps the single most important lesson to be drawn was that the results are viewed with more confidence, especially by policy-makers who are unfamiliar with these geographical issues and techniques, if a range of scenarios are presented. Handling uncertainty 'upfront' in this way was found helpful especially, of course, if the results did not vary too radically between the different scenarios. The research also proved, in practice, that uncertainty over policy shifts cannot be handled in a similar way.

Acknowledgements

We are grateful to the Lord Chancellor's Department who funded the research underlying this chapter, as part of a major research programme directed by Professor Jan Walker at Newcastle University's Centre for Family Studies (NCFS) (*see *Information Meetings and associated provisions within the Family Law Act 1996: Final Evaluation Report*, http://www.lcd.gov.uk/family/fla/fullrep.htm). We also fully acknowledge the inputs of numerous colleagues in the research programme, both within NCFS and elsewhere, whose contributions helped to ensure that the methodological developments briefly reported here were as appropriate as possible in meeting the overall objectives of the research programme. We also thank Dr Paul Densham (UCL) for his comments on the modelling work carried out.

All the GIS-based analyses conducted here depended on the use of ARC/Info software from ESRI(UK) who very kindly agreed to its use not only for the authors' academic research but also for this contract-funded research.

References

Beishon, S., Modood, T. and Virdee, S. (1998) *Ethnic Minority Families,* Policy Studies Institute, London.

Berrington, A. (1996) Marriage patterns and inter-ethnic unions, in Coleman, D. and Salt, J. (eds) *Ethnicity in the 1991 Census: Demographic Characteristics of the Ethnic Minority Populations,* ONS, London.

Böheim, R. and Ermisch, J. (2001) Partnership dissolution in the UK – the role of economic circumstances, *Oxford Bulletin of Economics and Statistics,* **63**: 197-208.

Clarke, L. and Berrington, A. (1999) Socio-demographic predictors of divorce, in Simons, J. (ed.) *High Divorce Rates: The State of the Evidence on Reasons and Remedies,* Research Series 2/99 (Vol. 1), Lord Chancellor's Department, London.

Coombes, M. and Raybould, S. (2001) Public policy and population distribution: developing appropriate indicators of settlement patterns, *Environment and Planning C: Government and Policy,* **19**(2): 223–48.
Densham, P. and Goodchild, M. (1990) Spatial decision support systems, *National Center for Geographic Information and Analysis Technical Paper 90:5* Santa Barbara, California.
Densham, P. and Rushton, G. (1996) Providing spatial decision support for rural public services that require a minimum workload, *Environment and Planning B: Planning & Design,* **23**: 553–74.
Fotheringham, A.S., Densham, P. and Curtis, A. (1995) The zone definition problem in location-allocation modelling, *Geographical Analysis,* **27**: 60–77.
Ghosh, A. and Harche, F. (1993) Location-allocation models in the private sector: progress, problems and prospects, *Location Science,* **1**: 71–106.
Gibson, C. (1996) Contemporary divorce and changing family patterns, in Freeman, M. (ed.) *Divorce: Where Next?,* Dartmouth, Aldershot.
Hodgson, M.J. (1990) A flow-capturing location-allocation model, *Geographical Analysis,* **22**: 270–9.
Kiernan, K. and Mueller, G. (1999) Who divorces?, in McRae, S. (ed.) *Changing Britain: Families and Households in the 1990s,* Oxford University Press, Oxford.
McCarthy, P. and Simpson, B. (1991) *Issues in Post-divorce Housing,* Avebury, Aldershot.
Murray, A.T. and Gerrard, R.A. (1997) Capacitated service and regional constraints in location-allocation modelling, *Location Science,* **5**: 103–18.
Thomas, R., Robson, B. and Nutter, R. (1991) Planning the work of county courts: a location-allocation analysis of the northern circuit, *Transactions of the Institute of British Geographers,* **16**: 38–54.
Townsend, P., Phillimore, P. and Beattie, A. (1988) *Health and Deprivation: Inequality and the North,* Croom Helm, London.
Walker, J. (2001) Divorce reform and the Family Law Act 1996, in Walker, J. (ed.) *Information Meetings and Associated Provisions within the Family Law Act 1996: Final Evaluation Report (Volume 1),* pp. 17–34. (http://www.lcd.gov.uk/family/fla/fullrep.htm)
Yeh, A.G.-O. and Man, H.C. (1996) An integrated GIS and location-allocation approach to public facilities planning – an example of open space planning, *Computers, Environment and Urban Systems,* **20**: 339–50.

17

New Methods for Assessing Service Provision in Rural England

Martin Frost and John Shepherd

Abstract

This chapter describes work undertaken on behalf of the Countryside Agency to use detailed information on the location of services and households to provide evidence that underpins policy development in two interrelated areas within England. The first is based on household access to services and facilities in rural areas while the second examines the role of small and medium-sized towns in acting as local centres of service provision. Commonly used geographical information system (GIS) procedures including the identification of nearest services and the use of point in polygon allocations are employed to measure levels of access to services and to generate service profiles of the towns. The results show that measuring access to services produces a generally more favourable view of rural service provision than the 'presence and absence' ratios traditionally used by the Agency. The profiles of small and medium-sized towns show high levels of variability suggesting a rather chaotic form for the concept of 'market towns' that in recent years has been a strong component of public policy.

17.1 Introduction

Although rural service provision has been a longstanding policy issue in England (for an early statement see, for example, Moseley, 1979), there has been an acceleration of interest in the 'state of the countryside' in recent years as the declining fortunes of farming, combined with the loss of many local shops and services in rural areas, has reinforced a recognition by the Government that problems of poverty and exclusion are found in both rural and urban areas of the country. The Government's view

Applied GIS and Spatial Analysis. Edited by J. Stillwell and G. Clarke
 ISBN: 0-470-84409-4

is captured in the introduction to the Rural White Paper, published in the Autumn of 2000 by the Department of Environment, Transport and the Regions, as follows: 'The challenge for rural communities is clear. Basic services in rural areas are overstretched. Farming has been hit hard by change. Development pressures are considerable. The environment has suffered' (DETR, 2000, p. 9).

The Rural White Paper (subsequently referred to as the RWP) goes on to set out a range of initiatives and targets relating to service provision, housing, transport, farming and the quality of the environment. In an era of 'evidence-led' policy development it is, perhaps, a little surprising that the level of empirical underpinning of many of these initiatives was, and still is, rather limited. In this chapter, we describe how the use of geo-referenced information analysed through GIS procedures has been used in research undertaken for the Countryside Agency to develop a more effective range of evidence in two key areas of Government initiatives: levels of access to rural services, and the role of market towns as centres of local service provision and economic development in rural areas.

In both areas of development, the work has been based on using information on the presence of individual service outlets in which the location of the outlets is either defined directly by a six-figure national grid reference (in the case of some commercial retail data such as that supplied by CACI) or has an address containing a unit postcode to which a grid reference can be attached. A range of data has been accumulated from both commercial sources and public agencies that locate many of the services identified by the Countryside Agency as of key importance in influencing the quality of life in rural areas of England.

The prime concern of the Agency, however, has been to progress from the identification of 'presence and absence' of services in two directions. The first is to assess, at the level of individual households, the geographical proximity that households have to their local services. The second is to use the range of services present within England's market towns to assess their 'vibrancy' in providing services and facilities to households living in surrounding areas.

In both developments, a combination of standard GIS procedures has been used to achieve these objectives. The analysis of household access to local services rests on the ability to identify an accurate geographical position for small groups of households using the postcode of their addresses and being able to locate (for each targeted service) the distance separating them from their nearest outlet. In the case of market towns, the emphasis is on being able to locate services within the built-up areas of towns that generally do not conform to recognised administrative boundaries (such as wards or local authority boundaries). This requires a point in polygon approach that has been supplemented in some of the analyses with the addition of buffer areas around the physical limits of each town that allow for recent expansion.

The chapter is divided into two parts. In the first part, we review the traditional ways of capturing information about the location of rural services as a basis for describing how this has been improved by adopting GIS approaches linked to digitally encoded national sources of information. In the second part, a discussion of the concept of market towns is followed by a demonstration of the profiling of these towns that is based on the capacity to locate key services within their boundaries.

17.2 Access to Services in Rural Areas: Identifying the Problems

The RWP generated no less than 15 headline indicators by which the state of rural areas within the UK was to be assessed. One of these was headed 'Geographical availability of services' and committed the Countryside Agency to maintain regular monitoring of the proportions of rural households living at specified distances from key services. The current list of key services monitored by the Agency includes banks and building societies, cash points, supermarkets, petrol stations, primary schools, secondary schools, doctor's surgeries, job centres, libraries and post offices. Although the notion of households' access to services may appear to be a simple one, producing measurements on a national scale that might be suitable to support the monitoring specified by the government has, traditionally, been far from easy. The problem has three components. The first is knowing where services are located, the second is knowing where households are, and the third is estimating the distances separating the two.

17.2.1 The Location of Services – The Rural Survey Approach

The traditional source of information on the location of services in rural areas has been the periodic surveys of rural services conducted by the Rural Development Commission (RDC). These were conducted in 1991, 1994 and 1997, and relied on parish clerks to complete questionnaires detailing the nature and extent of service provision within their parishes. The questionnaires were ambitious. They requested details on shops, pubs, post offices, banks, educational facilities, health facilities, medical facilities and recreational facilities, together with aspects of housing, local transport and welfare provision.

The basic data collection unit for all surveys was the civil parish which, in addition to its role in the reporting of other official data (notably from the Census of Population), constitutes an important focus of local democratic politics and policy action. The surveys covered only 'rural' parishes that were defined as those with a resident population of less than 10 000. The surveys recorded counts of the number of service outlets and facilities that lay within each parish. The results for the 1991, 1994 and 1997 surveys were published solely as paper documents (Rural Development Commission, 1992, 1995, 1998) with tabulations of the proportions of parishes containing a particular service or facility or a particular number of service outlets. Although this approach utilised the only comprehensive network of potential data collectors in rural areas, the parish clerks, its use in determining the location of rural services for the purpose of assessing levels of access was weakened by two key problems:

- *geographical coverage*: not all parish clerks filled in the forms or filled in the forms completely. Although considerable effort was put in by Rural Consultative Councils to achieve high response rates, in some counties response rates could be as low as 65%, even when an overall response rate close to 80% was achieved by the 1997 survey; and

- *urban/rural interaction*: limiting the surveys to parishes containing fewer than 10000 residents ignored the fact that many rural areas are close to towns or cities (which do not, in any case, contain defined parishes) and have access to the services and facilities contained in the urban areas.

Either one of these shortcomings means that it is not possible to use these records as a way of identifying consistently how far rural households have to travel to access their nearest service or facility. For many households the nearest facilities might lie in a neighbouring town or in a neighbouring parish which did not respond to the surveys.

17.2.2 The Problem of Locating Households

Using parishes – or indeed any administrative areas – as the sole basis for analysing information on services in rural areas also accentuates the problem of identifying the location of households accurately enough to permit sensible analysis of their distance from local services and facilities. Counts of residents and households within parishes are only generated by the decennial Censuses of Population. Even then, the value of these counts is diminished by the size and irregular shapes of many parishes. There is great variation in parish sizes (about 30 current parishes are of fewer than 100 hectares while 15 are larger than 10000 hectares) and, for many parishes, their elongated form makes it problematic to distinguish a meaningful centroid.

17.2.3 Assessment of Access from Survey Evidence

The combination of parish survey-based evidence on the location of services in rural areas, combined with the use of parish populations to assess the distribution of households, has severely limited the degree to which previous rural surveys have been used to measure levels of access to services. This is clear in the terminology used in the report on the 1997 Rural Services Survey (RDC, 1998) in which the term 'provision' of services is used consistently instead of any measure of 'access'. All tables within the report show the numbers of service outlets cross-tabulated against the population size of the parishes in which they fall, with many of the key headlines within the report summarising the proportions of parishes that do *not* contain a particular service or facility. A good example to illustrate some of the difficulties that may arise from this is to consider post offices as a key rural service. In 1997, 43% of responding parishes in the rural services survey did not contain a post office (RDC, 1998). By 2000, this figure had risen to 46%, reflecting the progressive closure of small rural post offices (Countryside Agency, 2001). Yet, using the GIS procedures outlined in the next section of this chapter, it was estimated that nearly 94% of households living in rural areas had a post office within 2 km of their homes.

The key variable in explaining the difference between these two sets of figures – one based on provision, the other based on access – is the uneven distribution of population across parishes. In the 1991 Census of Population, about 13% of all parishes had 100 or fewer residents, rising to 27% of parishes if the threshold is increased to 200 residents. In 2000, 88% of parishes with fewer than 100 residents did not contain a post office with an equivalent figure of 78% for those with between 100 and 200 residents. A concentration on provision alone is clearly weighted by the significant number of parishes that were defined many centuries ago but now contain few residents and correspondingly few services.

17.3 Using GIS Procedures to Measure Access to Rural Services

The development in recent years of a greater availability of digitally encoded lists of services and households that permit accurate geo-referencing, together with the software to link the two together, has transformed our ability to measure access to services at detailed spatial scales throughout the country. This can be seen in the contents of the Rural Services Survey of 2000, the first to integrate a parish-based survey with GIS-based calculations of household access to services. The key services used in these calculations are shown in Table 17.1.

All of these lists are supplied with the unit postcode of the service outlet. Where necessary, the Postcode Address File (PAF) was used to attach national grid references to each record (all CACI-sourced records are already geo-referenced). Additional data cleaning, using the Royal Mail's Address Manager to locate individual addresses where postcode details were inaccurate, was necessary to maximise the number of successfully matched postcodes. Address Manager software allows users to enter portions of an address and identifies an accurate matching unit postcode where an initial postcode has been wrongly recorded within the original digital lists.

Table 17.1 *Sources of digital lists of service locations within the Rural Services Survey, 2000*

Service	Data source	% successfully grid referenced (after cleaning)
Primary schools	DfES (Register of Educational Establishments)	100
Secondary schools	DfES (Register of Educational Establishments)	100
Post offices	Postcode Address File	98.8
Benefit offices	Benefits Agency	99.7
Job centres	Employment Service	99.5
Doctors' surgeries	Binley's Register of General Practitioners	99.5
ATMs (cash machines)	LINK Network plc	90.0
Banks and building societies	CACI Ltd	Already referenced
Hospitals	Yellow Pages and Dept. of Health	100
Petrol stations	CACI Ltd	Already referenced
Supermarkets	CACI Ltd	Already referenced

Source: RSS2000 Technical Report.

From Table 17.1, it can be seen that the rate of success in attaching grid references was generally high. Only ATMs posed any significant problems at this stage of the analysis.

Once grid-referenced, software was constructed to identify, for each unit postcode in England, the location of its nearest outlet for each service. This was assessed on the basis of the straight-line distance connecting the centroid of the postcode (using Royal Mail's Address Manager with 1 m centroid accuracy) to the grid reference of the service outlet. Whilst there are some limitations with this methodology over road lengths, straight-line calculations were used as the most practicable way of computing distances for about 1.3 million residential postcodes within England. After each unit postcode had been assigned a distance to its nearest service outlets, two avenues of development were used in the Rural Services Survey of 2000. In the first, the pattern of access was presented visually through plotting individual postcodes as points to generate maps at national and regional scales showing areas of relatively high or low access. An example showing the shortest distance to a hospital for all non-commercial postcodes in England is contained in Plate 7 (taken from Countryside Agency, 2001). Each dot on the map (there are about 1.2 million dots) represents a unit postcode with its colour indicating distance to the nearest hospital.

The real purpose of the map is to permit visualisation of those areas that are relatively remote from hospital services, picked out in dark blue in this example. Key areas in this category include North Cornwall and West Devon, the Welsh borders together with Lincolnshire and East Yorkshire. In addition, the sparsely settled highland areas stand out as blank, focusing attention towards those areas that are both relatively remote *and* relatively intensively settled.

In the second avenue of development, the fact that the number of households within each unit postcode can be estimated from its counts of residential delivery points was used to estimate, for standard administrative units (wards through to counties, regions and England as a whole), the proportions of households within each area having specified levels of access to key services.

An example of one of the products of using this approach is shown in Table 17.2. Developing the example of access to post offices used earlier in the chapter, this shows the breakdown by Government Office Regions of the proportions of rural households living within specified distances from their nearest post office.

The households used in this calculation are all those living within wards defined as 'rural' by the Countryside Agency. The results further confirm the arguments made earlier in the chapter that considering provision of facilities at a parish scale does not provide a meaningful view of the degree to which the great majority of households within rural areas live quite close to their nearest post office.

In these examples, using GIS linked to the increasing availability of machine-readable lists of services that can be, or already are, geo-referenced, combined with an ability to locate unit postcodes and their constituent households, has greatly improved our ability to address the three key issues of estimating access to services outlined earlier. It has provided a direct and generally cost-effective route to determining where services are located, where households are located and of estimating the distances separating the two. It has achieved this with comprehensive national

Table 17.2 Proportions of rural households living within a specified distance from their nearest post office

Government Office Region	0–2 km (%)	2–4 km (%)	>4 km (%)
East	92.3	7.0	0.7
East Midlands	93.8	5.7	0.5
North East	93.2	5.6	1.1
North West	95.0	4.4	0.6
South East and London	93.4	6.3	0.3
South West	93.9	5.8	0.4
West Midlands	91.9	7.7	0.4
Yorkshire & Humber	94.4	5.0	0.6
England	93.5	6.0	0.5

Source: Rural Services in 2000 (Countryside Agency, 2001).

coverage eliminating the problems of non-response to surveys and the problems of limited geographical coverage omitting services contained in the larger urban areas. The approach outlined here parallels that used in the construction of the access to services domain within the Indices of Multiple Deprivation (IDM) 2000, defined over a smaller range of services, further confirming the degree to which GIS methods are becoming central to the construction of monitoring of an increasing number of key Government policies and performance targets.

17.4 The Role of Market Towns

17.4.1 What are Market Towns?

The debate over access to services in rural areas has developed a further dimension over the last two years with the recognition that the bases for efficient and sustainable provision of a wider range of services in the countryside are likely to be found among the small and medium-sized towns. Compared with the need for provision of certain basic services within, or accessible to, villages and hamlets, market towns are perceived as having a wider economic and social role in rural areas, that is 'as a focus for growth in areas which need regeneration, and more generally as service centres and hubs for surrounding hinterland, exploiting their potential as attractive places to live, work and spend leisure time' (DETR, 2000, p. 73).

However, whilst the general perception of a market town may be deeply ingrained in the English psyche – viz. a free-standing, small to medium-sized place with a market square, a reasonable provision of professional, business and retail services serving a rural catchment area and, probably, a tourism function based upon the presence of historic buildings – in reality, the so-called 'market towns' are now a far more varied set of elements in the settlement system. In recent years, the profound changes that have taken place in the rural economy and its demographic and social structure have been partly driven by and partly mirrored in the functional evolution of 'rural' towns (Newby, 1985; Champion, 1989; Hodge and Monk, 1987; Marsden *et al.*, 1993). Moreover, the direction and pace of that evolution has been different in different parts of the country, depending on such contextual circumstances as levels of regional prosperity, location relative to larger urban places, access to motorways and other major roads and national planning policies (Shepherd and Congdon, 1990; Errington, 1997).

The RWP itself acknowledges the inherent variety suggesting that market towns are 'small rural and coastal towns, many of which serve a rural hinterland whether or not they have ever had traditional rural markets. Some may have grown up around a canal or railway junction or as a coastal resort while continuing to be an important commercial and leisure focus for a rural hinterland' (DETR, 2000, para 7.2.1). However, the first statement of a formal definition is given, with no further rationale, in the comment that 'there are over 1,000 towns in England with populations between 20,000 and 2,000' (ibid.). In particular, there is no hint here of a relationship between population size and the functional 'endowment' serving to help define market towns for policy purposes.

17.4.2 Problems of Estimating Indicators of Market Town Prosperity

Capturing the diversity of market towns in empirical terms is neither easy nor always totally convincing but, in attempting to do so for policy development and monitoring purposes, GIS have an indispensable role to play. In general terms, GIS have two main contributions to make in relation to the study of smaller urban places:

- they facilitate the integration of point and boundary data that make it possible to identify service levels available within and around market towns; and
- they permit the estimation of market town characteristics based upon the linkage of data captured at different geographic scales and for different types of data collection unit.

It is these contributions that we focus on here. A third area of analytical contribution, that of distance analysis, which is important in characterising the location (distance) of market towns relative to larger towns and cities to which they may look for jobs and major services/leisure opportunities and smaller places which they themselves supply with 'lower order' goods and services, is not covered.

In practical terms, and in the context of data specifically for England and Wales, the key problems of conducting empirical analyses of small and medium-sized urban places qua 'market towns' are as follows:

- the key dataset on urban places in England defines such places in terms of contiguous urban land uses (producing 'Urban Areas') and not, as in the case of the USA, for example, in terms of urban functions or the provision of jobs or services to the surrounding area (ONS/GRO, 1997; Office of Management and Budget, 1998);

- the boundaries of Urban Areas, which were last defined in 1991 are, at the time of writing, significantly out of date whilst, in some parts of the country in particular, it was known from other sources that there had been extensive residential development on the outskirts and within the catchment area of market towns (Bibby and Shepherd, 1997);
- those datasets which were available to characterise market towns (principally, though not exclusively, the Census of Population), are reported in spatial units (wards, districts, etc.) that, in terms of geographic scale, are better suited to analytical work on larger urban places rather than small ones and this difficulty is considerably magnified for those places in the lowest range of the market town 'hierarchy', i.e., in the region of 2000 residents;
- problems of the way market towns are defined as urban places and of the 'fit' of data reporting units to this definition make it very difficult, on the basis of quantitative data alone, to distinguish between 'true' local service centres and places with a significant or overwhelming dormitory function; and
- finally, no readily available information exists upon which to estimate the catchment areas linked to each town, a task which is important not only for 'free-standing' places, but to distinguishing or estimate overlapping market town catchments in areas with complex urban settlement patterns (ERM, 2000).

What follows relates entirely to the study of English (and possibly Welsh) market towns. In Scotland, different datasets are available which render certain aspects of studies of rural towns more straightforward.

17.4.3 Market Town Size and Regional Distribution

Although the 20 000 to 2000 population range for the definition of a market town is implicitly accepted as an indicator rather than a strict definition, its existence has nevertheless entered into the vocabulary of policy on market towns and has resulted in the list of places for consideration in strategic analyses of market towns within this range. The RWP gives a template for service availability in the East Midlands which defines 'Larger Market Towns' as having populations between 10 000 and 25 000 and 'Smaller Market Towns' as those with populations between 2000 and 10 000 (DETR, 2000, Table 7.1). However, the Countryside Agency's market town headline indicator of economic health focuses on Urban Areas in the 20 000 to 2000 range. This endemic flexibility in definition and the lack of certainty as to the character of market towns due to data problems, both imply the need for equally flexible (and powerful) approaches to policy analysis of which GIS is one such tool. Analytical flexibility is inherent, for example, in the need for awareness of:

- towns larger than 20 000 population which clearly form 'market town' functions in relation to their catchment area but which, because of their larger size, may be inappropriate for market town policy initiatives;

- places that are around the 2000 population level which, because of lack of services or the nature (lack of) settlement in their periphery, may not have an important catchment area function; and
- places that were (perhaps some way) below the 2000 population level at the time of definition but which nevertheless do serve a local catchment area.

There are two main ways in which small market towns may be omitted from current analyses: places which have grown significantly since the base population data (from the 1991 Census) were derived and places which were not originally defined for technical reasons related to the geographical 'capture' of population data within urban area boundaries. Bishop's Castle in Shropshire is a good example of the latter, a small, even classic market town, which fails to meet the population criterion because a 100 m buffer around the urban area fails to capture a sufficient number of census enumeration districts (ONS/GRO, 1997, p. 3).

As noted above, the identification of a place as a market town depends upon a boundary dataset for urban areas that relies upon a strict land-use definition of urban places. Using these data, there are 1210 market towns defined strictly on the '2000 to 20 000' principle within England. They range from Stow-on-the-Wold (with a 1991 population of 1999) to Thetford, in Norfolk, with a population of 20 058. This range excludes some towns which clearly serve a rural hinterland but which are only moderately larger, e.g., Newton Abbott (23 801), Chippenham (25 961), Sevenoaks (24 489) and Bishop's Stortford (28 403). On the other hand, it includes places which have either an extremely restricted 'catchment area' function (e.g., Westhoughton, population of 23 307, essentially a part of Greater Manchester) or are an appendage of a larger place (e.g., Bessacar, population of 19 907, now predominantly a suburb of Doncaster).[1] For the purpose of most analyses reported here, we have included essentially free-standing urban places in the population range 2000 to 30 000. Note, however, that this now incorporates some evident non-market towns such as Addington, an 'out-county' estate just outside Greater London, but just excludes Bury St Edmunds in Suffolk, a very clear candidate for true market town status.

The distribution by population size of market towns in this range is shown in Table 17.3. Over half the candidate market towns have fewer than 5000 residents and four-fifths (80%) have less than 10 000 population. In view of the dominance of market towns in the lower end of this distribution, there is a strong case, at the policy level, for closer scrutiny of the impact of size on market town type, service provision and catchment area range. At the level of empirical analysis, it further emphasises the need for awareness of the technicalities of definition (especially those, as noted above, in the linkage of census and other areas to urban land use boundaries) and especially any peculiar local instances of the fit of such areas to boundaries. Here, painstaking perusal of GIS mappings becomes *de rigueur.*

The absolute and proportional distributions of the size of market towns by region are shown in Tables 17.4 and 17.5. There are significant differences between regions in the proportional table, which indicates different regional settlement patterns based

[1] Bessacar (and no doubt others like it) is included on a technical point relating to the way discontinuous parts of larger urban areas are defined.

Table 17.3 *The distribution of market towns in England by population size*

Population size	Number	%
Over 20 000	64	5.0
15 000–20 000	80	6.3
10 000–15 000	111	8.7
5 000–10 000	309	24.3
2000–5000	710	55.7
Total	1274	100.0

Table 17.4 *English market towns by region and population size, 1991*

Region	2–5000	5–10 000	10–15 000	15–20 000	>20 000	Total
East of England	138	45	17	12	10	222
East Midlands	95	42	11	9	4	161
North East	41	23	7	5	6	82
North West	42	35	11	4	11	103
South East	147	49	18	23	15	252
South West	105	53	22	12	10	202
West Midlands	57	22	7	5	6	97
Yorks & Humbs	85	40	18	10	2	155
Total	710	309	111	80	64	1274

Table 17.5 *The proportion of market towns by population size and region, 1991*

Region	2–5000 (%)	5–10 000 (%)	10–15 000 (%)	15–20 000 (%)	>20 000 (%)	Total (%)
East of England	62.2	20.3	7.7	5.4	4.5	100.0
East Midlands	59.0	26.1	6.8	5.6	2.5	100.0
North East	50.0	28.0	8.5	6.1	7.3	100.0
North West	40.8	34.0	10.7	3.9	10.7	100.0
South East	58.3	19.4	7.1	9.1	6.0	100.0
South West	52.0	26.2	10.9	5.9	5.0	100.0
West Midlands	58.8	22.7	7.2	5.2	6.2	100.0
Yorks & Humbs	54.8	25.8	11.6	6.5	1.3	100.0
Total	55.7	24.3	8.7	6.3	5.0	100.0

upon different economic circumstances. Such differences have significant implications for market town policy (especially in relation to regeneration and equity issues), the interpretation of indicators and the evaluation of individual market town initiatives. Tables 17.4 and 17.5 indicate a generally even distribution of market towns by population size across regions, with the exception of the North West which is under-represented in the category of towns of 2–5000 and over-represented by towns with more than 20 000 population. However, this may have as much to do with the underlying definition of a town using land use criteria as with functional attributes and town–catchment area interactions.

The analyses that follow are the first stages in an attempt to identify some order in this group of settlements as a step towards identifying effective indicators of their condition and of identifying towns which are of particular importance to a rural hinterland. It is essentially experimental work that is trying to close the gap between policy and understanding.

17.5 Using GIS to Estimate Service Provision in Market Towns

The fact that the Urban Area boundaries and associated populations defining market towns were 10 years out of date at the time of the analyses was a cause for some concern. As previous studies had shown, small and medium-sized towns were the fastest growing places between 1971 and 1981 and the forces that had generated that growth were unlikely to have abated by much, if at all, in the subsequent decade (Champion, 1989; Shepherd and Congdon, 1990). The problem in gaining some idea of the rate of change among this particular set of towns lay in the scale difference between the areas for reporting inter-censal population estimates (i.e., wards with populations assigned from pro-rating local authority district estimates), and the smaller market towns which formed the majority of the overall size distribution of these places. One simple way to obtain population change information would have been to assign wards to market town urban areas on the basis of a boundary intersection/coincidence rule, for example, incorporating those wards which had any part of their boundary shared with or crossing that of an urban area. In practice, however, intersecting ward boundaries are not only a very poor fit to Urban Areas, but can also lead to complex overlapping of ward boundaries in locations where there are geographic clusters of the smallest market towns, thus making it difficult to allocate population growth to particular places.

An illustration of some of these difficulties can be seen in Figure 17.1. This shows both urban areas and ward boundaries centred on the Braintree urban area in Essex that in 1991 had a resident population of just over 30 000 people. Even at this scale, right at the top of the market town range, Census wards divide the town unevenly and often contain significant areas outside the built up area. In addition, the detailed rules by which the 1991 Urban Area boundaries were fixed often lead to rather ragged boundaries, particularly where urban sprawl has taken place. This can be clearly seen in the ward to the north of the centre of Braintree where a straggle of urban development protrudes into what would otherwise be an undeveloped ward. On the west side of the map another case can be seen where a smaller urban centre is divided into two more or less equal parts by the boundaries of two large rural wards.

Our initial approach, therefore, was to focus on the immediate environs of market towns by defining a consistent 500 metre buffer for all towns in the dataset and to use this to capture residential delivery points (RDPs) from Royal Mail's Address Manager database. This procedure had been used in the development of the Rural

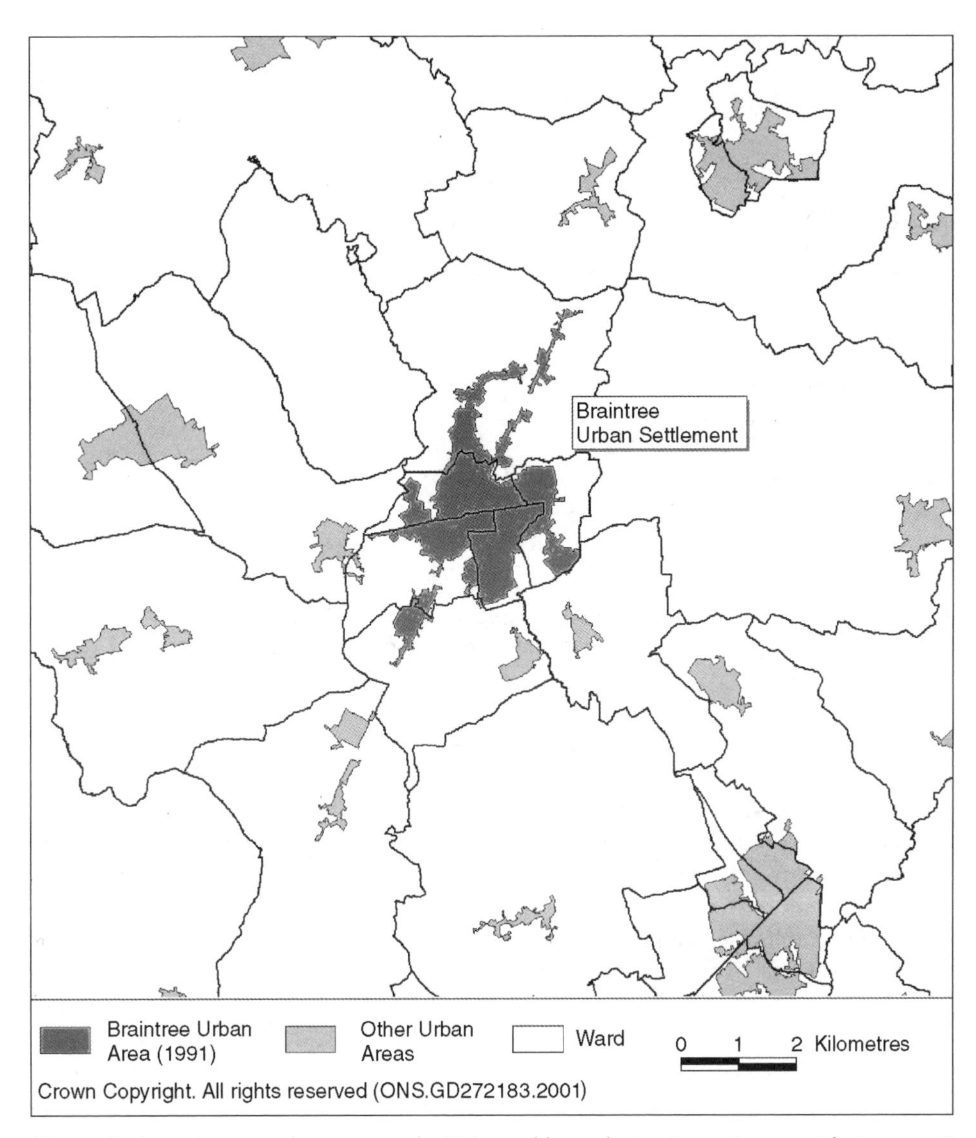

Figure 17.1 *A large market town and 1998 ward boundaries. From Countryside Agency. © Crown Copyright*

Settlement Gazetteer (RDC/HC, 1998) and in earlier studies using postcoded information to track population change (Raper *et al.*, 1992, pp. 220–33). In this case, however, there was no attempt to transform RDPs into population on the basis of regional or local differences in household size since, again, information on the latter, being derived from the 1991 Census was also likely to be inaccurate. In addition, there was no attempt, though it could be done, of differentiating between total household

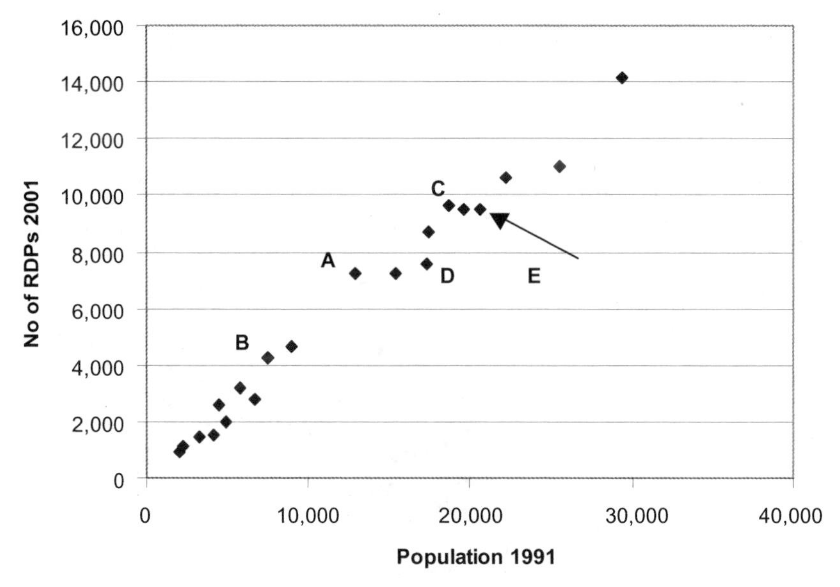

Figure 17.2 *Selected market towns; population 1991 (Urban Area) and number of Residential Delivery Points 2001 (Urban Area plus buffer)*

change (i.e., RDPs within Urban Area + buffer) and purely urban edge change (i.e., RDPs within the buffer).

Figure 17.2 shows a close relationship between population in 1991 and the number of RDPs in 2001. Of potentially more interest here, however, are the apparent departures from a 'straight-line' relationship reflecting, in some part at least, differences between market towns in levels of household growth in the period 1991–2001. These departures emphasise the importance of the economic structures of market towns and their regional contexts in accounting for population change. Stronger than expected growth can be seen in some of the small and medium towns in the south (e.g., East Dereham (A), Norfolk; Sherborne (B), Dorset; Spalding (C), Lincs) while those falling short of expectations are predominantly northern towns with a strong industrial or mining base (e.g., Spennymoor (D), Durham; Workington (E), Cumbria).

The first test rests on the well-established concept of urban hierarchy. We would expect certain medium to high level key services to be located in some of the larger towns in this group. The services selected from those where digital lists were available were banks (and building societies), secondary schools, supermarkets, large DIY outlets, Benefit Offices and Job Centres. To allow for fringe growth around the DETR urban boundaries an additional 500 metre buffer was added to each settlement.

These results demonstrate the considerable size ranges over which these services are found (Table 17.6). Only Benefit Offices and principal DIY outlets seem to have any power as threshold indicators of key centres and even for these it is limited. Most services are found throughout the size hierarchy, although with a tendency to be more completely and intensively present in the larger settlements.

Table 17.6 Presence and absence of selected services

	Largest town without service	Smallest town with service
Bank	15514	2000
Secondary school	18336	2012
Supermarket	15514	2052
DIY chain	19907	4591
Benefit Office	19907	6466
Job Centre	19907	2245

Sources: DfEE, CACI, Countryside Agency.

Given this range of occurrence, the initial approach to identifying the more important local service centres was based on overall relationships between population size and the number of service outlets found in each town rather than pursuing the concept of population thresholds. For a selection of important local services (secondary schools, banks and building societies, and supermarkets), the number of services present in each town was predicted using simple regression equations based on population size. The results were as follows:

$$Y_{bi} = -0.32^{**} + 0.00042^{**}P_i + e_{bi}\ (R^2 = 0.58) \text{ for banks} \tag{17.1}$$

$$Y_{si} = 0.02 + 0.00018^{**}P_i + e_{si}\ (R^2 = 0.59) \text{ for supermarkets} \tag{17.2}$$

$$Y_{di} = 0.45^{**} + 0.00013^{**}P_i + e_{di}\ (R^2 = 0.41) \text{ for doctors' surgeries} \tag{17.3}$$

where Y = count of services present, P = resident population in 1991, i = 1:1272 market towns and ** represents $p < 0.01$.

The residuals from these equations were then summed across all three services to derive an indicator of consistent over- or under-endowment of services. This is then divided by the population of each town to assess the intensity of these deviations.

$$I_i = \sum e_{ti} / P_I \tag{17.4}$$

where I = the intensity indicator and t = 1,3 service categories.

The purpose of this approach is to identify towns where there are significantly more services than would be expected from their local populations, suggesting a broader service centre specialisation. Table 17.7 shows the 12 towns with the strongest signs of positive specialisation in service provision.

Most of these towns appear as reasonably free-standing local service centres. The exception is Hemingford Grey which is part of a cluster of Urban Areas around Huntingdon/St Ives. Overall these suggest that there might be some path here to identifying the elusive market towns. This view is reinforced by considering the 12 towns with the lowest scores (Table 17.8). Many of these towns are either close to, or part of, larger urban or suburban settlements while some like Hatfield and New

Table 17.7 Towns with strong service centre functions

Town	Over-representation score
Stamford	2.30
Uttoxeter	2.41
Evesham	2.54
Tiverton	2.73
March	2.75
Tewkesbury	2.76
St Austell	3.16
Beverley	3.41
Diss	3.99
Wetherby	3.99
Hemingford Grey	4.03
Windermere	4.25

Table 17.8 Towns with weak service centre functions

Town	Under-representation score	Comments
New Addington	−3.84	near Croydon
Maghull	−3.74	outskirts of Liverpool
Hatfield (Yorks)	−3.55	close to Doncaster
Westhoughton	−3.08	close to Wigan
Golborne	−2.88	close to St Helens
Borehamwood	−2.75	fringes of London
Hoylake	−2.56	close to Birkenhead
Winsford	−2.49	more difficult to explain
Nailsea	−2.35	close to Bristol
New Rossington	−2.28	close to Doncaster
Trowbridge	−2.17	fairly close to Bath
Aylesford	−1.94	close to Maidstone

Rossington were developed partly as colliery settlements in addition to being close to a regional service centre.

Finally, we use this approach to examine the 'service level endowment' scores for a selection of classic free-standing small and medium-sized market towns. These are shown in Table 17.9.

It will be noticed that few of these scores are particularly strong but all are positive. Of the 1275 market towns defined in simple population terms, 735 show negative scores on this indicator, so to achieve 10 positive scores from the selection of towns is an interesting result. It confirms a little further that this approach has potential value in sifting the mixed group of towns that are now the focus of considerable policy effort.

Table 17.9 Scores for free-standing towns

Town	Town scores
Ely	0.07
Framlingham	0.08
Harleston	0.48
Bicester	0.24
Malton	1.51
Market Drayton	0.65
Bourne	0.21
Horncastle	0.56
Market Rasen	0.40
Thirsk	1.24

17.6 Conclusions

This chapter has demonstrated that relatively simple analytical approaches involving standard GIS procedures such as nearest neighbour measures and point in polygon allocations can, in combination with digital lists of services and households, be used to underpin the development and implementation of rural policies in England. At the moment, this work is in its early stages of development and will continue to be developed in coming years to provide a firmer base from which to specify and operate plans for greater rural support. At this stage, possibly the most interesting conclusions are not contained in the work itself but relate more to its timing.

In spite of public agencies widely proclaiming the virtues of evidence-led policy development, the political momentum of measuring access to services and, particularly, the use of the market town label preceded any systematic and thoughtful investigation of the concepts underlying the terms. All of the evidence points to the fact that little initial thought was given to how these terms might be effectively operationalised to produce measurements that would provide policy development with meaningful support.

In this setting, the application of GIS methods has provided a flexible way of responding to an empirical need generated by the speed of policy announcements and development of political agendas outstripping the capacity of public agencies to explore and comprehend relevant conceptual frameworks. We would argue that there is a risk in this flexibility. The power of modern software, linked to the increasing availability of digitally encoded information, can easily come to be seen as offering a quick fix to any empirical need thrown up in the policy development process. On the evidence of current practice, not all of these needs will be amenable to sensible analysis. The chaotic conception of market towns comes very close to this in bundling together towns with dissimilar functions, characters and histories into a common grouping linked only by their population size. If only some geographical analyses for exploring the nature of the current settlement and service-based urban hierarchy had been undertaken before, rather than after the term was coined, things might have been different – and better!

Acknowledgement

The research contained in this chapter was conducted by the South East Regional Research Laboratory, Birkbeck College London under contract to the Countryside Agency. The views expressed are those of the authors alone.

References

Bibby, P. and Shepherd, J. (1997) Projecting rates of urbanisation in England, 1991–2016: method, policy application and results, Town Planning Review, **68**(1): 93–124.

General Register Office (Scotland) (1995) *Census Key Statistics for Localities in Scotland*, HMSO, Edinburgh.

Champion, A.G. (1989) *Counterurbanisation in Europe Vol 1: Counterurbanisation in Britain*, Royal Geographical Society, London.

Countryside Agency (2001) *Rural Services in 2000*, Countryside Agency, Cheltenham.

Department of the Environment, Transport and the Regions (2000) *Our Countryside: The Future*, Cm 4909, HMSO, London.

Environmental Resources Management (ERM) (2000) *Indicators for Assessing the Character and Vulnerability of Market Towns in North West England*, for the Countryside Agency.

Errington, A.J. (1997) Rural employment issues in the peri-urban fringe, in Bolman, R. and Bryden, J. (eds) *Rural Employment an International Perspective*, CAB International.

Hodge, I.D. and Monk, S. (1987) Manufacturing employment change within rural areas, *Journal of Rural Studies*, **3**: 65–69.

Marsden, T. J., Murdoch, P., Lowe, M., Munton R. and Flynn, A. (1993) *Constructing the Countryside*, UCL Press, London.

Moseley, M.J. (1979) *Accessibility: the Rural Challenge*, Methuen, London.

Newby, H. (1985) *Green and Pleasant Land? Social Change in Rural England*, Wildwood House, London.

Office of Management and Budget (1998) Alternative approaches to defining metropolitan and non-metropolitan areas, Federal Register, **63**(244), December 21.

Office for National Statistics website (www.statistics.gov.uk).

Office for National Statistics/General Register Office Scotland (1997) *1991 Census, Key Statistics for Urban and Rural Areas*, HMSO, London.

Raper, J., Rhind, D. and Shepherd, J. (1992) *Postcodes the New Geography*, Longman, London.

Rural Development Commission/Housing Corporation (1997) The Rural Settlements Gazetteer, *Research Paper 26*, RDC, London.

Rural Development Commission (1992) *1991 Survey of Rural Services*, RDC, Salisbury.

Rural Development Commission (1995) *1994 Survey of Rural Services*, RDC, Salisbury.

Rural Development Commission (1998) *1997 Survey of Rural Services*, RDC London, Salisbury.

Scottish Executive, Ministerial Committee on Rural Development (2000) *Rural Definitions*.

Shepherd, J. and Congdon, P. (1990) *Small Town England: Population Change Among Small and Medium Sized Urban Areas 1971–1981*, Progress in Planning, **33**(1).

18

Forecasting River Stage with Artificial Neural Networks

Pauline Kneale and Linda See

Abstract

Forecasting models are required to determine continuous river stage for drought and water resource management, and for real-time forecasting of extreme flood events. The models use telemetred data from raingauges and river stage gauges located at key points in the catchment. Stage forecasts are traditionally made using physical models, but these large-scale systems have not proved to be accurate enough to provide timely flood warnings. Artificial neural networks (ANNs) provide an alternative modelling technology for building flood forecasting models. Artificial neural network software has been developed and installed for operational use by the Environment Agency in Northumbria, UK. This chapter describes its use in practice and compares the results using two feedforward neural network algorithms. Results from three experiments on three different sizes of catchment are compared. Overall the results suggest that ANNs are useful for operational purposes and represent an improvement over ARMA and regression modelling schemes.

18.1 Introduction

There is a continuing risk of flooding in the UK from rivers and from the sea. Some five million people live in two million properties that are situated in flood risk areas in England and Wales. Although there are over 36 000 kilometres of flood defences, the risk of flooding is too low in some areas for further physical defences to be considered cost effective. Therefore, timely flood warnings are required to give businesses and residents the opportunity to protect themselves and their properties. This

Applied GIS and Spatial Analysis. Edited by J. Stillwell and G. Clarke
© 2004 John Wiley & Sons, Ltd ISBN: 0-470-84409-4

requires real-time forecasting models to be in place which are capable of providing accurate river stage forecasts. Stage is the depth of flow above the river bed which is then used to calculate flow rates and volumes. The UK Environment Agency is responsible for providing flood warning services on main river channels, and they, in conjunction with DEFRA and local and regional authorities, are responsible for giving advice to planners.

Forecasting river stage has proved problematic. There are many examples of the continuing development of statistical and deterministic models but accuracy and reliability remain issues in a forecasting area where the events of primary interest, i.e., extreme floods, are infrequent and non-linear (Cameron *et al.*, 2000; Horritt and Bates, 2001; Mellor *et al.*, 2000; Stewart *et al.*, 1999). Artificial neural networks (ANNs) provide a potentially useful alternative to traditional flood forecasting approaches with rapid development times relative to large-scale physical models. ANNs are biologically inspired computational models that map a set of inputs, in this case rainfall and river stage data, to a set of outputs, such as river stage at downstream stations. The network achieves this task by learning the patterns in an historical dataset through a training or calibration process. Once trained the neural network can be utilised in a predictive capacity with unseen, independent input data and tested for accuracy using independent validation data. The results should be reliable provided the training dataset is representative of all the relationships being modelled, which is always an issue with flood forecasting. The largest flood event may be the next one and occur tomorrow.

There are many opportunities for neural network use in hydrology, and its application is becoming increasingly widespread (Dawson and Wilby, 2001). River stage forecasting is just one hydrological situation where ANNs are being adopted, as exemplified by Abrahart *et al.* (2001), Campolo *et al.* (1999), Chang and Hwang (1999), Dawson and Wilby (1998), Golob *et al.* (1998), Openshaw *et al.* (1998) and See and Openshaw (1999, 2000). In an early study, Hsu *et al.* (1995) showed that the non-linear ANN model provided a better representation of the rainfall–runoff relationship of the medium size Leaf River basin near Collins, Mississippi than the linear ARMAX (AutoRegression Moving Average with eXogenous inputs) time-series approach or the conceptual SAC-SMA (Sacramento Soil Moisture Accounting) model. They acknowledged one key difference in modelling philosophy; ANNs do not provide models that have physically realistic components and parameters and so are no substitute for conceptual watershed modelling. But they point out that ANNs provide a viable and effective alternative to the ARMAX time-series approach for developing input–output simulation and forecasting models in situations that do not require modelling of the internal structure of the watershed. In forecasting river stage there is no absolute requirement for physically realistic variables or mirroring of the catchment structure. Thus, ANNs are particularly useful as forecasting tools in situations where the relationships between the key variables are non-linear or cannot be adequately described by dynamic equations, or where the data are limited and more complex process models cannot be adequately calibrated.

Elsewhere in hydrology ANNs have been used to forecast water quality: phosphorus levels in 927 catchments across the USA (Omernik, 1997), river pH (Moatar *et al.*, 1999), Cryptosporidium and Giardia concentrations (Neelakantan *et al.*, 2001) and PAHs in water samples (Cancilla and Fang, 1996). Other examples include the prediction of fish numbers and habitat preferences in rivers (Lek and Baran, 1997; Mastrorillo *et al.*, 1997; Gozlan *et al.*, 1999) and algal numbers by Maier *et al.* (1998), Recknagel *et al.* (1997) and Whitehead *et al.* (1997).

While the range of hydrological applications using ANNs is encouraging, there is a pressing need to trial models cited in the literature, many of which are implementations in experimental or historical mode, to operational real-time settings. This work represents one of the first attempts at making this technology transfer. This chapter describes the development and application of operational neural network flood forecasting software, which was undertaken in collaboration with the Northumbria office of the UK Environment Agency (EA). Neural network theory and a description of the software tools are provided. Three case studies illustrating the application of the software highlight the ability of the networks to forecast events from limited records and beyond the range of the training data. Models were built using data from a seven-month winter period and tested using data from three further periods of continuous river stage data. The results were also compared with more conventional statistical models, multiple linear regression and autoregressive moving average models (ARMA).

18.2 Client Requirements

River stage forecasting, whether for engineering design or flood warning purposes, is one element in a wider hydrological modelling framework (Figure 18.1). A forecasting model sits between the data sources and a decision-making process. Although hydrological modelling has been developing since the 1920s, and numerical modelling has brought major advances since the 1960s, the accuracy of results is not yet satisfactory. Flood forecasting systems in current use are generally based on either distributed modelling ideas or on mass transport equations, which essentially move blocks of water down the river (HEC-RAS, 2001; ISIS, 2001; MIKE 11, 2001; NOAH 1D, 2001; RFFS, 2001; SSIIM, 1997; van Kalken *et al.*, 2001). The complexity of natural systems and the number of processes that continuously interact to influence river levels make traditional modelling based on mirroring natural processes with process-based equations very difficult. Where processes can be modelled numerically, there are constraints imposed by the spatial availability of data and the cost of collecting this data in real time.

Any modelling system is a considerable financial issue for an agency or company. Investment is required in the numerical model, the acquisition of historic and real-time input data and trained analysts to interpret the results. For real-time forecasting this involves a team of analysts available 24 hours a day, to interpret results and implement flood-warning procedures. Once a forecasting system is established within an organisation, moving to alternative systems will only happen if there are seen to be clear advantages, preferably in both financial and forecast accuracy. The cost of moving to a new system and the different modelling approach of ANNs are barriers to the acceptance of the technology.

At present the EA uses a number of conventional hydrological models. In a report, Marshall (1996) stated that 'the development/calibration of the models has, with few exceptions, not yet reached the stage where warnings can be reliably based on the

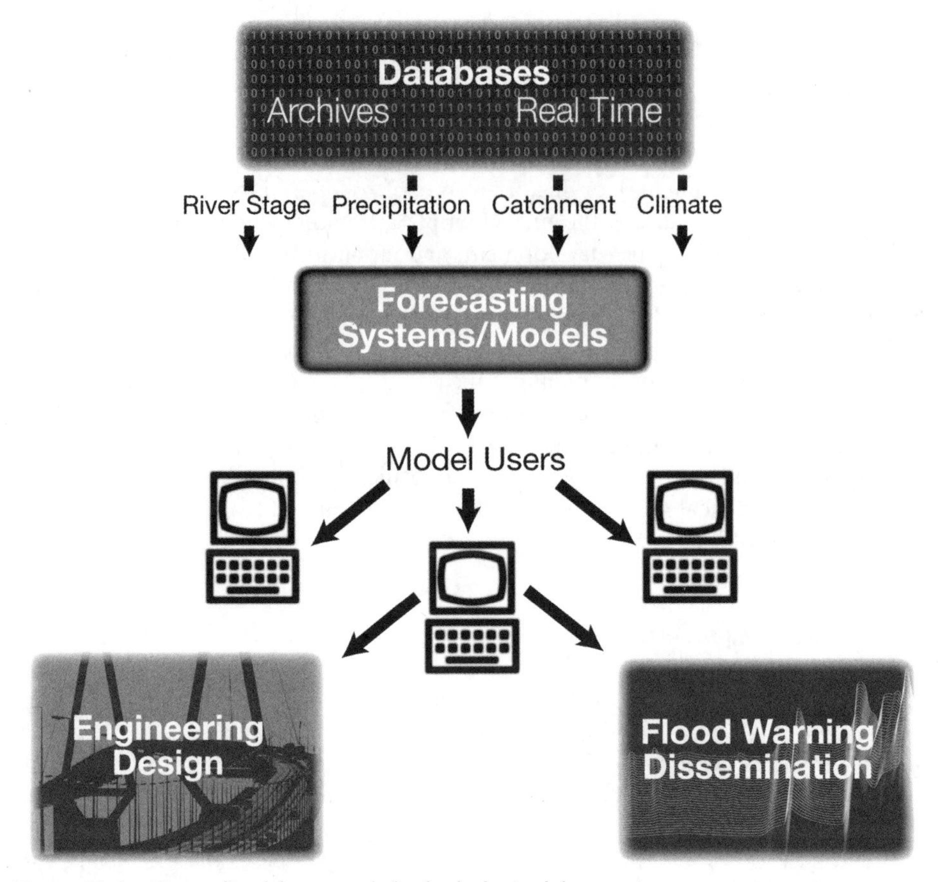

Figure 18.1 Generalised framework for hydrological forecasting

forecasts produced. The systems provide considerable help to duty officers by collecting, processing and displaying telemetry and radar data on a region-wide basis, but their operational performance as forecasting systems cannot yet be measured' (p. 7). Although none of the models in the regional offices were considered fit for operational purposes they are still in use at present. This leaves a gap in the market for alternative technologies that have the potential to provide affordable, accurate forecasting that is operationally straightforward. The neural network software developed for the EA represents an initial attempt to fill this market gap.

18.3 Methodology and Software Tools

18.3.1 Artificial Neural Networks

The basics of ANNs are discussed in detail by many authors (see e.g., Bishop, 1995; Gurney, 1997; Haykin, 1998); only a brief description is therefore provided here.

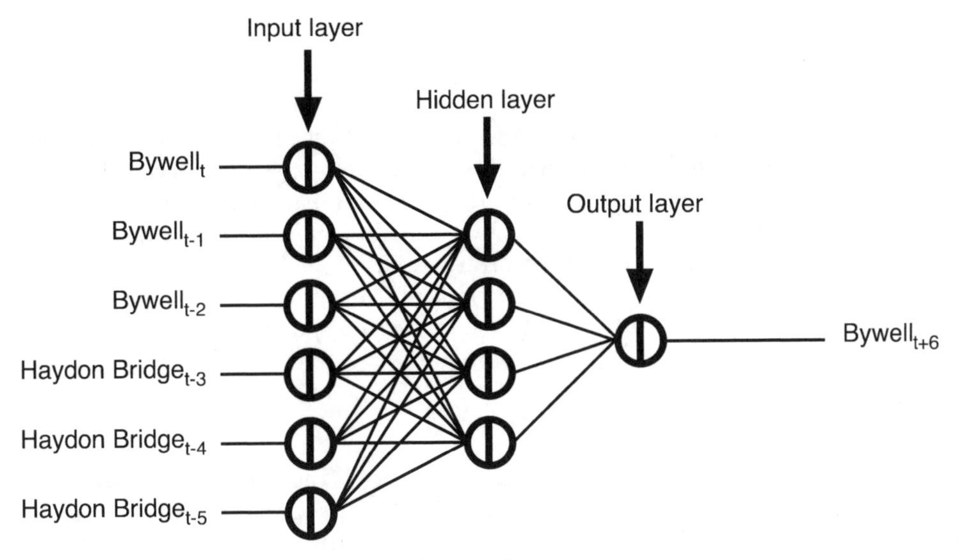

Figure 18.2 *Basic structure of a neural network*

ANNs perform a mapping of inputs to outputs via a series of simple processing nodes or neurons. The ability to map a function is derived via the configuration of these neurons into a set of weighted, interconnected layers as shown diagrammatically in Figure 18.2, where the neurons in the first and last layers have a one-to-one correspondence to the input and output values, respectively. Bywell and Haydon Bridge are two of the stage gauging stations on the River Tyne. Bywell is the lowest gauge on the non-tidal river where a forecast is required, and Haydon Bridge is upstream on the tributary River South Tyne (Figure 18.3). In this neural network mapping, the inputs are river stage at the two stations at the current time t, and t_{-1}, an hour before, and t_{-2}, two hours before. The inputs can be any combination of variables that are thought to be important for predicting the output for a specific case, in this case river levels at the forecast and upstream stations and rainfall would be appropriate. Thus, ANNs are capable of incorporating a large number of variables in a flexible manner. The input layer is connected to the hidden layer, which is further connected to the output layer where the forecast is produced. It is the programme-defined weights that are updated in the network learning phase (Beale and Jackson, 1990).

The task of any neuron, k, is to integrate the n inputs, x_i, from an external source or from other neurons within the network as follows:

$$S_k = \sum_{i=1}^{n} w_{ik} x_i \tag{18.1}$$

where w_{ik} are the weights between neurons i and k, and S_k is the sum of the weighted inputs. The neuron then further processes this sum via an activation function, such as the sigmoid:

$$f(x) = \frac{1}{1 + e^{-S_k}} \qquad (18.2)$$

To learn the relationships in the data the network is trained by iteratively adjusting the network weights in such a way that the network converges to an acceptable solution. Developed by Rumelhart and McClelland (1986), backpropagation is the first reliable algorithm for achieving this task and its development represented a major historical breakthrough in neural network research.

Backpropagation is a variation of the gradient descent optimisation algorithm. It is also known as the generalised delta rule for weight updating and is given by the following equation:

$$\Delta w = -\eta g \qquad (18.3)$$

where η is a learning rate parameter greater than zero. The gradient, g, is calculated in two passes through the network. In the forward pass, the network output for pattern p is calculated. The error between the calculated output, x_k, and the desired value, z_k, is defined as:

$$E_p = \sum_k (x_k - z_k)^2 \qquad (18.4)$$

In the backward pass, the partial error associated with the pattern is backpropagated through the network to update the weights. Weight updating can take place after each pattern or after each training cycle in which all the patterns are presented to the network. The network is trained until a stopping criterion has been satisfied such as an increase in error in the validation data set. When this situation occurs, further training will result in an overtrained network and an inability to generalise to independent data. The backpropagation training algorithm can be slow to converge so a momentum term can be added to the weight update equation to speed up the training process (Bishop, 1995).

A second, faster training algorithm incorporated into the forecasting software is Quickprop, developed by Fahlman (1988). In contrast to backpropagation, Quickprop is a second-order method because it uses information provided by the second derivative of the error surface to take larger steps down the gradient of the error function. The method assumes that the shape of the error surface is parabolic and it tries to move from the current position to the minimum of the parabola at each step. Quickprop computes the derivatives in the direction of each weight; firstly, the gradient is computed with regular backpropagation and then a direct step to the error minimum is attempted. The weights are updated as follows:

$$\Delta w(t) = \frac{g(t)}{g(t-1) - g(t)} \Delta w(t-1) \qquad (18.5)$$

where $g(t)$ is the partial derivative of the error function at the current time step and $g(t-1)$ is the last partial derivative.

Despite significant computational and methodological advances, there are fundamental problems involved in the selection of an optimal neural network architecture. As Abrahart *et al.* (1999) show, the choice of algorithm is a subject of lively debate. To simplify the choice of algorithms, only backpropagation and Quickprop are currently incorporated into the flood forecasting software developed for the EA, as outlined in the next section. The networks are also provided with default settings so that the users can apply the software with minimal knowledge of ANNs. In the future, additional algorithms will be built into the software.

18.3.2 Description of the Neural Network Software

The PC-based neural network model was developed for operational flood forecasting purposes by the University of Leeds and implemented by the Northumbria Environment Agency's flood defence and water resources department in December 1999 (Kneale and See, 1999; Kneale *et al.*, 2000a and b).

The software was designed so that the EA operators can develop and maintain their neural network models independently. It is written in Visual BASIC v.6 as a standalone program. The program can either use historical or real-time rainfall and stage data to forecast stage, and was programmed so that the EA staff could use it with data for any of the main rivers in the Northumbria region. The interface is designed to be user friendly. It contains a series of interactive lists that guide the operator through each stage of the forecasting procedure. There are three main components: an interface for specifying the data, a neural network component and a graphing feature.

The Data Component

The data interface provides a diagrammatic river network which permits the user to specify the forecasting station, the forecasting time horizon, station inputs to the network and the location of data files. A list of steps is provided on the right-hand side of the screen to guide the user through the process. Once the input data and files are specified, the user can build a training data set for the network. The following functions are available within the data component:

- *Data pre-processing*: There are two processes: (a) creation of hourly averages from the 15-minute stage data available from the historic records as used here, but some users might choose (b) first-order differencing of the data series prior to building the training data set in order to render it stationary.
- *Station selection and forecasting horizon*: The forecasting station is selected along with the forecasting horizon, ranging from 1 to 24 hours.
- *Build the training data set*: This function builds the training dataset from each of the station inputs specified by the user.
- *Travel time specification*: This allows the user to change the average travel times between stations.

The Neural Network Component

The neural network component allows the user to set up a neural network, choose an algorithm and specify training parameters. The trained network can be saved to a file or loaded and used to make forecasts. The functions available within the neural network component include the ability to:

- *Configure the network*: Specify the number of inputs and hidden nodes. Set training parameters if defaults are not desired. Choose one of two available network training algorithms: backpropagation or Quickprop.
- *Specify data set*: Indicate the name of the training or testing data set.
- *Select mode*: Real-time operational forecasting or training on historic data.
- *Train the network*: Train the network for the number of iterations specified by the user.
- *View the network performance*: Examine the training or testing statistics.
- *Save/load a network*: Save the trained net or load an existing one.
- *Forecast stage*: Use a trained network to forecast the stage at the user specified station.

As mentioned previously, the software has two neural network algorithms: standard backpropagation (Beale and Jackson, 1990) and Quickprop (Fahlman, 1988). In practice a researcher exploring the use of neural nets will look at alternative algorithms and choose the one which is most efficient for each investigation. Including two algorithms gives users the opportunity to experiment with different solutions and explore relative performance accuracy.

The Graphing Component

The graphing component provides the user with an interface to graph any input data set or to view network forecasts on a historical or real-time basis. The results required for hydrologists are a graph of the actual and forecast stage in metres on a common time frame. In some hydrological modelling the forecasting procedure checks the actual stage against the forecast value and then calculates the next step ahead based on the updated, real stage. In ANN modelling there is no step ahead updating, the forecast is continuous.

18.3.3 Use of the Software

There are two main ways to use the software. In the first mode, a single model is created with a set forecasting horizon. For example, networks can be built to predict river stage from 1 hour ahead up to 24 hours ahead. Predictions are always made from t_0 to produce a single forecast x hours ahead, where x is the forecasting horizon chosen by the user. To produce a continuous set of forecasts at three-hour intervals, for example, the user would need to build eight networks with increasing forecasting horizons. The second use is the more conventional forecasting model in which a single network is created with a short forecasting horizon, such as 1 hour, and stage and rainfall information are used to produce continuous forecasts over the next 24 hours.

It is important to recognise that a neural network must be developed for each specified forecasting station, so errors are not propagated downstream as occurs in a flow routing model. The programmes do not use any of the forecasted information as inputs to create additional forecasts. All the graphs in this chapter are validation forecasts created using independent data, and there is no step updating and correction as one would expect in a conventional hydrological forecasting model. In real-time operation the model can be run to make a new forecast every 15 minutes or whenever further real-time stage or rainfall data become available. The neural network forecasting model runs in seconds and can be retrained with new data in minutes. The retraining time is governed entirely by the time taken to access the real-time input data. It has the potential to be a very useful operational forecasting tool.

18.4 Case Studies

18.4.1 Background

The Tyne catchment in Northumberland, north east England, UK, has two main channels: the North and South Tyne. The River Nent and the East and West River Allens are important, ungauged lateral inflows (Figure 18.3). The underlying geology of the catchment largely consists of Carboniferous limestone and Millstone Grit. The Pennine upper reaches respond very quickly to rainfall inputs, the time between rainfall and rising river stage can be minutes. There is an approximate lag time of 4 hours between the raingauge at Alston and the river gauging station at Haydon Bridge. Flood alerts are often issued to the flood-risk areas along the river. Bywell is the lowest non-tidal gauging station on the River Tyne.

In these examples forecasts of stage are made for flood events at Alston, Haydon Bridge and at Bywell. Two types of forecasting are undertaken: event-based and continuous. Event-based forecasting involves extracting flood events from the historical record and using these small datasets to develop neural network models. Continuous forecasting uses a dataset over a continuous time period such as the winter period. Alston and Bywell are used to assess the value of ANNs for event-based forecasting using different forecasting conditions. Alston is a good example of a flashy headwater catchment where runoff starts at almost the same time as it rains. Bywell is a lowland station downstream from the North and South Tyne confluence. It is critical in providing flood warnings for Newcastle upon Tyne. In the experiment using Bywell, stage is forecast using only data from stations on the River South Tyne. Haydon Bridge is used for long-term forecasting based on continuous input data rather than individual events from the training record.

For the event forecasts the digital river stage archives (1985–2000) were searched to find the highest stage events at the input stations and Bywell. For the Alston experiments the input data were stage at Alston and rainfall at the Alston raingauge. For the Bywell forecasts, data from Haydon Bridge, Featherstone, Alston and the

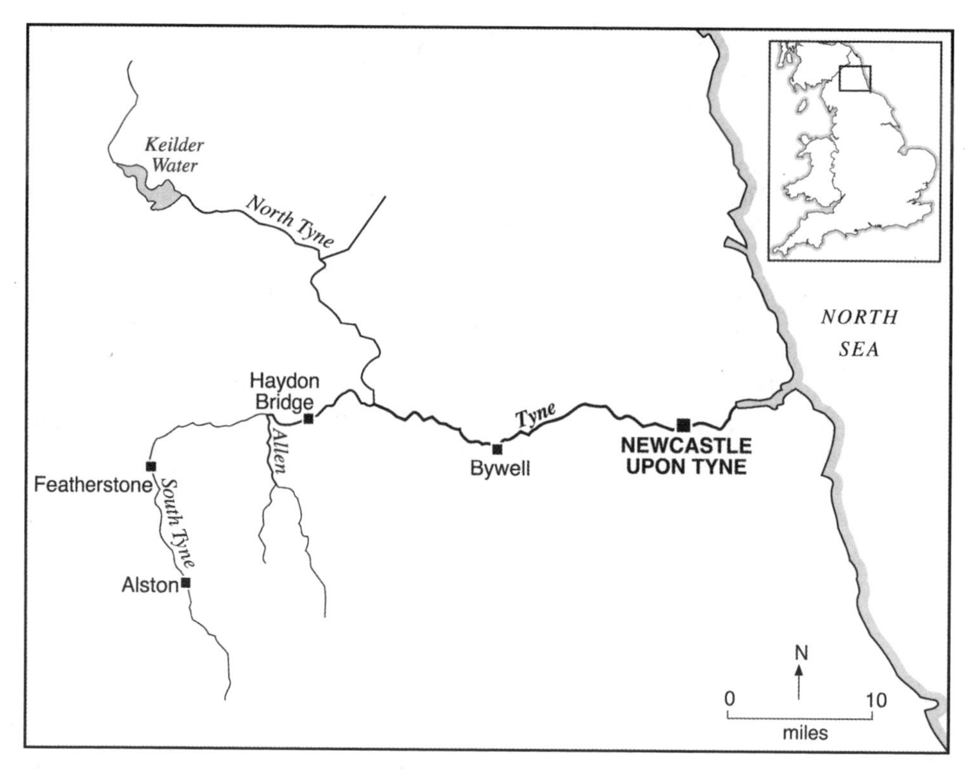

Figure 18.3 *Map of the catchment*

raingauge at Alston were used. In each case the 11 largest events were ranked. The odd-numbered events were used as the training series and the even-numbered events used for independent model validation. Networks were generated using both the backpropagation and Quickprop methods and for various lead times and training epochs (number of training iterations). The limited size of the event training data sets was deliberate to see how well the forecasts were made with small datasets.

For the long-term continuous forecasts at Haydon Bridge, the input data were stage at Haydon Bridge, Featherstone, Alston and the raingauge at Alston. Again backpropagation and Quickprop routines were compared.

Evaluation of the accuracy of forecasts by hydrologists is problematic. There is no single statistic that is uniquely helpful, primarily because the data of interest is frequently the rare high flow events in the record. For real-time on-line forecasting eyeballing the forecast stage is crucial, but where longer records are available statistical methods, root mean square error (RMSE) and the Nash and Sutcliffe (1970) index are standard hydrological procedures (Cameron *et al.*, 2002; Legates and McCabe, 1999). In addition to these measures we look at the accuracy of forecasts with respect to the real warning times which operators use on this river to notify the public of a flood event.

18.4.2 Event Forecasting for Alston

Alston is the headwater gauging station on the River South Tyne. It is a Pennine, upland moorland and rural catchment, with very flashy runoff patterns. The hydrographs (graphical representation of changing stage on a river) rise and fall within hours as water runs rapidly from the hillside to the river. In this type of catchment flood warning times are very short so any increase in warning capability is an advantage. For the EA flood warning team, the accuracy in forecasting the peak is not as significant as having a good forecast of the rising limb which triggers flood warning actions downstream.

A series of experiments have been undertaken using the approach outlined above, i.e., the flood events were extracted and ranked, the networks were trained on the odd-numbered events and validated on the even-numbered events, using both backpropagation and Quickprop algorithms to produce forecasts for 1 and 2 hours ahead. Two further variations were introduced. In the first case, models were developed to forecast stage at Alston using the stage data alone. In the second case, both the stage data and the rainfall records from a continuously recording, 'tipping bucket' gauge nearby, were input to the model.

The results showed that adding information from the raingauge significantly improved the forecast timing in this flashy catchment. Where a forecast is made using the gauging station data alone, the network cannot forecast a rise in stage until it observes a rise in stage, so the forecast of rise is delayed by at least the time step used in the calculations. The earlier inputs from the raingauge kick-start the forecast rise in river level. Examples of validation forecasts on the largest event in the record (Figure 18.4a) and a more typical event (Figure 18.4b) illustrate the forecasting ability of the networks. In these examples, the data resolution is 1 hour but these forecasts could be improved by using 15-minute stage and rainfall input data.

Similar results were also obtained in a parallel study for Kilgram, a comparable flashy Pennine station on the River Ure in the headwaters of the Yorkshire Ouse system. At both Alston and Kilgram, the Quickprop algorithm performs slightly better than backpropagation. While the peak is not always hit precisely, the warning levels on the rising limb were either hit on time or hit slightly ahead of time. This is an advantage because the sooner a warning can be given the longer local people have to respond. In future work, including rainfall data from real-time radar images could improve the quality of flood forecasts.

These results suggest that ANNs can be useful in flashy headwater areas with short response times and if further raingauge data are available, possibly from adjacent catchments, that could also improve the forecast. They also indicate that where data are limited, as in this example where only seven flood events were used in the training series, a neural network can provide a useful forecast.

18.4.3 Event Forecasting for Bywell

Two contrasting approaches were taken in this case study. Firstly, as in the Alston experiment, the highest 11 stage events for which there are data at the four input

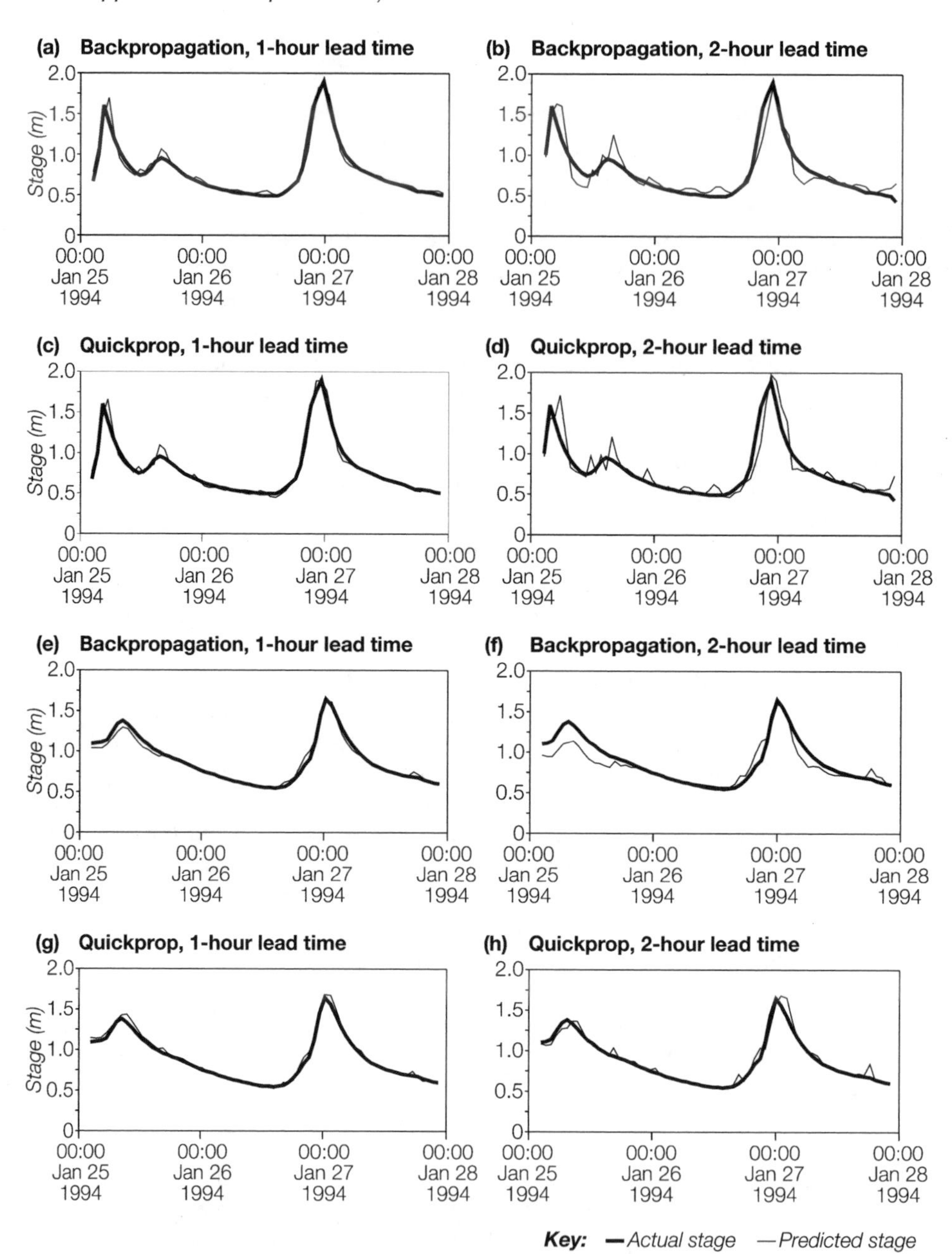

Figure 18.4 Validation forecasts for events 1 (a–d) and 2 (e–h) at the Alston stage gauge, using stage at Alston and the raingauge data from Alston as inputs, and trained for 3000 epochs using the odd events

stations and Bywell were ranked. The odd-numbered events were used as the training series and the even-numbered events were used for independent model validation using both the backpropagation and Quickprop methods for 1 and 2 hour lead times. The first flood warning alarm stage at Bywell is 3.5 m.

The peak forecast for the largest event in the record is acceptably modelled in both shape and timing (Figure 18.5a and b). The 3.5 m warning stage is forecast more accurately by the backpropagation algorithm for both 1 and 2 hours ahead. The lowest event on record (Figure 18.5b) is, by contrast, overpredicted at the peak and the warning times are forecast up to a couple of hours early. This may be in part due to the slower travel times of lower flood peaks down to Bywell and that this was the lowest event in the validation record. However, this peak height would not cause a flooding problem in the catchment.

This experiment was then extended to develop models for a longer lead time, i.e., 6 hours. However, the results proved less than satisfactory. The peaks and total runoff were underestimated and the forecast time of rise was delayed by 2 hours. While the hydrograph shape was correct, this model is operationally less useful, although under extreme circumstances it may be better than nothing. The fast response time of the catchment, where travel from Alston to Bywell is of the order of 6 hours, means that looking from 6 hours out there has not been enough significant rainfall and rise in stage to indicate that a significant storm runoff could be expected. The Quickprop algorithm performed better than backpropagation here. Under normal forecasting circumstances the network would be created using stations on both main rivers and from the reservoir at Kielder. If forecasts of this type are to be valuable then experiments that look at the radar rainfall in adjacent catchments should be integrated with the neural network data to provide longer lead time warnings.

In the second approach the event data were used to see how the neural network models would handle flood event data outside the training envelope. The critical question here is can neural nets cope when asked to model stage patterns that are higher or lower than the data used to develop the model? In this experiment the challenge is greater in that the model is given no guidance from stage levels at Bywell itself, or from the whole of the North Tyne system. The three highest flood events were excluded from the training series. The neural network was generated on events ranked 4–11 and used to forecast stage at Bywell for the three unseen events. Results for the third largest and largest unseen events are shown in Figure 18.6a and b. The training data has peaks ranging from 3.05 to 4.57 metres. The validation forecasts are for events peaking at 4.75–5.88 metres.

Encouragingly, Figure 18.6a shows a good forecast of both peak and timing for an event with stage 18 cm higher than seen in the training set. For the highest event on record with a peak of 5.88 metres, and without knowledge of the North Tyne inputs which contribute 40–50% of the stage, the forecast is very respectable for the rising limb but not surprisingly in three of the four forecasts underestimates the peak (Figure 18.6b). The results suggest that the network can cope with forecasting rising limb responses for extreme stage events. The accuracy of the peak and the modelling of alarm levels would seem to be appropriate for routine forecasters' purposes.

In modelling terms, it was to be expected that given a limited number of flood events and a situation where the validation data are outside the range of the

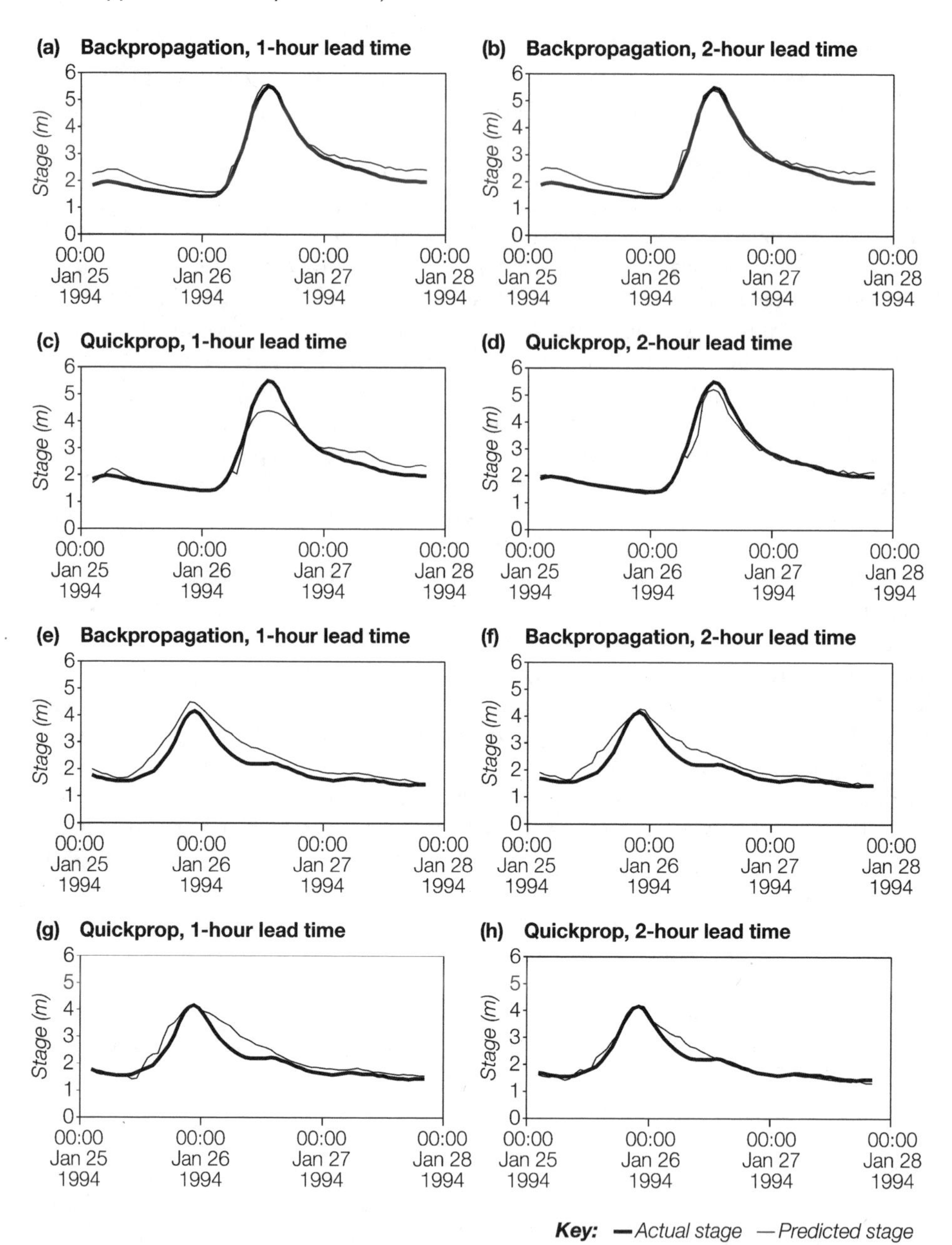

Figure 18.5 Validation forecasts for events 1 (a–d) and 2 (e–h) at the Bywell stage gauge, using stage at Alston, Featherstone and Haydon Bridge and the raingauge data from Alston as inputs, and trained for 3000 epochs using the odd events

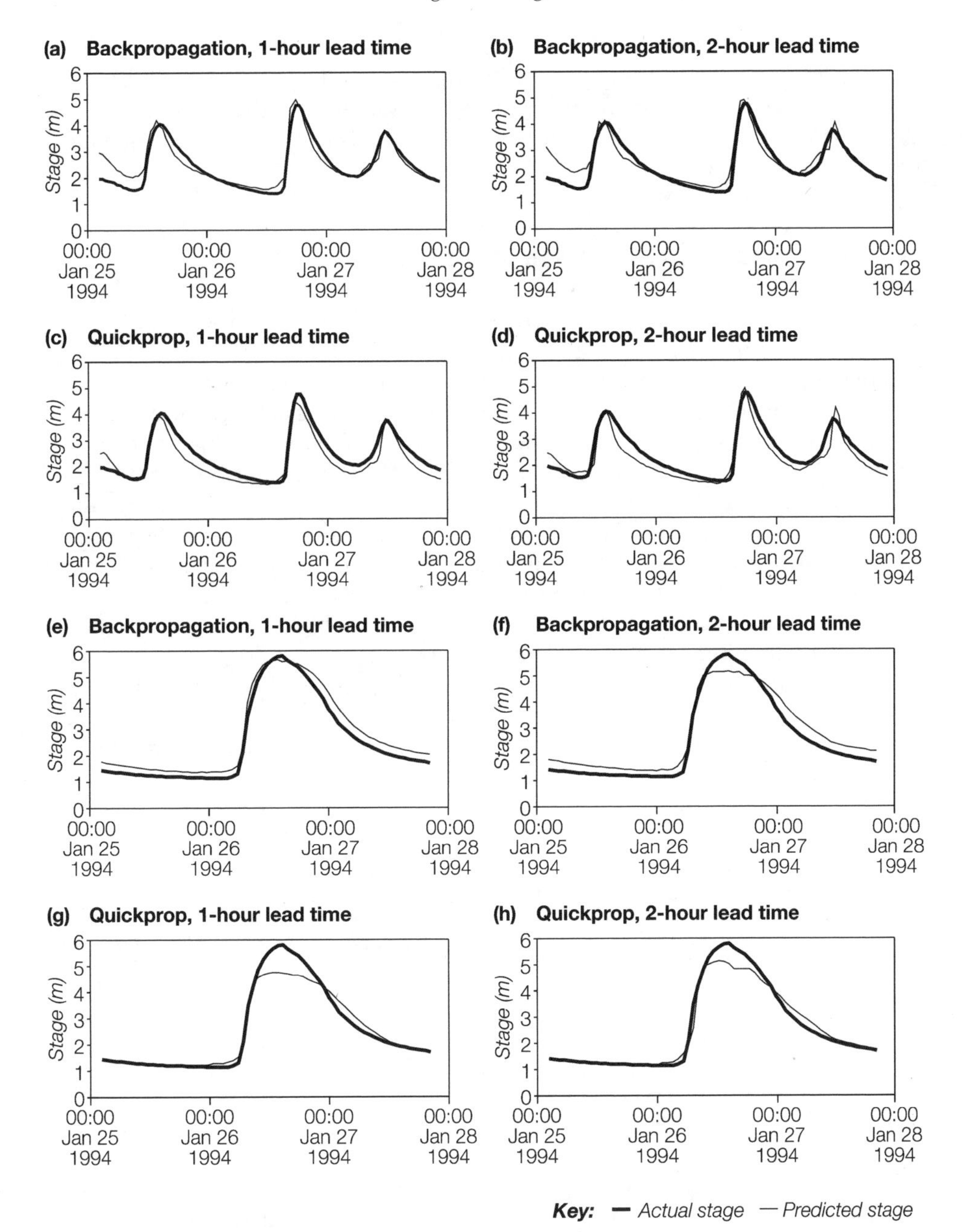

Figure 18.6 Validation forecasts for events 1 (a–d) and 2 (e–h) at the Bywell stage gauge, using stage at Alston, Featherstone and Haydon Bridge and the raingauge data from Alston as inputs, and trained for 3000 epochs with the events ranked 4–11

training data set that the model would underpredict very high events (Figure 18.6b) and overpredict a low event (Figure 18.5b). Ideally models would be developed with longer training sets, but where new gauging stations are installed and records are short, these results indicate that neural network models are useful even with limited data.

18.4.4 Continuous Forecasting for Haydon Bridge

In this set of experiments, models were developed to forecast river levels at Haydon Bridge. The inputs used were stage at Haydon Bridge and Featherstone, and stage and rainfall at Alston. The ANNs were trained for 3000 epochs using both the backpropagation and Quickprop algorithms, various experiments having shown that there was minimal improvement in the networks' ability to forecast at larger numbers of epochs. The networks were trained on data from October to April of 1996–97 and validated using data for October to April of 1994–95, 1995–96 and 1997–98. Table 18.1 shows the summary evaluation statistics for all four years for the two ANNs. The networks are compared with results from models developed using multiple linear regression and ARMA (AutoRegressive Moving Average) techniques.

The Nash–Sutcliffe and R^2 statistics are encouragingly high for the neural network models. The multiple linear regression and ARMA models do slightly less well. RMSE values, indicating the absolute error in stage between the actual and forecast levels, are of the order of centimetres for all models. Quickprop produces the best model followed by backpropagation, multiple linear regression and then the ARMA model.

Table 18.1 *Evaluation measures for long-term calibration of neural networks, multiple linear regression and ARMA models*

Algorithm	Historical period	Nash–Sutcliffe coefficient	*R*-squared	RMSE (m)
Backpropagation	Winter 1994/95	0.914	0.966	0.097
	Winter 1995/96	0.891	0.912	0.071
	Winter 1996/97	0.965	0.971	0.056
	Winter 1997/98	0.939	0.953	0.065
Quick propagation	Winter 1994/95	0.981	0.984	0.052
	Winter 1995/96	0.962	0.967	0.043
	Winter 1996/97	0.971	0.971	0.054
	Winter 1997/98	0.979	0.979	0.041
Multiple regression	Winter 1994/95	0.918	0.941	0.108
	Winter 1995/96	0.843	0.876	0.086
	Winter 1996/97	0.916	0.916	0.092
	Winter 1997/98	0.936	0.937	0.071
ARMA	Winter 1994/95	0.784	0.805	0.176
	Winter 1995/96	0.813	0.830	0.093
	Winter 1996/97	0.717	0.751	0.168
	Winter 1997/98	0.776	0.798	0.132

Table 18.2 *Accuracy and timing in predicting the operational level of 2.0 metres*

Event	Observed stage	Backpropagation	Quickpropagation	Regression	ARMA
1994/11/13	2.07	2.20 (+1.0)	2.18 (+1.0)	2.12 (+2.0)	2.09 (+3.0)
1994/12/08	2.21	2.48 (0.0)	2.17 (0.0)	2.04 (+1.0)	2.21 (+2.0)
1994/12/11	2.03	2.16 (+3.0)	2.03 (+1.0)	2.02 (+4.0)	2.32 (+4.0)
1994/12/31	2.26	2.28 (−1.0)	2.06 (0.0)	* (N/A)	2.22 (+2.0)
1995/01/31	2.04	2.14 (+1.0)	2.10 (+1.0)	2.23 (+2.0)	2.27 (+3.0)
1995/02/12	2.29	2.15 (0.0)	2.27 (0.0)	* (N/A)	2.34 (+2.0)
1995/02/19	2.10	2.19 (+1.0)	* (N/A)	* (N/A)	2.06 (+2.0)
1995/02/22	2.61	2.08 (0.0)	2.52 (+1.0)	2.08 (+1.0)	2.74 (+3.0)
1995/03/01	2.13	2.06 (0.0)	2.18 (+1.0)	* (N/A)	2.30 (+3.0)
1995/03/26	2.01	2.09 (−1.0)	2.02 (+1.0)	* (N/A)	2.09 (+2.0)
1995/10/03	2.02	2.34 (+0.0)	2.00 (+1.0)	2.12 (+0.0)	2.25 (+2.0)
1995/11/15	2.37	2.25 (+1.0)	2.40 (+1.0)	2.14 (+2.0)	2.02 (+2.0)
1996/02/18	2.18	2.31 (0.0)	2.30 (+1.0)	* (N/A)	2.03 (+2.0)
1998/01/01	2.09	2.24 (0.0)	2.08 (+1.0)	* (N/A)	2.05 (+2.0)
1998/01/08	2.17	2.17 (0.0)	2.39 (+1.0)	2.19 (+2.0)	2.36 (+3.0)
1998/03/07	2.28	2.58 (0.0)	2.21 (0.0)	2.10 (+2.0)	2.37 (+3.0)
1998/04/03	2.12	* (N/A)	2.12 (0.0)	* (N/A)	2.06 (+2.0)

Note: *Forecast did not reach the operational stage at all.

Table 18.1 summarises the results for the entire seven-month winter series, but in event forecasting the accuracy with which the networks predict the operational alarm stage of 2.0 m on the rising limb of the hydrograph is also very important; Table 18.2 details the results from 17 independent, validation events. The Quickprop neural network proved to be the most accurate of the four models in terms of predicting the timing of the alarm levels. The ARMA model consistently predicted the operational stage 2 to 3 hours late.

These results show that the multiple linear regression model did not reach the alarm levels on eight of the 17 events and was generally 2 hours late at other times.

If there were no significant inputs between the gauging stations then one might expect a linear relationship between stage at Featherstone and Haydon Bridge, and a linear routing model or regression equation should work well. However, the River Allens join the Tyne between Featherstone and Haydon Bridge, as does the River Nent just downstream of Alston adding significantly to the discharge. The ability of ANNs to handle non-linearities, for example from ungauged tributaries, is its strength over alternative linear approaches.

In addition to improved accuracy, a clear advantage of the neural network software is that model development times are rapid, even over a long continuous time series. In these experiments the training runs took approximately 12 minutes for 3000 epochs, and validation of the models with the three years of winter data was virtually instantaneous. Multiple linear regression equations were also very fast to develop but their use appears to be limited. The ARMA models, on the other hand, required a great deal of experimentation and the results were not very satisfactory.

18.5 Conclusions

The incidence of river flooding is increasing and forecasts of global warming indicate that this trend is likely to continue. Building new flood defences to contain higher flows, defending vulnerable sites and domestic and commercial insurance claims all have serious financial implications. Accurate forecasting systems are required for real-time day-to-day emergency forecasting and for the planning and design of new defences. ANNs are flexible mathematical structures capable of identifying complex non-linear relationships between input and output data sets. Their use in hydrological forecasting is still very new but it has been shown here that they can provide a viable alternative approach to statistical models.

This software has been in use since December 1999. Operators report that it is easy to use and straightforward to manage in both training and real-time operations. The experiments described here are designed to test the system response given difficult hydrological conditions with the emphasis on testing to see whether the alarm levels on the rising limb of the hydrograph can be satisfactorily forecast. The results would give the operators time to generate warnings at the first alarm stages.

Efficient use of a neural network model requires a model trainer who builds appropriate networks and selects the best algorithm. To find the lowest error solution may take many runs, but training runs are quick to do, typically less than 10 minutes per run. Given this high operational speed, a series of experiments can be defined to ensure that alternative algorithms and stage regime models are fully investigated. Forecasters can therefore make a mature selection of networks for operational use on the basis of wide experience. Real-time forecasts with established networks are effectively instantaneous once the data are downloaded. Forecasters must therefore take the time to investigate the effects of model architecture to ensure the validity of the results. However, having established a suite of models for operational use, these may be used in the flood room by additional operators after a short period of training (one day at most). Moreover, the experience with the software to date indicates that it is a viable technology that should be considered as an additional tool for hydrological forecasting.

The software was developed in response to a growing demand for user-friendly access to ANNs but integrated within a single, user-customised software environment. These approaches are already used commercially in areas such as database mining and financial market prediction, but this software represents the first commercially available, operational hydrological forecasting tool that allows for the development of river stage forecasting models. In addition to incorporating other neural network algorithms, it is envisaged that the software will eventually be linked to a GIS in order that forecasters can visualise the flood wave as it travels down the catchment, both from a plan and a 3-D perspective. The spatial element will also allow forecasters to calculate flood inundation. The potential for constructing spatial decision support systems that incorporate the expert knowledge that is currently embedded in human forecasters remains an exciting area for future development.

References

Abrahart, R.J., See L. and Kneale, P.E. (1999) Using pruning algorithms and genetic algorithms to optimise network architectures and forecasting inputs in a neural network rainfall-runoff model, *Journal of Hydroinformatics,* **1**: 103–14.

Abrahart, R.J., See, L. and Kneale, P.E. (2001) Investigating the role of saliency analysis with a neural network rainfall-runoff model, *Computers and Geosciences*, **27**(8): 921–8.

Beale, R. and Jackson, T. (1990) *Neural Computing: An Introduction*, Adam Hilger, Bristol.

Bishop, C.M. (1995) *Neural Networks for Pattern Recognition*, Clarendon Press, Oxford.

Cameron, D., Beven, K., Tawn, J. and Naden, P. (2000) Flood frequency estimation by continuous simulation (with likelihood based uncertainty estimation), *Hydrology and Earth System Sciences*, **4**(1): 23–34.

Cameron, D., Kneale, P.E. and See, L. (2002) An evaluation of a traditional and a neural net modelling approach to flood forecasting for an upland catchment, *Hydrological Processes,* **16**: 1033–46.

Campolo, M., Soldati, A. and Andreussi, P. (1999) Forecasting river flow rate during low-flow periods using neural networks, *Water Resources Research,* **35**(11): 3547–52.

Cancilla, D.A. and Fang, X. (1996) Evaluation and quality control of environmental analytical data from the Niagara River using multiple chemometric methods, *International Association Great Lakes Research*, **22**: 241–53.

Chang, F.J. and Hwang, Y.Y. (1999) A self-organization algorithm for real-time flood forecast, *Hydrological Processes,* **13**(2): 123–38.

Dawson, C.W. and Wilby, R. (1998) An artificial neural network approach to rainfall-runoff modelling, *Hydrological Sciences Journal,* **43**(1): 47–66.

Dawson, C.W. and Wilby, R. (2001) Hydrological modelling using artificial neural networks, *Progress in Physical Geography,* **25**(1): 80–108.

Fahlman, S.E. (1988) An empirical study of learning speed in backpropagation networks, *Technical Report, CMU-CS-88-162*, Carnegie Mellon University.

Golob, R., Stokelj, T. and Grgic, D. (1998) Neural-network-based water inflow forecasting, *Control Engineering Practice*, **6**: 593–600.

Gozlan, R.E., Mastrorillo, S., Copp, G.H. and Lek, S. (1999) Predicting the structure and diversity of young-of-the-year fish assemblages in large rivers, *Freshwater Biology*, **41**: 809–20.

Gurney, K. (1997) *An Introduction to Neural Networks*, UCL Press, London.

Haykin, S. (1998) *Neural Networks*, Prentice Hall, London.

HEC-RAS (2001) HEC-RAS water surface profile model, Boss International, online http://www.bossintl.com.hk/html/hec-ras_details.html.

Horritt, M.S. and Bates, P.D. (2001) Predicting floodplain inundation: raster-based modelling versus the finite-element approach, *Hydrological Processes,* **15**(5): 825–42.

Hsu K.L., Gupta, H.V. and Sorooshian, S. (1995) Artificial neural-network modeling of the rainfall-runoff process, *Water Resources Research,* **31**(10): 2517–30.

ISIS (2001) ISIS Flow / Hydrology, Wallingford Software Ltd and Halcrow Group Ltd, online http://www.wallingfordsoftware.com/products/isis.asp.

Kneale, P.E. and See, L. (1999) *Developing a Neural Network for Flood Forecasting in the Northumbria Area of the North East Region, Environment Agency: Final Report*, University of Leeds.

Kneale, P.E., See, L. and Kerr, P. (2000a) Developing a prototype neural net flood forecasting model for use by the Environment Agency, *Proceedings 35th MAFF Conference of River and Coastal Engineers 2000*, 11.03.1–11.03.4.

Kneale, P.E., See, L., Cameron, D., Kerr, P. and Merrix, R. (2000b) Using a prototype neural net forecasting model for flood predictions on the Rivers Tyne and Wear. British Hydrological Society, 7th National Hydrology Symposium, 3.45–3.49.

Legates, D.R. and McCabe, G.J. (1999) Evaluating the use the 'goodness-of-fit' measure in hydrologic and hydroclimatic model validation, *Water Resources Research*, **35**: 233–41.

Lek, S. and Baran, P. (1997) Estimations of trout density and biomass: a neural networks approach, *Nonlinear Analysis: Theory, Methods and Applications,* **30**(8): 4985–90.

Maier, H.R., Dandy, G.C. and Burch, M.D. (1998) Use of artificial neural networks for modelling cyanobacteria *Anabena* spp. in the River Murray, South Australia, *Ecological Modelling*, **105**: 257–72.

Marshall, C.T. (1996) Evaluation of River Flood Forecasting Systems, R&D Technical Report W17, Environment Agency, Bristol.

Mastrorillo, S., Lek, S. and Dauba, F. (1997) Predicting the abundance of minnow *Phoxinus phoxinus* (Cyprinidae) in the River Ariège (France) using artificial neural networks, *Aquatic Living Resources,* **10**: 169–76.

Mellor, D., Sheffield, J., O'Connell, P.E. and Metcalfe, A.V. (2000) A stochastic space–time rainfall forecasting system for real time flow forecasting II: Application of SHETRAN and ARNO rainfall runoff models to the Brue catchment, *Hydrology and Earth System Sciences*, **4**(4): 617–26.

Mike 11 (2001) MIKE Flood watch, A management system for real-time flood forecasting and warning, Danish Hydraulics Institute, online
http://www.dhisoftware.com/mike11/index.htm.

Moatar, F., Fessant, F. and Poirel, A. (1999) pH modelling by neural networks. Application of control and validation data series in the Middle Loire river. *Ecological Modelling,* **120**(1): 141–56.

Nash. J.E. and Sutcliffe, J.V. (1970) River flow forecasting through conceptual models, Part 1: a discussion of principles, *Journal of Hydrology,* **10**(3): 282–90.

Neelakantan, T.R., Brion, G.M. and Lingireddy, S. (2001) Neural network modelling of Cryptosporidium and Giardia concentrations in the Delaware River, USA*, Water Science and Technology,* **43**(12): 125–32.

NOAH 1D (2001) NOAH 1D Modelling System, Newcastle Object-oriented Advanced Hydroinformatics, online http://www.ncl.ac.uk/noah/.

Omernik, J.M. (1997) *Non-point Sources Stream Nutrient Level Relationships: A Nationwide Study*, EPA-600/3-77-105, Corvallis Environmental Research Laboratory, Office of Research and Development, US EPA, Corvallis, Oregon.

Openshaw, S., Kneale, P., Corne, S. and See, L. (1998) *Artificial Neural Networks for Flood Forecasting: Final Report*, MAFF Project OCS967P, University of Leeds.

Recknagel, F., French, M., Harkonen, P. and Yabunaka, K. (1997) Artificial neural network approach for modelling and prediction of algal blooms, *Ecological Modelling,* **96**(1): 11–28.

RFFS (2001) Real-Time Flood Forecasting System, Institute of Hydrology, Wallingford, online http://www.nwl.ac.uk/ih/www/products/iproducts.html.

Rumelhart, D.E. and McClelland, J.L. (eds) (1986) *Parallel Distributed Processing*, MIT Press, Cambridge, MA.
See, L. and Openshaw, S. (1999) Applying soft computing approaches to river level forecasting. *Hydrological Sciences Journal*, **44**: 763–78.
See, L. and Openshaw, S. (2000) A hybrid multi-model approach to river level forecasting, *Hydrological Sciences Journal*, **45**(4): 523–36.
SSIIM (1997) SSIIM 1.4, SINTEF Group, online http://www.sintef.no/nhl/vass/ssiim.html.
Stewart, M.D., Bates, P.D., Anderson, M.G., Price, D.A. and Burt, T.P. (1999) Modelling floods in hydrologically complex lowland river reaches, *Journal of Hydrology*, **223**(1–2): 85–106.
Van Kalken, T., Huband, M., Cadman, D. and Butts, M. (2001) Development of a Flood Forecasting Model for the Welland and Glen Rivers in East Anglia, UK, online http://www.dhisoftware.com/uc2001/Abstracts_Proceedings/Papers01/122/122.htm.
Whitehead, P.G., Howard, A. and Arulmani, C. (1997) Modelling algal growth and transport in rivers: a comparison of time series analysis, dynamic mass balance and neural network techniques, *Hydrobiologia*, **349**: 39–46.

CONCLUSION

19

Undertaking Applied GIS and Spatial Analysis Research in an Academic Context

Robin Flowerdew and John Stillwell

Abstract

This is the final chapter in a book that has aimed to demonstrate the utility of applying quantitative methods to solve real-world problems in human geography. The collection clearly exemplifies the wide range of applications and flags up the potential for future endeavour. Yet doing applied research within the confines of an academic environment is neither easy nor straightforward. This chapter concludes the book by describing the setting and some of the applied quantitative work done at the North West Regional Research Laboratory and at the Yorkshire and Humberside Regional Research Observatory during the 1990s. Some of the constraints under which the organisations operate are discussed, along with more detailed examples of four applied projects. Finally, some comments are made on the future scope for commercial applications of geography and their implications for the discipline.

19.1 Introduction

The chapters contained in this volume have reported on a spectrum of applications of GIS, spatial analysis and geographical modelling for a variety of clients in a range of different contexts. Whilst certain contributions refer to projects undertaken by individuals working as consultants for companies in the private sector (such as PricewaterhouseCoopers, David Simmonds Consultancy) the majority of the applications have been undertaken by researchers employed by organisations created

Applied GIS and Spatial Analysis. Edited by J. Stillwell and G. Clarke
 ISBN: 0-470-84409-4

within their universities specifically to provide an environment in which applied research can flourish (such as CURDS, Nexpri, SERRL), or by academics working from within their respective university departments. These university-based studies reflect the extent to which 'applied geography' has emerged within the discipline of geography and how quantitative and computer-based approaches have been at the forefront of this development over the last 20 years in particular.

In the sections of this chapter that follow, we concentrate on three aspects of applied research. Firstly, we review the rationale and structure of two different institutional models for applied research established in neighbouring regions in the north of England, the North West Regional Research Laboratory (NWRRL) and the Yorkshire and Humberside Regional Research Observatory (ReRO). Secondly, the types of projects undertaken by each of these organisations are reviewed and selected GIS or spatial analysis-based applications are detailed for exemplification. Finally, based on our personal experiences of working for the NWRRL and ReRO respectively, we identify some of the major benefits arising from applied research but set these positive factors against a number of constraints or difficulties that have been associated with the conduct of this type of research.

19.2 New Regional Research Institutions for Applied GIS and Spatial Analysis

19.2.1 Introduction

Government funding for the social sciences, including human geography, comes via the Economic and Social Research Council (ESRC), renamed from the Social Sciences Research Council because of Government scepticism in the 1980s about their findings (which were often at variance with the underpinnings of Government policy). Then (and now) the ESRC received a tiny proportion of the total science budget, and had to argue its case to maintain this share. It saw its best chances of obtaining funding through emphasising that some parts of the social sciences were relevant to business and industry, and that social sciences too could adopt the technological 'laboratory' style of the physical sciences. GIS came along at just the right time and met the criteria. Arguably at least GIS was part of human geography and hence social science; requiring 'equipment' and 'laboratory facilities' (just like 'proper science'); seemingly objective and free from questions of values and interpretations; and with the potential to be relevant to a wide range of clients and 'customers'.

At the same time, the ESRC also attempted to exert more influence on the future of the social sciences, increasing the extent to which it decided what areas to put funding into, as opposed to merely being responsive to the ideas of practising social scientists. This was done through creating research initiatives or programmes, whose general topics were announced in advance with a request for proposals, several of which were to be funded. In addition to those proposals that were funded, this strategy had the effect of encouraging other research groups to think about the issues and in some cases to appoint relevant staff. Unsuccessful ESRC research programme proposals have often been funded through other ESRC research money or by other funding agencies, so the research programme approach can be regarded as levering more research in the specified area than is actually funded.

19.2.2 The Regional Research Laboratories

It was in 1987, prior to the publication of the Chorley Report on the development of a national GIS policy, that the ESRC began the trial phase of its Regional Research Laboratory (RRL) initiative, moving in 1988 to the main phase. Eight RRLs were created in the main phase:

- the South East RRL (based at Birkbeck College London and the London School of Economics);
- the Midlands RRL (based at the Universities of Leicester and Loughborough);
- the North East RRL (based at the University of Newcastle upon Tyne);
- the North West RRL (based at Lancaster University);
- the Urban and Regional Policy Evaluation RRL (based at the Universities of Liverpool and Manchester);
- the Wales and South West RRL (based at the University of Wales Institute of Science and Technology and the South West Regional Computing Centre at the University of Bath);
- the Scottish RRL (based at the University of Edinburgh); and
- the Northern Ireland RRL (based at the University of Ulster).

In addition, the RRL initiative provided support for a Co-ordinator (Ian Masser, at the University of Sheffield) and a Technical Advisor (Mike Blakemore, at the University of Durham). The individual RRLs were constituted by a team of academics with permanent contracts with their universities, supplemented by appointed research staff. In five cases, the RRL was run by geographers, who also played a major role in a sixth, and people whose disciplinary affiliation was to geography, although sited in a planning department, dominated another. Over time, the composition of these RRLs has changed, as a result of moves of key people from one site to another and changes in degrees of involvement. Another trend was for the multi-site RRLs to split, so that the roles of the LSE, Loughborough and Bath diminished, and the Urban and Regional Policy Evaluation RRL split into two separate operations at Liverpool and Manchester. Other groups doing GIS research have also been granted 'RRL status', so groups from Glamorgan, Glasgow and York, among others, are part of RRL.net, the organisation that exists to coordinate the former RRLs in their post-ESRC situation.

The intention of the RRL Initiative was to provide resources for the groups receiving funding to allow them to become 'well-founded' laboratories for GIS work

(Masser, 1988). Money was provided for equipment and for staff, but the RRLs were always expected to become self-supporting through working on a commercial basis with clients in government and industry. At the end of the main RRL funding in 1991, the RRLs remained in existence but have followed different paths, some effectively disappearing while others have grown in size through a mixture of strategies.

There were always some ambiguities and contradictions in the roles that they were expected to play, reflected in part in the name. Part of the original idea was that each RRL would be particularly responsible for collecting and providing access to datasets about its own region, and for liaison with potential clients in that region. This function worked reasonably well in the RRLs with responsibility for Scotland, Wales and Northern Ireland, where there were government and commercial offices with devolved responsibilities with which they could work. The English RRLs, however, tended to work with datasets and organisations at the national level, and although they may have built up closer relationships with local government or other organisations in their own regions, their roles quickly became demarcated in terms of subject specialisation rather than region.

The term 'Research' also relates to an ambiguity in the role of the RRLs. The requirement to become self-supporting appeared to demand a concentration on commercial and consultancy work. However, the potential 'users' of the research, especially those from the private sector, were very concerned that the RRLs might become subsidised competitors for their own companies. Hence there was also a requirement on the RRLs to conduct basic research in GIS, which might lead to innovations that could be developed commercially by private sector organisations. Many or most of the RRLs applied successfully for GIS-related funding for what were clearly research projects, and indeed ESRC funded several projects in this general area, including a collaborative research programme with the Natural Environment Research Council (NERC).

It is clear that the RRL designation 'laboratory' was part of the ESRC's attempt to provide the social sciences with a more scientific image. Although RRL staff seldom wore white coats and handled test tubes, the RRLs did usually possess large rooms where several people worked together at a range of computers. For many in human geography this may have been a change from the model of the single scholar working in isolation in his or her own office with occasional trips to the library or archive. In the next section we outline the ReRO model which provided an alternative approach to bringing together academics from differing institutions to work together on applied projects.

19.2.3 The Yorkshire and Humberside Regional Research Observatory

No RRL status was awarded to a university in Yorkshire and Humberside. This was surprising given the large group of applied geographers at the University of Leeds and the international status of the modelling work carried out there under the leadership of Alan Wilson. In fact a consultancy business named GMAP had already been set up within the University and had begun to flourish, using spatial interaction modelling to advise companies on locational decisions and building intelligent GIS (Clarke *et al.*, 1995; Birkin *et al.*, 1996). However, in 1990, separate from its commercial consulting activities under the GMAP banner, the Yorkshire and Humberside Regional Research Observatory (ReRO) was created to provide clients in the region with access to a region-wide data and analysis service.

Initially, ReRO was a joint venture between the School of Geography at the University of Leeds and the Centre for Urban Development and Environmental Management (CUDEM) at Leeds Polytechnic (later Leeds Metropolitan University). The aim was to create a network of research centres, or 'Associate Research Units' (ARUs) as they became known, each offering a range of skills and expertise in data collection and processing, problem-solving, strategy formulation and policy analysis. Such a network was regarded as being advantageous in that research teams might be assembled at relatively short notice from different ARUs with the skills and local knowledge required to respond to any particular request for research from the public or private sector. Two further units, the Policy Research Unit at Leeds Polytechnic, and GMAP Ltd, became part of the network at the outset and research units from other universities in the region joined the network subsequently, including the School of Geography and Earth Sciences at the University of Hull, the School of Social and Professional Studies at the University of Lincolnshire and Humberside, the Regional Office, University of Sheffield, the Policy Research Centre (PRC) and the Centre for Regional Economic and Social Research (CRESR) at Sheffield Hallam University, the Institute for Transport Studies at the University of Leeds as well as groups of researchers at the Universities of Bradford, Huddersfield and York. Thus, through its network of ARUs, the Observatory was able to bring together not only subject specialists with skills in data processing, spatial analysis and policy evaluation, for example, but also academics and researchers with a wealth of knowledge and understanding of their own local geographical areas, thus creating a rich base of experience and expertise and a unique resource with which to address the range of strategic, developmental and research issues related to Yorkshire and Humberside.

The general structure of ReRO (Figure 19.1) indicates that the ARUs represented the formal members and, through a Committee of Members, constituted the strategic development body for the Observatory through its annual meeting. ARUs paid an annual subscription for the benefits of being part of the research network. The executive function of the Observatory was managed through the Board of Directors consisting of five senior academics from the ARUs. The day-to-day management and operation was handled through the ReRO Office housed in the School of Geography. In the early 1990s, the Director was supported by a secretary, a research associate, a graphics design technician and a data manager, all of whom were appointed on a part-time basis.

Research Council funding was never secured by ReRO, whose staff, running and development costs were financed entirely from subscriptions from the member ARUs, overhead earnings (20%) from research contracts and consultancy work, sales of ReRO publications and a limited amount of sponsorship from companies such as Coopers & Lybrand Deloitte, Barclays Bank, Halifax Building Society and Northern Foods.

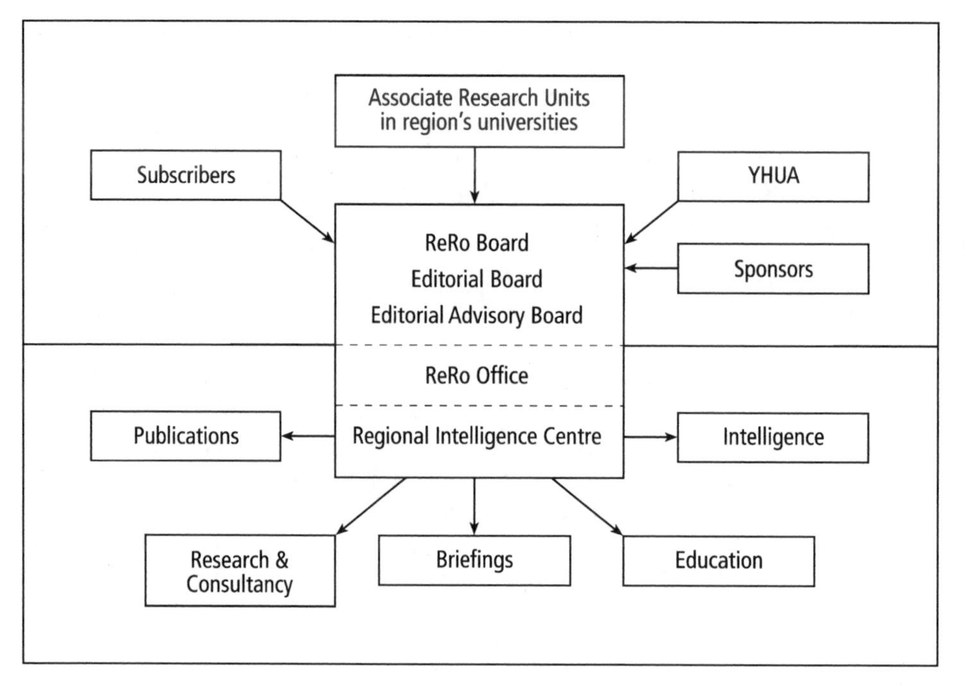

Figure 19.1 *The structure of the Yorkshire and Humberside ReRO*

19.3 Applied Research Activities

In the following two sub-sections, we discuss the range of research activities by the North West RRL and the Yorkshire and Humberside ReRO and present two more detailed examples of projects undertaken at each of the two centres.

19.3.1 Projects at the North West RRL

One very important category of research activity at the North West RRL has been the larger and longer term projects funded by research councils, some of which have been theoretical in nature. They are particularly valuable because they have tended to run for quite a long time, promoting financial stability and job security for the staff employed as researchers. They have usually brought in a substantial amount of money and should have led to opportunities for publication in the academic literature (though the need to chase the next project has sometimes meant that fewer papers reach publication than should be the case). Work in this area has been initiated by RRL staff and allows the RRL to build alliances with others around the university (or outside). The NWRRL has worked on projects involving spatial statistical models, areal interpolation and the modifiable areal unit problem. There have also been research council funded projects that have been more specific, such as evacuation modelling in the context of emergency planning.

In the more applied context, work has been done with Eurostat on GIS software for EU applications, with the European Environmental Agency on environmental databases, with the Home Office on developing GIS software for the Safer Cities Programme, and with the Department of Environment, Transport and the Regions (DETR) on pollution modelling using the critical loads approach. Much of the RRL's work, however, is made up of much smaller and shorter term projects, often based on the employment of relatively straightforward GIS capabilities for practical purposes. Such projects may only bring in a few hundred or thousand pounds, but are relatively simple to accomplish. A common type of project involves the mapping of data of various types, usually involving the conversion of unit postcodes to grid references. Sometimes the work involves aggregation of the data to areal units and production of choropleth maps. Another type of small project is the construction of a spatial database, for example the Lancaster Urban Archaeology Database which involved creating and linking georeferenced records for archaeological finds in the city of Lancaster. Sometimes the work is basically data analysis, not necessarily involving GIS skills. The RRL has also been asked to assist in the design of spatial sampling, involving the collection of data on household waste and the analysis of rural poverty among older people. Another frequent task has been to solve routing problems, for example through examining the extra travel distances or times if a particular facility is closed down or another one is opened. This type of analysis has been important in the study of health care and leisure facilities as well as commercial service delivery systems. A particularly important set of tasks has been the study of visibility, especially in relation to the location of wind farms, where the RRL has developed close links with a number of companies, providing technical expertise for planning enquiries as well as relevant information for siting issues.

Two examples of applied research provide an indication of some of the advantages and disadvantages of being an RRL in particular contexts. The first is an ESRC-funded project on using GIS to measure remoteness in the context of issues related to funding for public services, and the second is a large public financed project on the problems for North West England of changing demographic trends.

Remoteness of Services

This project was primarily academic rather than commercial, though its genesis was very much applied. The original beginning of interest in this area was a set of consultancy projects with local authorities relating to the Standard Spending Assessment (SSA) system. The latter was designed to provide a fair basis for central government to disburse funds to local government with the objective of providing standard levels of service across England. The formulae used were open to question on several grounds (Flowerdew *et al.*, 1994). In particular, several rural authorities felt that the SSA calculations did not adequately reflect the additional costs they incurred because of the time taken to deliver services to clients. The project was based on the idea that GIS provided a methodology for calculation of the actual distances that service deliverers would have to travel to reach the entire population. If assumptions were made about how service delivery centres were located, estimates of the costs in service delivery could be derived. Calculations of this type could be done for all local authorities, on the basis of several sets of assumptions about the locations of service delivery points. These estimated costs could then be compared with existing allocations. The ESRC agreed to support the project, which employed one research assistant half time for a year.

In the event, the project suffered somewhat from being carried out alongside several other commercially funded projects. It was difficult for the researcher concerned to spend 50% time on this project when there were other shorter term projects requiring significant input to a tighter deadline. More pre-processing of data was required than anticipated before the calculations could be started, and the latter proved to be more time-consuming than expected. Although the availability of experienced staff may have been an advantage of being conducted within an RRL, the necessity of carrying out the work at the same time as other commercial projects proved to be a problem. A further problem arose in writing up the results of the project, which conflicted with the necessity for all staff members to proceed with tendering for new projects to maintain the RRL's income flow, and to carry out those projects for which tenders had been successful. This was exacerbated by the Research Council's policy of paying overheads at rates much lower than commercial ones. Such projects are still worth doing for an RRL, however, because they are worthwhile for the group's academic profile, even if they do not bring in much additional money.

Ageing Population

Another project undertaken by the RRL involved studying the implications of changing population structures for North West England. This project was sponsored by a consortium of regional-level bodies, including the North West Regional Assembly, the North West Development Agency, the Government Office for the North West, the National Health Service Executive North West and Age Concern. The main aims of the project were to produce population projections by age at the district level for the North West and to investigate the consequences of these projections for a range of economic, social, health and planning issues.

Population projections are available from the Office of National Statistics for the population as a whole until 2025. However, for the purposes of this project it was important to project conditions at the local level for a longer period, and also to project the populations at the local level for population subgroups, including ethnic groups and the disabled population. These projections raised a number of interesting academic issues as well as being highly relevant to policy questions. A crucial issue in the reliability of projections turned out to be the effect of students, whose migration behaviour, before and after moving away from home for higher education, is particularly difficult to estimate given available data (Simpson, 1995). It seems likely that the use of census data may underestimate student population in university cities and the use of National Health Service registration data may overestimate it. As a result, the methodology used in this context produced wildly disparate projections for places with large student populations.

Producing local-area projections by ethnic group and disability status also raised methodological issues. Arbitrary decisions had to be made to produce these projections and the unavailability of all the desired data meant that simplifications and unrealistic assumptions were necessary. For example, it was difficult to make projections for places that had a very small population in one ethnic group but had experienced migration of ethnic group members in or out. If migration for the ethnic group was assumed to be proportional to the population of that ethnic group, this could result in a very rapid and totally unrealistic rise (or a rapid disappearance) in the numbers of that group. Accordingly, we assumed that the numbers of migrants rather than their proportion would remain constant. Whereas government has, quite understandably, refused so far to make projections by ethnic group, we felt constrained to produce the best projection we could, even though we knew it was unrealistic in several respects.

In any exercise of this nature, it is vital to draw a distinction between projection and forecast. The nature of a projection is that it is based on the continuation, in some sense, of current trends. This is quite unlikely in many situations and, in some cases, it is more than likely that an undesirable projection will generate some sort of policy response to prevent it actually coming true. Of course there are many other factors, predictable or otherwise, which will mean that the assumption of conditions remaining constant will not be accurate.

19.3.2 Applied Research through Partnership

In summary, ReRO developed a portfolio of activities that can be grouped broadly under five headings:

- research and consultancy: through the assembly of multi-institutional and multi-disciplinary teams;
- intelligence: through the interpretation and analysis of data rather than mere cataloguing of current data sources;
- education and training: through a European initiative to provide a pan-European course in training of regional managers;
- publication: through a regular journal, *The Regional Review*, and a periodic bulletin; and
- networking: both amongst its member ARUs but also with relevant bodies in the region and beyond.

The majority of these services are provided through collaboration between one or more of the ARUs and examples of research contracts are listed in Table 19.1. In most cases, the applied work was focused at the regional scale and was carried out at a time when the regional tier of agencies and administrative institutions was embryonic. Work has been commissioned by central government departments, national agencies, regional agencies, local government bodies, charitable organisations and private sector companies.

The view at ReRO was that urban and regional planning and policy-making had substantial needs for information and intelligence. Good regional data capture, maintenance and analysis were considered imperative as the basis upon which to

Table 19.1 Examples of ReRO research contracts

Title of research project	Funding body
Regional Textile Database Analysis: Annual Report	Barclays Bank and six textile companies
Yorkshire and Humberside Universities: A Potential for the Region	Yorkshire and Humberside Universities Association
Profiles of English Regional Arts Boards	The Arts Council
The Impact of Inward Investment into Yorkshire and Humberside	Department of Trade and Industry and the European Commission
The Role of Leeds in the Regional Financial System	Leeds Development Agency and Leeds TEC Ltd
Research into the Application of Bio-technologies in North Yorkshire	Yorkshire and Humberside Development Association
Regional Technology Plan	The Regional Technology Network and Yorkshire and Humberside Partnership
Research into School Performance	Government Office for Yorkshire and the Humber
Developing a Strategic Framework for Yorkshire and Humberside: Review of Existing Policies and Strategies	The Regional Assembly for Yorkshire and Humberside

understand problems and to evaluate feasible solutions. Yet it was evident that few guidelines were available as to which particular demographic, economic, environmental or social variables should be used to measure regional performance. This vacuum has been highlighted by the debates on measuring 'sustainability' that preoccupied various organisations in the 1990s. ReRO always placed a high premium both on the collection of data and on 'adding value' to that data. On the one hand, this involved maximising the availability of computer technology to manipulate, analyse and display data. On the other hand, it also meant using expert local knowledge to provide reliable interpretation of the cross-sectional and time series datasets that have been assembled. As with the North West RRL, several of the smaller projects undertaken by ReRO involved collecting or assembling data, geocoding postcode-based records and mapping spatial distributions of phenomena such as inward investment establishments, poorly performing schools or organisations receiving funding from the Prince's Trust. Further details of two projects are provided to exemplify the nature of the applied GIS-based and spatial analytical work.

Adjustment of Population Projections

In the early 1990s, the data available from the 1991 Census provided local authorities with the opportunity to evaluate the accuracy of the 1989-based sub-national population and household projections for 1991 produced by the Office of Population Censuses and Surveys (OPCS, 1991) and the Department of the Environment (DoE, 1991) respectively. ReRO was commissioned to carry out this assessment for the region by the Yorkshire and Humberside Regional Planning Conference (representing the interests of all the local authorities in the region) and the Regional Office of the DoE. In addition, the ReRO research team, comprising staff from the Universities of Leeds and Sheffield Hallam, were asked to utilise the findings in the formulation of a range of alternative population and household projections which could subsequently be used to inform the assignment of land for the future development of housing, part of the process of formulating Regional Planning Guidance (RPG).

It was found that, overall, the 1989-based projections for the region as a whole underestimated the mid-1991 estimates by 0.5% of its 4.9 million inhabitants but that differences between projections and estimates varied appreciably by local authority and by age group. In North Yorkshire, for example, overprojection of the total population was due to significant overprojection of those living in the county of young working age, offsetting an underprojection of those aged over 40. In Humberside, on the other hand, the population projections were lower than the population estimates for mid-1991 in almost all age groups, providing a substantial underprediction of the population of the sub-region as a whole. One of the key factors in explaining the differences was net migration. A methodology was developed to update the existing 'official' population and household projections for local authorities rather than constructing a new projection model (ReRO, 1993). The procedure for adjusting the population projections involved applying revised population base factors and adjusting the net migration assumptions on the basis of three scenarios about the range of possible futures. A modelling methodology was then used to convert the population projections into a revised set of household forecasts and involved subtracting the institutional populations from the projections, applying marital status factors (the proportions of a given population group that are either single, married or widowed/divorced) to the male and female projected populations by age group, and finally applying headship rates (the proportion of members of a population group defined by age, sex and marital status that are heads of households) to the relevant age–sex–marital status group projections. The details of the methodology and results of the project have been reported in Gore and Stillwell (1994).

The experience of undertaking the project was both rewarding and frustrating for the researchers involved. A huge database was created, a large amount of new demographic information about the region was uncovered, and an innovative solution was derived to a difficult problem. However, it became very clear to the research team at an early stage in the project that what was required to do a proper job was a full multi-regional population projection model linked to a regional model for household projections which would generate outputs that could be converted into projections for a variety of different uses. Frustration arose from the inability, due to time and cost constraints of the project, to deliver a fully fledged regional model that would become an immensely useful tool to support this type of analysis in the future. Further experience of undertaking applied quantitative research in population geography for local and regional authorities has been reported in Stillwell and Rees (2001).

Educational Performance

In the mid-1990s, ReRO was contracted by the Government Office for Yorkshire and the Humber to undertake a study of the educational performance of primary and secondary schools throughout the region. Recent studies by Gordon (1996) and

Clarke and Langley (1996) had shown the extent of spatial variation in school performance and preliminary work at the Local Education Authority (LEA) level had indicated huge variations in performance in Yorkshire and Humberside at Key Stages 2 (age 11) and 4 (age 16). At one end of the spectrum, the average numbers of pupils achieving level 4 in English, Mathematics and Science at primary schools in North Yorkshire were 66.4%, 66.5% and 73.5% respectively whereas equivalent figures for Kingston upon Hull, at the other end of the performance league table, were 55.8%, 54.5% and 63.1%. Similarly, the average percentage of pupils attaining five or more A* to C GCSE grades was 52.7% in North Yorkshire and only 24.1% in Kingston upon Hull (Stillwell and Langley, 1999). This geographical disparity is explained by a number of factors collectively summarised as levels of social disadvantage. These factors include children's personal characteristics, family structures, socioeconomic features, educational levels and ethnicity.

The focus of the project was to examine the geographical variations in achievement in more detail by looking at the performance of individual schools. This involved filtering out special schools, non-state schools and schools for which sets of examination results were incomplete and then geocoding the schools remaining in the database according to their unit postcodes. The locations of the schools were mapped against a set of ward boundaries and the best and worst performing schools were overlaid on a ward-based choropleth map of deprivation as measured by the Department of Environment's Index of Local Conditions, a measure based on seven deprivation indicators from the 1991 Census (unemployment, children in low earning households, overcrowded housing, housing lacking basic amenities, households with no car, children in unsuitable accommodation and educational participation at age 17). The spatial coincidence between schools performing badly and higher levels of deprivation was particularly significant. However, one of the key difficulties in linking examination performance and social deprivation is the use of ward-based socioeconomic characteristics as measures of school catchment area, when it is well known that social conditions vary within wards and that the size of school catchment areas differs widely from location to location. Consequently, one of the most important recommendations of the project was to use GIS buffering techniques to identify enumeration districts within a defined radius of each school and to compute new socioeconomic indicators based on the areas located within the buffer (see Stillwell and Langley 1999, for details). Gender-specific data was also made available to the researchers to look at the extent to which girls across the region were performing better than boys. Remarkable variation exists in the gender-gap, even at LEA level, where the East Riding was ranked 18th out of all 129 LEAs with 12.5% difference between the proportion of girls and boys getting five or more grades at A* to C, and North East Lincolnshire was ranked 124th, where the difference was only 3.9% (Bright, 1998). The project found no spatial relationship between the size of the gap and level of examination performance.

19.4 Cultural Differences, Constraints and Opportunities

The overviews of selected projects presented in the previous section exemplify some of the work undertaken within two different research institutions specifically established to conduct applied research. In this section, some generalisations regarding the relative nature and priorities of applied work in contrast to academic work are offered and some of the benefits and problems associated with applied research are considered. Here we focus primarily on the role of the RRL although many of the points made relate to applied research in an academic context more generally.

19.4.1 Differences Between the Applied and Academic Research Cultures

There can be little doubt that operating within the framework of an RRL had a major impact on the nature of the work done and on the balance between theoretical and applied work. This is partly because an RRL has a particular remit to work in the GIS field, but largely because of the financial arrangements under which it has had to operate since the conclusion of RRL funding. The key issue of course is that the RRL employs a number of staff funded to work on GIS projects in general rather than on any one specific project. The first priority in considering possible work is therefore to provide enough money to pay the staff, and if possible, to give them a reasonably long-term degree of security. One major difference between RRLs and commercial consultancy companies is that the director of the RRL will usually have job security through a continuing contract with a university. The negative aspect of this is that he or she will also have academic responsibilities, and must be prepared to fit consultancy work around teaching commitments, marking examinations, and other non-applied research.

The differences in time scales between academic and commercial research are well known. Academic projects have a long lead time and last for a considerable period, sometimes for several years. Commercial projects are likely to come up with little prior notice, insufficient time to prepare the tender (especially when it may involve staff in a number of different institutions), no lead time between winning the project and starting work, and usually a very tight time scale to final report. Academic time scales allow posts to be advertised so that a new employee can be brought in with appropriate skills. For commercial projects, it is essential to have staff available already with the right sorts of skills and with time available to work on the project intensively and immediately. Generally universities find it difficult to operate in this way, but in an RRL it is easier, provided that staff have the willingness and the ability to try their hand at a wide range of projects. Certainly, the more commercially orientated environment makes it very hard for any on-the-job training to be accomplished or for RRL funding to be used for supporting research students.

The academic environment still tends to be fairly casual about punctuality. In the commercial world, however, clients are more concerned that work be produced on time. In many circumstances, their business needs mean that a sub-optimal answer provided on time may be more useful than a better answer which does not arrive until two or three weeks after it was needed. Sometimes it may in fact be more important to say something than to say the right thing. Decisions have to be made to fixed deadlines, and sometimes a client may be worse off if no decision is made than a decision that is flawed. Whereas in the academic world, few people would want to argue for their point of view unless they had carefully considered the arguments and the available data; in the commercial world there is seldom time for reflection.

The *modus operandi* of applied research creates difficulties because work is often concentrated at certain times of the year and absent at other times. Also, because most commercial work is on a tight timetable, it is relatively hard to schedule it to iron out the peaks and troughs. This causes difficulties in that RRL staff may be overworked at one time and at a loose end at another. The latter should not be a problem, in that slack periods can be used for writing up work that has been done already (though this seems surprisingly difficult to do for commercial projects), or for seeking out new business. It may also be possible to bring in casual labour to help out at busy periods. Often a university is well equipped to do this because it contains well-trained students who may be prepared to work part time and short term on an interesting project. On the other hand, if RRL staff have different skills and interests, there may be times when one staff member is needed to work on several different projects at the same time while another has nothing to do, only for the situation to be reversed a few months later.

The establishment of the RRLs coincided with the development of Masters level courses in GIS, and these courses have been fruitful sources of researchers for the RRLs. This is particularly so where, as at Edinburgh and Leicester, successful MSc courses are taught in a department where an RRL is located. It has been relatively easy to persuade many of the best MSc students to stay on to work on RRL projects, or to become research students in the department. Even in other RRLs, such as the North West at Lancaster, finishing MSc students have been a good source of able and well-trained researchers, who have often been keen to acquire experience of the mixture of theoretical and applied research projects characteristic of RRL research. At Leeds, whilst GMAP recruited new graduates, masters' students and PhD students (particularly from geography) in considerable volume from year to year, the philosophy of ReRO was to assemble teams of established researchers from institutions within the region.

The best solution to recruitment and staffing problems is through establishing economies of scale. The more people there are, the easier it will be to accommodate changes in the workload, and to find somebody suitable to work on a short-term short-notice contract. Unfortunately, however, this requires a bigger overall turnover and one which is not sufficiently guaranteed to allow permanent appointments to be made. Over time, and with a consistent record of high-quality work, it becomes possible to build up the volume of work, but there is still going to be a workflow problem. Furthermore, writing tenders and submitting proposals is a difficult and time-consuming activity, which will normally fall to the leader of the RRL, whose task becomes increasingly difficult as the size of the operation increases.

The working environment for applied research has often in the past been at odds with a university's normal expectations of security for research staff. The researchers themselves need the promise of job security to commit themselves to working in an RRL, but university finance offices are reluctant to promise such security unless money is in hand or contracts have been won sufficient to pay the staff for whatever length of contract is specified. In a situation where income is irregular and projects may appear at very short notice, RRL directors may have to convince their institutions to take the risk that sufficient money will be brought in to pay for new appointments.

Another implication of the ambiguous position of an applied research community that sits between the academic and commercial worlds is that there may be a conflict between staff's research interests and the portfolio of funded projects. Sometimes the most interesting ideas may arise from unfunded or underfunded work; developing these ideas may only be possible by neglecting income-generating projects. Often, the most socially worthwhile projects are sponsored by organisations with very limited resources. Applied research staff may also feel an obligation to colleagues and students; a lot of time at the North West RRL, for example, has been spent in helping students with GIS problems, and in exploring possible GIS aspects of the research of colleagues with other interests. Although it is important to show colleagues in the same university how GIS and spatial analysis may help them with their research, there is sometimes a risk of setting up a free and unresourced GIS consulting service.

Academics are trained to make sure they provide proper acknowledgement of other people's ideas, and that any claims not substantiated within their papers must be referenced, so that other academics may verify their truth or trace their intellectual provenance. Nobody in commerce or government will have time to do this, and referencing sources is usually unimportant. It may even produce a negative impression, as it may be taken to emphasise the academic nature of the project, using the word 'academic' in its sense of 'not leading to a decision; unpractical' (Onions, 1972). The style of reporting on a project is also an important difference between academic and commercial requirements. Academics are used to hedging their conclusions with all manner of reservations and *caveat*s, always aware that what they are saying may depend on context and may be matters of probability, nuance or weight of evidence. Such mealy-mouthed conclusions are not popular in the outside world, where a plain and definite statement is required and a short executive summary of any major report is paramount since this is all that the key decision-maker is likely to read. The conclusions, and indeed the description of the project, must also be short, succinct and definite. When working on issues related to government policy, the 'taxi ride principle' applies; apparently there is little point in saying anything that is too long or complex for a civil servant to explain to a minister during the taxi journey from the Ministry to the House of Commons!

A further simplification is the tendency for decision-makers and the media to ascribe work not to the individuals who conducted it, nor to the RRL or ReRO, but to the university with which they are affiliated. Academics may be taken aback or overwhelmed with responsibility if their analyses and policy recommendations are ascribed not just to them, but to their universities. This is especially so if they are aware that many of their colleagues would not share their views about the issues involved.

From the academic's point of view, publication, especially publication in refereed academic journals, is an important part of career development. In fact this is much more important than dissemination of research that may be of practical relevance. Academics are notoriously bad at writing about the practical application of their work for the layperson to understand. This issue is particularly familiar to one of the authors after 12 years of editing a publication (*The Yorkshire and Humber Regional Review*) trying to inform policy-makers of regionally relevant research. Equally, applied projects may be difficult to write up for academic journals for a variety of reasons. In terms of the points just discussed, the academic may need to reinsert the nuances and the qualifications that were taken out to meet the client's needs for clarity and definitiveness. There may also be problems relating to confidentiality concerns. A client may not want his or her company's problems or operating policies published to the world in an academic journal, and anonymising the client may be insufficient if describing the situation makes the identity obvious. If the academic analysis indicates problems with the client's activities, the client will not usually want these known. Equally, if the research has identified new and better ways of doing something, considerations of commercial confidentiality may mean that the new methods cannot be described explicitly lest competitors adopt them, destroying any commercial advantage that the clients may have gained.

There may also be problems from the academic side. Academic research is usually best regarded if its findings are general. Results specific to one place or one company are less likely to be regarded as important by journal editors and referees. If important details are withheld or anonymised for reasons of commercial confidentiality, the academic article is further weakened. The perspective may also need to be different. In most fields, academic work would be expected to present a neutral review of situations and problems, and an even-handed discussion of what can be done about them. In contrast, the commercial report will necessarily be based on the position of one specific actor, the client, and will be primarily concerned with what the clients can do for their own benefit rather than for the good of society as a whole.

19.4.2 Benefits and Constraints

Despite the differences and problems discussed above, most participants in the research efforts of the RRLs and ReRO would probably agree in seeing many benefits to involvement in commercial projects. First, the wide range of applications is highly stimulating and the requirement to produce solutions is very challenging. Certainly some commercial problems are repetitive and formulaic, but very frequently there are interesting differences which require variation of a standard approach, and in some cases require revisions of an existing method which may shed new light on the problem from an academic perspective. A very common situation occurs when a fairly standard situation is encountered but there are limitations in the data available to develop a solution, and interesting intellectual problems arise in trying to make the optimal use of whatever data can be found. Sometimes, new and very valuable datasets are unearthed which can be used in other research projects or contexts.

Second, some intellectual satisfaction can be gained from the problem-solving nature of much applied research. Admittedly the problems must be solved with limited time and sometimes limited data, and some academic researchers may find this frustrating because they recognise what might be possible with greater resources. However, there are compensations in the finite nature of the problems. A researcher can feel he or she has reached the best possible solution within the constraints available. In contrast, much academic research is inevitably open-ended; one always feels slightly unsatisfied, in the knowledge that further research or further thought could produce a fuller or more sophisticated understanding of the point at issue.

For both the authors, quantitative geographers trained in the 1970s, applied research in GIS and spatial modelling has often involved a journey back in time, to some of the operational research and statistical models that were then at the cutting edge of geography. Though fashions have changed within the discipline and not many geographers get excited by location-allocation, spatial interaction models or multivariate classifications, they do have a commercial value. So do some of the skills that quantitative geographers have, less valued perhaps within academic geography than they were, but able to command a considerable price in the commercial marketplace, particularly if combined with some more advanced computer programming skills.

The use of the very term 'applied geography', or of 'applied GIS and spatial analysis' more specifically, begs some important questions. If geography is to be applied, what is it to be applied to? Who is to be doing the applying? Who will benefit from the application? If geography is to be applied for money, for the support of geographers working in higher education, then the answer must be that it will be applied for the benefit of those who have the resources to pay for it. This will usually be business or government, and it will usually be for their own benefit. It is rare for applied research to be performed in the interests of poor or excluded groups within society, simply because they are poor and excluded and hence unable to commission such work. The Joseph Rowntree Foundation is an exception to the general rule. Even within commercial organisations, it is the larger and more aggressive who are most likely to be clients. If geography is applied to help such companies with store location or market penetration, it will inevitably be at the expense of small businesses and corner shops with which we might otherwise tend to sympathise. In the RRL and ReRO context, however, ethical considerations of this type must be set against the ethical imperative to provide work and earn funds for the institution concerned.

It could also be argued that elements of applied geographical methods may themselves have questionable ethical effects. In constructing market areas or accessibility measures, for example, it is usual to adopt drive times as measures of accessibility. If facilities are located on this basis, their location pattern will no doubt make sense as far as car users are concerned, but there is a risk that those reliant on bus or rail travel, or other transport modes, will not be served adequately by the facilities. Further, for environmental reasons, we may not be happy to recommend a system that relies on even more universal usage of private cars with the well-known negative social and environmental externalities that they generate.

A further critique that applies to some aspects of applied geography in the commercial sector concerns invasions of privacy (Goss, 1995). Geodemographics and related marketing tools increasingly require large databases, including information about many individuals and local geographical areas. Some of this is collected in unobtrusive ways, such as recording address and purchasing information from store loyalty cards, or from participants in competitions and promotional activities. Apart from the invasion of privacy such activities can be regarded as constituting, there may be practical consequences, such as the refusal of credit, or the insistence on high insurance premiums, to somebody purely on the basis of where they live.

In terms of the concerns of contemporary geography, moreover, it can be argued that commercial applications lead to a focus on the tangible and perhaps on the economic. There may be little scope for studies in this area to engage with questions about values, interpretations and the taken-for-granted world. Arguably, a concern with the mappable and the unambiguously locatable tends to distract our attention from the ambiguous perceptions, interpretations and projects of those we are studying. The concern with the interests of clients, often largely their commercial interests, detracts from the full humanity of the people who are reduced to a geocode and a record of purchasing behaviour.

Other problems with a concentration on applied geography within an academic setting are institutional rather than ethical. Many of these have already been mentioned in the previous sub-section, such as the funding treadmill which requires a constant need to bring in more funding to keep the organisation going and to keep research staff in post. University administrations are not designed to promote commercial activities, however much current funding systems have forced them to try. Academic goals and commercial goals are not the same. Universities cannot ignore their finances and their need to bring in revenue, but to an important extent they still value truth and the search for knowledge. Commercial companies and government, quite rightly, have their own different goals and values, and an academic trying to combine both sets of goals must inevitably struggle to satisfy the latter without abandoning the former.

19.5 The Future for Applied Research

In conclusion, and despite the comments in the preceding section, our experience makes us generally optimistic about the future of applied geography and the use of the computer-based systems and methods that are now emerging for spatial analysis and forecasting. The geographical perspective, geographical data and geographical skills are highly relevant to many aspects of contemporary society. The activities of the RRLs, ReRO and companies like GMAP have done an enormous amount to make more people aware of how quantitative geography can be applied, and have helped some geographers understand better how to market their services and to meet the needs of outside clients. There is plenty of scope for applying geography.

Geographers may be in competition with other academics, especially economists, many of whom may be better integrated into the commercial and policy worlds. Certainly potential clients have a clear idea of how their organisations may benefit from applied economics, while geography still suffers from a 'capes and bays' or 'geomorphology and climate' image. Fiercer competition may come from consultancy firms employing the most recent generations of graduates or postgraduates. These firms have the economies of scale and the knowledge of client expectations to attract a lot of business. In contrast, academics may offer lower rates and more specialised knowledge, but often may be perceived as less likely to deliver on time, and more likely to produce work that is unhelpful to the client because it does not offer clear-cut solutions and recommendations. Our experience suggests that geographers in commercial work in order to compete effectively must learn the importance of presentation, the importance of meeting deadlines, and the importance of presenting their work in a manner which the clients can most easily use.

Quantitative skills are necessary for much of what is done in the name of applied research. The quantitative revolution left geographers in the 1970s with degree courses that included statistical analysis and some practical work with maps and numerical data. Compared with many other social scientists, they could be effective in applying geography because they possessed practical and quantitative skills as well as the ability with write coherently about their results. In the current intellectual environment, some of these gains may be threatened. It is becoming easier to graduate in human geography without acquiring much quantitative expertise, and at the dissertation and postgraduate levels, there may now be more prestige in conducting a qualitative survey or a discourse analysis than doing quantitative analysis or modelling work. The discipline may lose out if we do not continue to teach and to apply quantitative skills. We must also be prepared to work to real-world rather than academic standards. The work must recognise the importance of punctuality rather than perfectionism. A willingness must be shown to commit to ideas and policies without having all the evidence.

We do not wish to argue that all geography should be applied, but there are huge opportunities for applying the work we do and it would be a great shame if we do not take further advantage of them, especially those that make use of GIS, analysis methods and modelling techniques. The examples contained in the earlier chapters of this book are testament to what has been achieved. Much more is possible in the future.

Acknowledgements

This chapter represents the fusion of ideas presented in earlier papers by Robin Flowerdew at the Annual Conference of the Royal Geographical Society with the Institute of British Geographers in Plymouth in January 2001 and by John Stillwell at a seminar on regional observatories at the University of Complutense in Madrid in November 2000.

References

Birkin, M., Clarke, G., Clarke, M. and Wilson, A. (1996) *Intelligent GIS: Location Decisions and Strategic Planning*, GeoInformation International, Cambridge.

Bright, M. (1998) Boys performing badly, *The Observer*, 4 January, p. 3.

Chorley, R. (1987) *Handling Geographic Information. Report to the Committee of Enquiry*, chaired by Lord Chorley, HMSO, London.

Clarke, G. and Langley, R. (1996) A review of the potential of GIS and spatial modelling in the new education market, *Environment and Planning C*, **14**: 301–23.

Clarke, G., Longley, P. and Masser, I. (1995) Business, geography and academia in the UK, in Longley, P. and Clarke, G. (eds) *GIS for Business and Service Planning*, GeoInformation International, Cambridge, pp. 271–83.

DoE (1991) *Household Projections England 1989–2011: 1989-based Estimates of the Numbers of Households for Regions, Counties, Metropolitan Districts and London Boroughs*, Housing Data and Statistics Division, DoE, London.

Flowerdew, R., Francis, B. and Lucas, S. (1994) The Standard Spending Assessment as a measure of spending needs in nonmetropolitan districts, *Environment and Planning C: Government and Policy*, **12**(1): 1–13.

Gordon, I. (1996) Family structure, educational achievement and the inner city, *Urban Studies*, **33**(3): 407–23.

Gore, A. and Stillwell, J. (1994) Updating population and household projections for Yorkshire and Humberside: an input into the strategic planning process, *Planning Practice and Research*, **9**(4), 381–93.

Goss, J.D. (1995) We know where you are and we know where you live, *Economic Geography*, **71**(2): 171–98.

Masser, I. (1988) The regional research laboratories initiative: a progress report, *International Journal of GIS*, **2**, 11–22.

OPCS (1991) *Subnational Population Projections: Population Projections by Sex and Age for Standard Regions, Counties, London Boroughs, Metropolitan Districts and Regional and District Health Authorities of England from Mid-1989*, Series PP3 no 8, HMSO, London.

Onions, C.T. (1972) *The Shorter Oxford English Dictionary on Historical Principles*, Third edition, revised with addenda, Clarendon Press, Oxford.

ReRO (1993) *An Analysis of the 1991 Census Results for Population and Household Projections in Yorkshire and Humberside*, Final Report for the Yorkshire and Humberside regional Planning Conference and the Department of the Environment, ReRO, Leeds, p. 259.

Simpson, S. (1995) Using the special migration statistics: examining ward migration and migrants' ethnic group, in Simpson, S. (ed.) *1991 Census Special Statistics on Migration, Workplace and Students*, Local Authorities Research and Intelligence Association, London, pp. 49–55.

Stillwell, J. and Langley, R. (1999) Information and planning in the education sector, Chapter 17 in Stillwell, J., Geertman, S. and Openshaw, S. (eds) (1999) *Geographical Information and Planning*, Springer, Heidelberg, pp. 316–33.

Stillwell, J. and Rees, P. (2001) Applied population projection for regional and local planning, Chapter 7 in Clarke, G. and Madden, M. (eds) *Regional Science in Business*, Springer, Heidelberg, pp. 115–36.

Author Index

Subject Index